AMERICAN NEUROSCIENCE IN THE TWENTIETH CENTURY
Confluence of the Neural, Behavioral, and Communicative Streams

AMERICAN NEUROSCIENCE IN THE TWENTIETH CENTURY

Confluence of the Neural, Behavioral, and Communicative Streams

BY

HORACE WINCHELL MAGOUN, Ph.D.

Brain Research Institute, University of California, Los Angeles, USA

EDITED AND ANNOTATED BY

LOUISE H. MARSHALL, Ph.D.

Brain Research Institute, University of California, Los Angeles, USA

Louise H. Marshall

Good luck!

A.A. BALKEMA PUBLISHERS / LISSE / ABINGDON / EXTON (PA) / TOKYO

Cover design: Studio Jan de Boer, Amsterdam, The Netherlands
Typesetting: Charon Tec Pvt. Ltd, Chennai, India
Printed in The Netherlands by Krips, Meppel

Published by: Swets & Zeitlinger Publishers
www.szp.swets.nl

ISBN 90 2651 938 9

Contents

Preface and Foreword

Natural History of the Manuscript

Origins

Few committed scientists attain their senior years without succumbing to the urge to "autobiograph" their careers and mature thoughts. A few examples among many easily come to mind: Wilder Penfield's *No Man Alone* (1977), Charles Sherrington's *Man on His Nature* (1941), and that wonderful prototype series, *Psychology in Autobiography*, suggested in 1928 by the discipline's consummate scientist-historian, Edwin G. Boring and inaugurated 2 years later under Carl Murchison's editorial hand. The young domain of neuroscience history is grace-noted by similar autobiographies, perhaps the first of which is the collection celebrating the 70th anniversary of Francis O. Schmitt titled *The Neurosciences: Paths of Discovery*, published in 1975 Other, less known autobiographical sources in neuroscience are the oral histories collected since 1975 by the Neuroscience History Archives of the Brain Research Institute of the University of California, Los Angeles. More recently, the Society for Neuroscience initiated a series of autobiographical volumes and video interviews with its prominent members.

Horace Winchell Magoun (1907–1991) in his later years did not share that self-conscious characteristic so conspicuous among his peers. When asked to contribute to the Schmitt celebratory volume he detoured around recording his own career and submitted an elegant description of the rise of institutes and organized research units committed to studies of brain and behavior. The chapter written by the conceptualizer of one of the most powerful theories of how the brain works, the ascending reticular system, was relegated to the rear of the collection, where it joined other essays in the impersonal fields of organizational history and neuroscience theory.

In keeping with his self-effacing persona and harking back to his love of history inspired by a high school teacher, Magoun agreed to write on neuroscience for the two-volume *Advances in American Medicine: Essays at the Bicentennial* (Bowers & Purcell, 1976). In that chapter lay the seed of the idea of a final contribution to the advancement of the discipline – to write the history of American neuroscience in the 20th century – and the preliminary resources were at hand. There were several distractions, however, before reaching that objective: A series of large posters prepared for the American

Academy of Neurology in 1983; preparation of a monograph (1984) on the first quarter-century of UCLA's Brain Research Institute for which he had been the prime mover; and several published articles on historical topics of interest to him.[1] With the accumulation of about 400 pages of typescript, Magoun applied to the National Library of Medicine for continuation of support. Unhappily, a newly appointed study section of professional historians (after a long lapse in rotation of members) did not find this scientist's treatment of history in line with their "external" view and the proposal received the dreaded "pink slip." The 73-year-old author laid down his pen, and the manuscript rested undisturbed in a low cabinet drawer until after his death, when it became part of the Horace Winchell Magoun Papers in the Special Collections of the UCLA Research Library.[2] The final twist of events was a casual mention of the manuscript's existence to the then-science editor of Swets and Zeitlinger, Publishers, which elicited a contract to prepare the first draft for publication. At that point, Magoun's written words commenced a life of their own.

Revival of the Manuscript

On close scrutiny, the draft sheets consisted of an elaborate Table of Contents and a partial, looseleaf text that had been rearranged during archival processing, plus reference slips stuffed into a shoebox. By combining typescript and handwritten sheets from his working papers, it was possible to reconstruct Magoun's grand plan, including a reordering of sections composed at different times, to achieve a coherent whole. The rearrangement revealed gaps in the text which had to be filled for continuity (they are identified in the pages that follow), a process that entailed finding and absorbing the missing historical sources.

Magoun's pattern of acknowledgement of the foreign background of the American work has been preserved, but without detailed citing of the literature. Parallel past events cannot be interwritten and a strict chronology is impossible, therefore this overall history is presented as topics from their origin to the later years of the century; there are flashbacks and confluences that become ever more complex as the inter- and multidisciplinary aspects of neuroscience acquired momentum. Magoun's grasp of their directions ceased in the mid-1980s and some extensions from then were added to his text as indicated.

Another of the author's propensities was to quote extensively from his reading. To stay within the publisher's agreed length and to prevent the reader losing the thread, most of the quotations have been paraphrased; when

[1] A bibliography is available on request to the Neuroscience History Archives, Brain Research Institute, University of California, Los Angeles 90095-1761 or lmarshall@mednet.ucla.edu

[2] A finding aid is available on request.

retained, they were checked for accuracy as Magoun had the quaint habit of unconsciously improving on any quotation. His emphasis on the primacy of support systems in advancing a discipline or an organization was apparent in his earliest historical writings. It was manifest in his chapter for the Schmitt volume mentioned above, and in the present text it threatened to overshadow the narrative of research and investigators. For those reasons, a second volume is planned on support systems in neuroscience, to include such topics as the roles of primate centers, freshwater and marine laboratories, journals and translations, foundations, and many more sources specifically in North America that have contributed to the explosion of neuroscience. The history of those support systems will be contributed by selected authors to complete the outline as Magoun envisioned it.

As a scientist, Magoun harbored no arguments over historical issues that he needed to defend. His trained eye for detail and his conceptual powers discerned relationships not seen by other chroniclers of antecedent events. An example is the spread of the evoked potential technique from the peripheral to the central nervous systems. His version may not be useful to externalist historians but it will afford a reminiscence to mature neuroscientists and serve as a source of new knowledge to those entering the field.

This "Magnum Opus"

The first of seven chapters introduces the history of the neural sciences with three prodromal events highlighting the early portals of neurological knowledge from Europe to America: Gall's cranioscopy at Boston, organ size at Philadelphia, and New York's "brain watchers." The inspiration kindled in research-minded men during their scientific "grand tours" abroad fascilitated the transition of neural sciences into the 20th century, epitomized by the elder Herrick, Clarence, who inserted neuroanatomy into psychology through the comparative study of nervous systems. The younger Herrick, Judson, was a pioneer in the study of animal forebrains and American neuroscience experienced a consolidation in the eastern seaboard and spread to the Midwest. Electrophysiology and its enablers, the oscilloscope, amplifiers, and stereotaxic devices, overtook morphology and the hegemony of European centers, as American studies on primate somatic localization and the excitable cortex multiplied. The turn to visceral and somatic physiology of the lower brain stem yielded theories of postural control, the "discovery" of neuroendocrinology, and hypophyseal control of homeostasis. Meanwhile, studies of peripheral and autonomic nervous systems in American centers were contributing to the seminal work in British laboratories on the action potential and neurotransmitters. After the usefulness of the EEG and evoked potentials as tools became clear, the spotlight moved to the central nervous system and reliable data were available from a single neuron deep in the brain of an awake, normally behaving laboratory (and later, human) animal. Soon after midcentury, the biological basis of mental health was recognized and the door opened to neuropharmacology and its sister, pyschoneuropharmacology. The American beginnings of tissue culture paved the road to tissue and stem-cell implants that revolutionized the therapeutic possibilities, as had chlorpromazine three decades previously.

Chapters 8 through 13 explore the behavioral sciences, originating on foreign ground in late 19th century with Wundt but quickly shifting to American soil as William James, Clarence Herrick, and many others returned from training abroad. It is interesting to note that two early-century American texts, Clarence Herrick's translation of Lotze and its neuroanatomical supplement and Margaret Washburn's textbook, *Animal Thinking*, were written for undergraduates, in contrast to the postdoctoral training available in European centers. Comparative psychology in America received impetus from detailed studies of beavers, nonhuman primates, and the Norway rat. The laws of effect and introspection were developed as well as many variants of the conditioned reflex with data and theoretical analyses and the prefrontal cortex of primates became a focus of interest with the establishment of primate colonies. The theory of emotions received contentious attention and was bolstered by elaboration of the "circuit of Papez" and subsequently the "triune brain." This combination of anatomic elements with behavioral components was spectacularly manifested by the Klüver-Bucy syndrome in which removal of specific brain regions produced functional reversals. Meanwhile, human sexuality cautiously tip-toed into a respectable science, and surveys were made of its physiology and biochemistry, producing the "Pill." At the top of the higher brain functions is the "root of consciousness," a topic inseparable from the broader "What is life?" A new light was cast on those questions by the consequences of a therapeutic procedure that effectively isolated the cerebral hemispheres. Subsequent detailed testing of sensory modalities and cognitive functions (memory and learning) revealed the exquisite precision of hemispheric functional asymmetry.

Chapters 14 and 15 concern the communicative sciences among nonhuman and human animals respectively. In each of those domains, only small packets of information are presented, a slighting necessitated by the vastness of the subject and the intricacies of their details; communication by means other than sound (with pheromones, pressure and touch, visual signals) has not been included. Chapter 14 summarizes the attempts to persuade nonhuman animals to communicate their inner lives in an understandable form, i.e., in human symbolism. Experiments and arguments about human-nonhuman communication dominate the foreground and obscure the basic questions on the nature of speech. The mammalian communicative strategies culminate in the second of the higher brain functions: human speech and its many variations. Chapter 15's perspectives include reading and the rude awakening to the extent of the American school population handicapped by various degrees of symbol reversals, or "developmental dyslexia." As the 20th century merged with the new millennium, a revival of research in human speech and animal vocalization becomes apparent. Finer analyses of animal communicative competence and human infants' developmental progress in information exchange position the communicative sciences firmly within the constellation overarched by neuroscience, where they embrace the neural and behavioral sciences and in turn are strengthened by them. Clues to human speech fundamentals are found in nonhuman communication and in the late 20th-century efforts are being made to organize such studies.

In a final, short postscript, a few major trends are drawn from this overview of a most complex, exquisitely fine-grained subject in an attempt to delineate a coherent profile. Both author and compiler set out to describe American directional signs along the path to a mature, international neuroscience that uses a multidisciplinary and interdisciplinary approach to how the brain works. The author's aim in attempting a history of neuroscience is perhaps best expressed by his own vivid analogy. In the preface to the neurophysiology section of the *Handbook of Physiology,* Magoun wrote: "Its goal, like that of chariot racing, has been to secure a balanced perch astride the rushing progress of investigative advance" (Magoun, 1959, p. xi). For those omitted individuals and topics we apologize and know that others will fill the lacunae. We will be pleased if our efforts afford new knowledge and appreciation to a few readers.

The Formalities

Acknowledgements

Warm gratitude is extended to four categories of assistance that endured throughout the preparatory years. First, the organized support systems that Magoun championed in all his projects: the Brain Research Institute and its successive directors, C.D. Clemente, A.B. Scheibel, and A.J. Tobin; the staff of the Special Collections of the Louise M. Darling Biomedical Library; and the Office of Instructional Development of the UCLA School of Medicine. Second, I am highly appreciative of the careful work of the unnamed reviewers who rendered essential constructive critiques and of the successive science editors of Swets and Zeitlinger: Martha Chorney, who recognized the manuscript as a contribution to the history of neuroscience, and Arnout Jacobs, the current, patient incumbent, who found the reviewers. Third, there was a succession of associates who have gone on to their own distinct careers but started with transcribing, typing, data entering, and library research: Joseph Dumit, Lilas Eastman, Caroline Pierce, Janesri Ranasinghe, Joan Thompson, and Zhan Xu. Special thanks go to Kalpana Shankar and Russell

Johnson without whose skillful help the manuscript would not have taken shape. And fourth are the volunteers who researched, interviewed, transcribed, or proofread, as the case may be: Marion Anker, Frances Brewer, Virginia Hansen, Shirley Lavenberg, Elizabeth Lomax, Ynez O'Neill, and Sally and Bennett Shaywitz.

Statement of Responsibilities

This editor claims responsibility for any lapses in paraphrasing, arranging, and referencing of text, and nonoriginal passages inserted into the Magoun manuscript. I join whole-heartedly the author's endeavor to produce an authoritative look at antecedent events that coalesced in a new angle of view in advancing knowledge of the to-and-fro relationship of brain and behavior.

Credits

During the early years, preparation of the manuscript was funded in part by grant LM 03069 from the National Library of Medicine/National Institutes of Health. The work of the later years had support from the Frances Margaret Keddie Trust of the Brain Research Institute. To both these sources we are most grateful.

The original source of the figures, and when appropriate, permission to reproduce, have been included in the captions. All figures not thus acknowledged are from the Neuroscience History Archives in association with the History and Special Collections Division, Louise M. Darling Biomedical Library, University of California, Los Angeles.

Louise Hanson Marshall, Ph.D.
Neuroscience History Archives
University of California, Los Angeles
October, 2001

Introduction: Early North American Science

During the century just completed, one of the most significant developments in the life and health sciences was the differentiation of an integrated neuroscience. The identity of this novel multidiscipline gradually took form in the convergence of three initially disparate yet interdependent subdisciplines: the neural, behavioral, and communicative sciences. Development of those earlier fields commenced principally in northern Europe, then spread westward, first to England, thence to the eastern seaboard of North America and from there successively to the Midwest and West; surprisingly, an unrecognized contribution from China merits mention. Translocation of centers of primacy in brain exploration was the result of three factors: First, the initial dependence on Europe for postdoctoral training in medicine and basic sciences was progressively severed as American centers became more numerous, productive, and scholarly. Second, increasing numbers of career scientists in the subdisciplines of anatomy and physiology chose to specialize in studies of the nervous system. And third, after a period of lackluster accomplishments, substantive discoveries began to emerge from the American laboratories.

Among the circumstances that contributed to the slow start of American neuroscience in the 20th century was the absence of an indigenous biomedical grounding from which to build. American medicine was in a "lamentable state" (Frank, Marshall, & Magoun, 1976, p. 553) and remained so until the Flexner Report of 1912 raised the standards of medical curricula and the requirements for granting of degrees. At the beginning of the 20th century there were few basic biomedical science departments or laboratories, and in those few (Harvard, Johns Hopkins) medically trained men were engrossed in studies of cardiovascular and respiratory systems and the biologists and zoologists were interested in primitive life forms. When attention finally shifted to the mammalian nervous system, several unproductive decades of consolidation with behavioral studies passed before a recognizable neuroscience developed. The discipline's relative newness, however, conferred a certain advantage: "The relative recency of developments enables personal recollection of many of the major investigators and programs, as well as their academic and governmental settings" (Ibid., p. 552).

Before the last third of the 19th century, the only significant American investigations related even remotely to the nervous system occurred as the result of accidents with explosives. In each case the victim luckily was in the hands of a physician blessed with scientific curiosity. In 1833, William Beaumont, a frontier doctor, conducted a series of experiments on the "fistulous" Alexis St. Martin who had sustained a shotgun blast into his abdomen. Among other trials, Beaumont reported "the establishment by direct observation of the profound influence on the secretion of gastric juice and on digestion of mental disturbances" (Beaumont, 1833, p. xxv). The second case was described by a Vermont physician, J.M. Harlow (1848), as the spectacular survival and change in character of a blasting foreman, Phineas Gage, after a destructive wound to his frontal brain. The report was of sufficient general medical interest to become the only American entry that year in a listing of the international current literature. The scientific void on the eve of the American Civil War was in striking contrast to the state of the neurological sciences in Europe and Great Britain where a flurry of laboratory studies on the anatomy and physiology of nervous systems was underway and the results published in long-established journals.

The first and second decades of the 20th century have been called the Progressive Era in U.S. history, and, as the medical historian John C. Burnham pointed out (1971, pp. 252–259), it was a time conducive to the belief that pure science could and would solve social problems. Advances in technology, the formation of philanthropic foundations, well defined professions, aspirations, and a positive public stance toward science in general finally nudged the inauspicious beginning into a period of rapid growth. The fledgling neuroscience was sparked in part by new discoveries about the nature of the nerve impulse, many from the newly awakened midwestern laboratories, and by technological advances generated by two World Wars.

Stemming from an expanded post-Second World War economy, new laboratories, centers, and buildings were dedicated to brain science. Training programs that crossed disciplinary boundaries were set in place and academic departments created. With the organization of an ever-increasing constituency, accelerating research on the brain has pushed outward the possibilities of discovery. The coining of a term – neuroscience – to describe the multidisciplinary experiments that had taken over the laboratory bench in the 1960s conveyed a sense of self to the new neuroscientists; the teaming of brain and behavior approaches to investigation of nervous systems at all levels of organization proved rewarding in terms of production of frontier research. Nourished by molecular biology and the fraternal twins of biochemistry and biophysics, neuroscience at the final decade of the 20th century became a mature and strong science devoted to exploration of new territory that may hold some answers to "What is Life?" (Schrödinger, 1944).

As in all scientific fields, the growth and spread of the subdisciplines destined to constitute neuroscience rested on the earlier achievements of outstanding individuals or groups of investigators and educators, now deceased, whose former distribution can be specified in time and location. Fortunately, the identity of the two aggregates of such figures in the neural and behavioral sciences has recently become available in published compendia. The first is *The Founders of Neurology* (1970), compiled and edited by Webb Haymaker and later with Francis Schiller, which includes 74 leading pioneers in neuroanatomy (39), neurophysiology (29), neurochemistry (3), and neuropharmacology (3). The general features of these data, and of similar information on individuals in the behavioral sciences, are depicted in Volume 2, and for the sake of moving on with history, only a summary is presented here.

Of the 74 selected founders of the neural sciences, seven were from the United States and all of these were among the most recently born, between 1870 and 1890, beginning their careers during the decades surrounding the turn of the century, and continuing their activities until 1940–1960.[1] This exceedingly limited representation from the United States (scarcely 10 percent of the group) bears witness to the relatively recent development of the neural sciences in North America and emphasizes the importance to American neuroscience of the contributions of founders in other countries through the nineteenth and early twentieth centuries. It also reveals a gap, for the largest portion of those foreign neural scientists had already made or were in the throes of making their contributions before the small number of American contributors were born or had begun their career pursuits. The preponderance of founders were from Germany, closely followed by France and Great Britain.

The relative maturity of the basic medical sciences and philosophy (the "new psychology") in Europe naturally attracted young Americans to undertake their advanced doctoral and postdoctoral study overseas, with gains in proficiency and research experience as well as in teaching for both sides. With the largest concentration of such figures in Germany, its leading universities became the mecca of those from many countries who could afford short or long sojourns abroad.

In the neural and behavioral sciences, the University of Leipzig and, more focally, the institutes of two of the professors of its remarkable faculty formed a major attraction. The first of these was the Institute of Physiology directed by Carl Ludwig. During his 30-year tenure (1865–1895), Ludwig

[1] In birth order, they were: Ross Harrison, Walter Cannon, George Coghill, Joseph Erlanger, Stephen Ranson, James Papez, and Herbert Gasser. They were joined by three émigrés from abroad: Brown-Séquard (France), Otto Loewi (Austria), and Dusser de Barenne from Holland.

attracted some 250 advanced students from all parts of the world. Associated with him in the life and health sciences were Wilhelm His in neuroembryology, Max von Frey in gross anatomy, and Paul Flechsig in neuropsychiatry. In the behavioral sciences at Leipzig, the Institute of Psychology, directed by Wilhelm Wundt, professor of philosophy, was a comparable magnet. During his 42 years there (1875–1917), Wundt attracted some 120 advanced students internationally and with his associates established the field of experimental or physiological psychology (see Chapter 8).

With the numbers showing so clearly that the U.S. endeavor was germinated abroad, each subsection of this history of American neuroscience pays introductory homage to its foreign roots that both nourished and bent the ground-breaking young discipline. Tracing the asymmetrical development of the branches of American neuroscience has meant following each theme in turn from its early decades to the end of the 20th century. Mindful that at any moment of history the developmental stage and speed of growth varies from branch to branch and new twigs are adding constantly to the arbor's complexity, the succeeding pages attempt to elucidate some of the antecedent events in one of the fascinating verdant domains of current knowledge.

Part 1
Neural Sciences

Chapter 1
Prodromal Nineteenth-Century Developments

SKULLDUGGERY ABROAD AND AT HOME

One of the major imports of European neuroscience to the American scene in the 19th century was a concept of brain organization that moved westward from France by way of Great Britain to the so-called "New World." Currently known as "phrenology," it gained entry to the United States through the nation's two centers of intelligentsia: Boston and Philadelphia. At Boston, John Collins Warren, Harvard's famous Professor of Surgery, on his visit to Paris in 1801–1802, had learned about a new system developed by Franz Josef Gall. On his return, Warren introduced the concept in lectures at Boston and Cambridge and, in 1820, made it the subject of an annual dissertation before the Massachusetts Medical Society. In Warren's view, phrenology led both to the development of knowledge of the anatomy and physiology of the nervous system, and in some measure, determined the intellectual power possessed by individuals. In Philadelphia, early interest in phrenology was introduced by Nicholas Biddle, a prominent banker who had attended Gall's lectures at Carlsruhe in 1806 and 1807 and, ever after, had kept a skull that Gall had personally marked for him.

Cranioscopy in France

Franz Josef Gall (1758–1828) was born in the German Rhineland, the son of a merchant of Italian extraction. The family name was Gallo – familiar to anyone acquainted with the California wine industry. He entered medical school at Strasbourg, where he contracted an unhappy marriage. Moving to Vienna, Gall obtained his medical diploma in 1785 and within a decade had developed such a large practice that he turned down an offer to become physician to the Emperor. He also began giving public courses on his doctrine of the brain, in which "[h]is main goal was … to establish an anatomy and physiology of the brain that would at the same time be a new psychology"

(Ackerknecht & Vallois, 1956, p. 8). For about 8 years he was closely assisted in anatomical studies by Johann Caspar[1] Spurzheim (1776–1832). The social climate, however, was contrarian and the Emperor interdicted Gall's courses because they advocated a doctrine of the brain that His Majesty and many of his subjects found materialistic, immoral, and antireligious. So it is not surprising that in 1805, Gall and Spurzheim left Vienna to lecture and collect materials at centers in Germany, Switzerland, Holland, and Denmark. After 2 years, they settled in Paris, and during 1810–1819, published in French a multivolume *Anatomy and Physiology of the Nervous System in General and of the Brain in Particular*, the first two volumes of which were prepared in collaboration. There was a falling out, however, and Spurzheim left for England and during the next decade developed the phrenological movement in that country.

Owesi Temkin (1947) has called attention to the important role that their conceptual organization of the nervous system played in the formation of their system, which Gall called "cranioscopy" and Spurzheim termed "phrenology". Steeped in comparative anatomy, they began with the primitive invertebrate pattern of disseminated ganglia, each of which operated autonomously in its focal region and function, but maintained some connections for interrelating with the others. In man, this pattern was replicated in the peripheral autonomic system where each of the ganglia was a focal neural center concerned with the independent and isolated action of its innervated viscus, though still maintaining some coordination with the other ganglia.

Gall was seeking the "psycho-physiology" of the brain, the neural substrates of the psychic faculties. He thought he had found 27 of them, 2 of which, both concerned with speech, he located in the frontal lobe, and associated them with some of the clinical signs of aphasia. Later, Paul Broca, who knew Gall's work, reported (1861) his belief that a center for language resided in the left inferior frontal convolution, an announcement that eventually brought praise to Broca. Gall's suggestion, on the other hand, was lost in the 25 other centers that have never been confirmed and for which he received only ridicule (Clarke & Jacyna, 1987). The fate of the organ of "amativeness" is described on p. 7.

Phrenology Imported

In the decades of the 1820s and 1830s, translations of the six-volume *Anatomie et physiologie* had become available in America and additional enthusiasm was aroused by the American tour of Spurzheim. Landing in New York on August 4, 1832, he went to New Haven, gilded the commencement exercises

[1] Also written as "Gaspard."

at Yale, and reached Boston on August 20. There a series of his public lectures at the Athenaeum were transferred to the Masonic Temple to accommodate the crowds. Additional courses at Harvard College, Cambridge and a series of presentations of the anatomy of the brain to the Medical Society, committed him to a strenuous schedule. Shuttling back and forth between town and gown, with additional visits to local schools and prisons, as well as exercises by the Phi Beta Kappa Society and attendance at Harvard's Commencement, Spurzheim's health broke under this exhausting time-table. Despite all possible care by his mentor, Dr. Warren, he died, only 3 months after arrival in this country.

His funeral was a great event: the bells of Boston tolled for this public calamity, and some 3,000 citizens, including members of the Boston Medical Society en masse, marched to the Old South Meeting House for the funeral ceremonies. Doctor Warren had previously conducted an autopsy on Spurzheim at Harvard Medical College, preceded by a lecture on his teaching. As medical mementoes, Spurzheim's skull and brain were preserved as he had requested and his other remains were buried in Boston's exclusive Mount Auburn Cemetery.

Flourens's Reputation in America

Although the evidence seems to show widespread popular acceptance of the tenets of phrenology, the harsh critique of Pierre Flourens was devastating to the theory of phrenology among the American scientific literati of the time, as it had been in Europe. The lofty reputation of the French physiologist had crossed the Atlantic some 20 years earlier when the Marquis de Lafayette sent a copy of the just-released *Recherches expérimentales*... to his friend, Thomas Jefferson. In grateful acknowledgment, Jefferson wrote: "I have never been more gratified by the reading of a book...." (quoted in Johnson, 1995, p. 3). He expounded on Flourens' ideas that matter might "exercise" thought, and that the organ of thought resides in the cerebrum, loss of which "removes all sense." Jefferson's "keen insight into the ramifications of the science he read and then proceeded to describe and advocate in an uncommonly literary voice" rendered his opinion unusually influential (Ibid., p. 2).

An American Version: Fowler and Wells in New York

Notwithstanding the intelligentsia's resistance to phrenology, the impact of Spurzheim's short visit on the popular growth of phrenology in America was spectacular.

> [I]n less than two months, [he] kindled the phrenological flame into a bonfire.... [His] new doctrine shedding light upon the dark recesses of the mind... seemed to have a special relevance for the new country.... [T]his

> doctrine that man's character could be read from the shape of the skull and improved by the exercise of various mental functions, introduced a simple and practical system of the mental philosophy of the mind .... The times were ripe and the soil receptive for the system.... (Stern, 1971, p. xiv).

There was a universal feeling of heads, lecturers went from town to town, explaining the new science, and giving public and private examinations. Periodicals were published to promulgate the new philosophy, and a library of phrenological books was issued rapidly. Many of the country's greatest men accepted its pronouncements: Horace Greeley affirmed them in his *Tribune*, Horace Mann and Samuel Gridley Howe[2] applied them to educational reform. Two of the great American authors of the century, Walt Whitman and Edgar Allen Poe, culled a new vocabulary from phrenology. But Spurzheim's short visit was only the match that lit the fire; fuel came from impressionable college students.

Among the many adherents who sprang up, at nearby Amherst College two undergraduates found phrenology a gripping subject. One was Henry Ward Beecher, a future Unitarian minister who delivered phrenology in his sermons. The other student was his classmate, Orson Squire Fowler, from upstate New York, whose family joined him in rising to prominence on Spurzheim's ideas. "Professor" Orson Fowler and his brother, Lorenz, undertook a lecture circuit and became practicing phrenologists. They examined the skull, its shape and size, and palpated the individual faculties on the cranium and numbered them from 1 to 27. The size of the organ of amativeness, for example, at the base of the skull, was designated by numbers: if small (1 to 3) it required cultivation; if large (7), it required control and repression and, if between the two, "vive la différence." The Fowlers' sister, Charlotte, joined the phrenological crusade, a Philadelphia office was established, and publication of the *American Phrenological Journal* was initiated. Charlotte married Samuel R. Wells and on moving to New York City in 1842, this "Headquarters of phrenology in America" occupied successive offices on Broadway as "Fowler and Wells." In 1866, the "Broadway Cabinet" became "The American Institute of Phrenology," incorporated under a New York State Charter, with a faculty of eight who offered six-week courses in phrenology, leading to a certificate or diploma for their graduates.

In their New York setting, the group continued to attract a wide range of clients. As a young man, G. Stanley Hall paid 5 dollars for an examination at the Fowler and Wells Emporium (Hall, G.S., 1923, p. 180) and Oliver Wendell

[2] Howe applied phrenology to the education of the blind, notably Laura Bridgman, at the Perkins Institution and Boston Asylum of the Blind, and would not allow a "woman servant to be engaged [there] until ... she had removed her bonnet, giving him an opportunity to form an estimate of her character" (Stern, 1971, p. 29).

Holmes, Sr. published a mirthful report in the *Atlantic Monthly* (vol. 4[22], August, 1859) of his experience at "Messrs. Bumpus and Crane's (Fowler and Wells) Physiological Emporium." Lorenz Fowler was Walt Whitman's analyst. Later, Lorenz established a British branch of the Phrenological Cabinet in London, where Mark Twain, remembering the itinerant phrenologists of his Hannibal, Missouri days, decided to make a "small test." It consisted of two visits to the phrenologist, the first under an assumed name and the second, his own. In the initial examination, Fowler

> found a *cavity*... where a bump would have been in anyone else's skull. That cavity represented the total absence of a sense of humor!.... After three months I went to him again, but under my own name this time. Once more he made a striking discovery – the cavity was gone, and in its place was... the loftiest bump of humor he had ever encountered in his life-long experience! (Neider, 1959, p. 66).

In his own seniority, Orson Fowler moved to Boston where for the remainder of his long life he "was driven to the point of eccentricity, by one consuming desire to provide sexual guidance to all who bared their heads... ." (adapted from Stern, 1971, p. 191). In the 1860s and 1870s, when Sigmund Freud was barely in his teens, this pre-Freudian Fowler devoted himself to advancing what he called "Right Sexuality," in which celibacy was regarded as an outrage to nature. In a tome of 940 pages, *Sexual Science... As Taught By Phrenology* (1870), Orson Fowler distilled his views on sex and sexuality, providing a work which, he claimed, 100,000 women rated "next to their bibles." Harking back to phrenology's founder, Gall, and his #1 faculty, Orson later republished another monograph: *Amativeness; Embracing the Evils and Remedies of Excessive and Perverted Sexuality, Including Warning and Advice to the Married and Single* (Fowler, 1848/1889). Jessie Allen Fowler, Lorenz's daughter and the last practicing member of the family dynasty, died in 1932. Almost to the day, a century earlier, Johann Spurzheim had delivered in Boston those lectures that had unfolded to this country a new philosophy of the mind. Phrenology had its origin with Gall in Vienna and a century later, a second Viennese physician founded another theory of mind called psychoanalysis.[3]

Evaluation

Gall and Freud invented, rather than discovered, a system or doctrine, the novelty of which attracted young disciples but led to their rejection by the

[3]This analogy had also occurred to Karl M. Dallenbach who in 1954 devoted his presidential address to the Southern Society for Philosophy and Psychology to "Phrenology versus Psychoanalysis," a title that may not have precisely conveyed his view that "the drama of psychoanalysis is similar enough to be a plagiarism of the one just told" (Dallenbach, 1955, p. 519).

academics; their activities were either ignored or roundly condemned by scientific groups and the medical profession. Not holding faculty status, the leaders instructed their followers privately at home or office, or by public lectures. Both groups established a specialty organization of their own: The Boston Phrenological Society was inaugurated in 1832, on the evening of Spurzheim's funeral; and the International Psychoanalytic Association was organized by Freud, with its first meeting in Vienna in 1910. Each was more like a religious cult than a scientific society and periodicals and publishing houses were established to promote its system. At about this stage of development, or before, the exuberant radicalism of the younger disciples began to exceed the Master's tolerance. Interpersonal relations were strained, schisms deepened and, finally, the radicals were expelled and expressed their independence by modifying the original doctrine or elaborating novel systems of their own. Further, the leaders were invited to visit and lecture in this country, Spurzheim in Boston in 1832, and Freud to help celebrate the 20th anniversary of Clark University, in Worcester, Massachusetts, in 1909; each planted the seeds of his doctrine in a foreign soil.

BRAIN WATCHERS

Concurrently with the often grotesque fascination in phrenology largely on the part of the American public, serious students of the brain were directing their attention to that organ's overall size and attempting to fashion correlations between its capabilities and the life-style of its departed owner.

Correlations of Human Brain Weights

Scientific interest in the human brain in America was kindled by John Call Dalton, Jr., M.D. (1825–1889), Professor of Physiology and later president of the College of Physicians and Surgeons in New York City. Considered by Weir Mitchell (1890, p. 179) to be America's first professional physiologist, earning his living by teaching and writing, Dalton's research interest lay in locating visceral centers on the cerebral cortex. During his retirement, Dalton assembled and published a three-volume *Topographical Anatomy of the Brain* (1885), consisting respectively of striking life-size photographs of the human brain in external views, serial horizontal (axial) sections, and serial vertical (coronal) sections.

Four years later, an American Anthropometric Society was established by leading anatomists and neurologists from Philadelphia and New York City. The chief objective was to preserve the brains of its members for scientific study in an effort to determine whether or not the brains of persons of superior intellectual capability could be distinguished from ordinary brains by special anatomic characteristics. That was a scientific, as well as a theoretical question in the attempt to show mankind as the culmination of evolutionary forces.

The first six brains acquired[4] were studied by Edward Anthony Spitzka (1907), then Professor of Anatomy at Jefferson Medical College, Philadelphia. He already had assembled data and constructed histograms of the heaviest brain weights recorded worldwide that showed a greater number of relatively heavier brains among the 100 leading "eminent men" compared with the distribution among 1334 "ordinary cases" along a scale of 1 kilogram.

In 1891 and 1892, Henry Herbert Donaldson (1857–1938), then at Clark University, had reported on the brain of the blind deaf-mute, Laura Dewey Bridgman, who had earlier been studied psychologically by G. Stanley Hall. The finding of modifications due to destruction of the sense organs at a specific age focused his research interest on quantitative studies of nervous system growth in frog, rat, and man (Conklin, 1939). Subsequently, at the Wistar Institute in 1928, Donaldson published "A study of the brains of three scholars: Granville Stanley Hall, Sir William Osler, and Edward Sylvester Morse." In that study, Donaldson collaborated with Myrtle M. Canavan, who three years earlier had published a small monograph on *Elmer Ernest Southard and his Parents. A Brain Study* (1925), and asked "Did the son inherit his parents' brain pattern? It does not appear that he did" (p. 29). Southard was Bullard Professor of Neuropathology at Harvard Medical School and director of the Boston Psychopathic Hospital, following its construction near the medical school in 1912. Before the First World War, he had tried to interest Harvard and the Rockefeller Foundation in establishing a research institute in that setting for the study of nervous and mental disorders. In 1919, Southard's proposal was supported by the State Board of Insanity, and he became founding director of the new Massachusetts State Psychopathic Institute. In less than a year, however, he died of pneumonia at the age of 43.

In Donaldson's study, after preservation in formaldehyde the three brains were each examined by the same methods and the results reported in the same manner. G. Stanley Hall, psychologist noted for having established the first experimental psychology laboratory in this country,[5] was at Clark University 1889 to 1919. William Osler was a celebrated Professor of Medicine at John Hopkins and Oxford, whose award-winning biography was written by Harvey

[4]In the order of their weights in grams, these were the brains of William Pepper (1593 g), Edward D. Cope (1545 g), Joseph Leidy (1545 g), Harrison Allen (1531 g), Philip Leidy (half-brother of Joseph) (1415 g), and Walt Whitman (1282 g). On the Spitzkas, father and son, and an interesting evaluation of their contribution to American neuroscience, see Haines, 1995.

[5]At Johns Hopkins University in 1881.

Cushing. Edward S. Morse, a protégé of Louis Agassiz at Harvard, had spent 1877 to 1880 as Professor of Zoology at the University of Tokyo and subsequently became director of the Peabody Museum at Salem, Massachusetts, and keeper of Japanese pottery at Boston's Museum of Fine Arts. Taking the brains of the three as a group, it appeared that they differed

> by being somewhat heavier, ... having a slightly greater extent of cortex, ... with deeper sulci, ... and also a greater extent of the cortex in the frontal and occipital areas .... From the data on hand, we conclude, therefore, that the scholars had brains that were somewhat better grown and therefore nourished than those with which they are here compared [the Southards'] and that this ... constitutes a fundamental condition favoring superior performance (Donaldson, 1928, pp. 83–84).

Thus, the concept of bigger is better was strengthened among those who were most blessed, the educated male.

Gender Differences

One of the best studies of differences in brain weights between men and women at the turn of the century was by biostatistician Raymond Pearl (1905) at John Hopkins. He analyzed four extensive North-European series, two from Germany and one each from Sweden and Czechoslovakia, reporting the brain weights of a total of 2,700 individuals, 1,747 male and 952 female, age 20 to 80 years, who were autopsied in large public hospitals. The overall average brain weight was 10 percent lighter for females than in males. Among other comparisons, Pearl matched selected groups of individuals of the same body stature and age[6] and concluded that they accounted, at most, for less than one-third of the observed gender difference in brain weights. To summarize, the average human male brain was about 140 grams heavier than the female brain, and about 100 grams of this was a true gender difference. Late 19th-century efforts to correlate brain weights with intellectual capacity have long since been abandoned.

Certain 19th-century studies had attributed the gender difference in brain weights, in part at least, to smaller frontal lobes in the female brain. In 1909, however, Franklin P. Mall, with the assistance of Florence Sabin, found in a large series of hemicerebra that the percentage of their frontal lobe fluctuated between 38 percent and 49 percent, with a mean of 43.5 percent. "[O]n an average," he wrote, "the percentage of the frontal lobe was the same in both sexes" (Mall, 1909, p. 15). And later: "It is often said that the brains of women are of a simpler type, but if their weight is not considered it is questionable

[6] Brain weight increases with body size and decreases with age.

whether a collection of brains could be assorted according to sex with any degree of certainty .... [The German neuroanatomist] Waldeyer states that to determine whether a brain came from a man or woman is much like identifying the sex of the individual from which a given skull came. I am not so optimistic," Mall added, "and would rather take my chances with the skull" (pp. 24–25).

These problems of comparison of structures and areas of the human brain have continued to the present day. In 1979, a two-day, NIH-sponsored workshop was held in Vienna on brain dissection with the goal of developing a consensus that would lead to worldwide standards for collection, processing, storage, and structural definitions of brain specimens. The workshop was organized because lack of standardization in dissection techniques – and even in the definition of key brain regions – had made reliable comparison of data from different laboratories difficult if not impossible; a summary was published later (Pope, 1983).

Oliver Wendell Holmes's Experiment

At the end of the 19th century, physiology as a discipline was firmly established in the United States (Fye, 1987), and studies of how the nervous system functions were joining the inquiries about brain size and its implications. One of the more ingenious experiments, appealing in its simplicity, was carried out by Oliver Wendell Holmes, Sr. (1809–1894), who measured the speed of the nerve impulse with no more apparatus than a professional chronometer (Holmes, 1870). He instituted experiments with two classes, each of 10 students, who joined hands in a circuit of about 66 feet; so that a hand-pressure transmitted ten times around the circle traversed 660 feet. The chronometer was a "horse-timer," marking quarter seconds. After some practice, the time of transmission ten times around, which had stood at 14 and 15 seconds, came down to 10 seconds, that is, one-tenth of a second for the passage through the nerves and brain of each individual.

With a baccalaureate and medical doctorate from Harvard (1836), Holmes was appointed the George Parkman Professor of Anatomy and Physiology at his Alma Mater in 1847, succeeding John Collins Warren, the man who had vigorously supported Spurzheim's phrenology (see p. 6) and had retired after 31 years.[7] Famous for his literary gifts, Professor Holmes's insights concerning the anatomy and physiology of the nervous system gained their most eloquent expression in his Phi Beta Kappa address, "Mechanism in Thought and Morals."

[7]The year before, in 1846, Dr. Warren had operated on two patients anesthetized with sulfuric ether by W.T.G. Morton, a dentist.

> The brain, he asserted, is an instrument, necessary ... to thought .... [A]ll thinkers ... should see that the study of the organ of thought, microscopically, chemically, experimentally, on the lower animals, in individuals and races, in health and in disease ... is just as necessary as if mind were known to be nothing more than a function of the brain ... as digestion is of the stomach ....

And he added:

> The central thinking organ is made up of a vast number of little starlike bodies embedded in fine granular matter, connected with each other by ray-like branches in the form of pellucid threads; the same which, wrapped in bundles, become nerves, – the telegraphic cords of the system. The brain proper is a double organ ... its two halves being connected by a strong transverse band, which unites them like the Siamese twins.

After comparing the brain with an English walnut, he continued:

> The brain must be fed, or it cannot work. Four great vessels flood every part of it with hot scarlet blood, which carries at once fire and fuel to each of its atoms. Stop this supply, and we drop senseless. Inhale a few whiffs of ether, and we cross over into the unknown world of death, with a return ticket .... Infuse a few drachmas of another fluid into the system, and, when it mounts from the stomach to the brain, the pessimist becomes an optimist; the despairing wretch finds a new heaven and a new earth, and laughs and weeps by turn in his brief ecstasy. But so long as a sound brain is supplied with fresh blood, it perceives, thinks, wills .... (Holmes, 1870, pp. 188–189).[8]

Jeffries Wyman, Comparative Anatomist

Brain studies at the Harvard University Medical School in Boston were the center of attention of Jeffries Wyman (1814–1874), who derived his baptismal name from the distinguished Dr. John Jeffries, with whom his father studied medicine. Moreover, in 1847, the son later succeeded John Collins Warren as Hersey Professor of Anatomy at Harvard, perhaps an indication of how small was the pool of eligible candidates and certainly an example of the "old boys syndrome."

Young Jeffries Wyman was prepared for college at Phillips Exeter Academy where he followed his bent for catching and dissecting bullfrogs and manifested a skill in rapidly drawing the specimens that he so closely observed. Wyman gained the degree of Doctor of Medicine from Harvard in 1837, and in the winter of 1840–1841, delivered a course of lectures at the Lowell Institute. With the proceeds, he spent a year in Paris, both in study of human anatomy and in

[8] As a culminating expression of his literary endowment, Holmes composed with the brain in mind, "The Living Temple." See Volume 2.

lecture rooms of the physiologists Flourens, Magendie, and Longet, as well as the masters of comparative anatomy in Paris and in London.

Studies of Gorilla Anatomy

After serving as chair of anatomy and physiology for the Medical Department of Hampden Sidney College at Richmond, Virginia, Wyman assured his position among eminent comparative anatomists in a communication to the Boston Society of Natural History (1847), in which the gorilla was first named and introduced to the scientific world. The distinctive structures of the animal were so thoroughly described that there has been little to add and nothing to correct. In this memoir, Professor Wyman stated that "The specific name, *gorilla*, has been adopted, a term used … in describing the 'wild men' found on the coast of Africa, probably one of the species of the Orang" (pp. 419–420).

Another subject in which Wyman and others of the period were interested was the homology between nervous systems. This was elaborated in his "Anatomy of the nervous system of *Rana pipiens*" (1853), in which his earlier penchant for dissecting bullfrogs finally found scientific expression. With admirable logic, he wondered about the function of the frog's undeveloped cerebellum and reasoned that contrary to the theory of Gall and Spurzheim the cerebellum in frog is not the seat of the sexual instinct.

Vertebrate and Invertebrate Differentiation

Turning to the relations between vertebrates and invertebrates, Wyman believed that in spite of frequent attempts to homologize the nervous system of vertebrates with articulates, there were serious contradictions. Those were, first, the brain and the spinal cord in vertebrates are enveloped in a common sheath, and are never in the same cavity with the viscera, even in the embryo [as they are in the articulates]. Second, in vertebrates the spinal cord is always dorsal and in the articulates the ganglionic chain is always ventral. Wyman participated also in discussion of the then popular "vertebrate theory of the skull" which rested chiefly on and, indeed, was a primary motivation for, the study of the number and differentiation of the cranial nerves. Wyman's conclusions were based on the cranial nerves of the frog, in which he identified (excluding the "sense nerves") only three pairs of cranial nerves, each of which had the characteristics of a spinal nerve, namely, motor and sensitive roots and a ganglion. This work, like that on the higher apes, placed Wyman's reputation on a level with that of the leading European comparative anatomists, in that it is quality, and not necessarily quantity, which makes a first-rate investigator. The high regard in which Wyman was held was attested by all his memorialists, among them S. Weir Mitchell and Mitchell's friend and fellow poet/scientist at Harvard, Oliver Wendell Holmes, Sr. Characteristic of the reverence for men of science that prevailed in that era, a prominent New

England poet, James Russell Lowell, wrote "To Jeffries Wyman" in which the emphasis was placed on sublime character rather than on the science (see Volume 2).

WHERE *IS* SEX IN THE BRAIN?

The essay, *Sex in Brain* (Gardener, 1893),[9] the title of which is paraphrased here, is perhaps one of the first American vindications of equality of the sexes. Widely read and translated, the book introduced a topic that reaped serious attention in the 19th century due in part to Franz Gall's effort to delineate, from his anatomical studies of the human brain, the foci concerned with "propensities, sentiments, and intellectual faculties." The location of each cortical brain area, he conceived, was demarcated on the overlying skull, and hence could be palpated on external examination.

Abroad: Gall, Flourens, and Leuret

The historians, Ackerknecht and Valois (1956, p. 23) pointed out that "For the search of the fundamental faculties and their organs, Gall recommended… [examining] people gifted with one talent." From his list of 27 fundamental faculties, fortunately we are concerned here with only #1, "the instinct of reproduction," to which Gall in 1838 devoted an entire publication, with the subtitle "The History of the Discovery that the Cerebellum is the Organ of the Instinct of Reproduction." That concept arose from observations first made on a young widow who experienced sexual ecstasy accompanied by tension and a feeling of heat in the nape of her neck. Later neurologists reported additional supportive cases and cranioscopy was renamed and spread as "phrenology," especially by his assistant, Johann Spurzheim, as described on pp. 4–5. Attacks were made on Gall's and Spurzheim's doctrines in general and, in particular, castigated their bizarre localization of sex in the cerebellum, the most sweeping of which were those of Flourens, who declared (1846) the cerebellum solely regulated the movements of locomotion.

Fifteen years after Gall's death, the eminent French physiologist, Pierre Flourens, published *Examin de la Phrénologie* (1843), a treatise that is generally assumed to have proven fatal to this pseudo science. In republications extending over 20 years, its substance was epitomized from the first:

> The entire doctrine of Gall is contained in two fundamental propositions, of which the first is, that understanding resides exclusively in the brain, and the

[9]This essay was first read before the International Council of Women in Washington, D.C. in 1888, where it made a great impression on the men and women who heard it and was well received by the newspapers.

second, that each particular faculty of this understanding is provided in the brain with an organ proper to itself.

Now, of these two propositions, there is certainly nothing new in the first one, and probably nothing true in the second one (Flourens, 1846, p. 18).

In conclusion, Flourens made a facetious, yet positive contribution to their program: "Gall and Spurzheim forgot to place curiosity among their primary faculties. They were wrong. But for the credulous curiosity of mankind, how could they have explained the success of their doctrine?" (Ibid., p. 128).

In his cool critique, Flourens twice made reference to François Leuret's "fine work on the convolutions of the brain", entitled *The Comparative Anatomy of the Nervous System* (Leuret & Gratiolet, 1839). In his elegant monograph, Leuret, dissatisfied with Gall's and Spurzheim's simplistic observations, concentrated his attack on the localization of sex in the cerebellum by using objective, quantitative data and initiated a study of the influence of sex on development of the cerebellum in horses. Comparing the relative weights of the cerebella and the brains of gelded horses, stallions, and mares,[10] he found that the stallions' cerebella were the least developed, those of the mares were next, and the geldings the highest developed, exactly the opposite of Gall's assertion.

American Studies

Burt Wilder at Cornell

A sequence of American studies on the influence of sexual activity on the human brain may begin with those at Cornell University, Ithaca, New York. When Cornell opened in 1868, Burt Green Wilder (1841 1925; Fig. 1.1) from Harvard was appointed Professor of Vertebrate Zoology and Neurology, a position he held for 42 years. Of New England stock, Wilder was born in Boston, gained his baccalaureate at Harvard's Lawrence Scientific School in 1862, and served as a surgeon-cadet in the Massachusetts Infantry during the Civil War. After the war he obtained an M.D. at Harvard Medical School and, for a time, collected brains for Louis Agassiz's Museum of Comparative Zoology at Cambridge, where his interest in the nervous system took root. On his retirement from Cornell in 1911, at the age of 70, his protégé, Simon H. Gage, wrote:

> In the almost universal interest in the nervous system at the present time few know that with a prophetic kind of insight Professor Wilder saw that in the progress of anatomy and physiology the nervous system was to play the most important part. In 1870–71 he gave lectures in comparative neurology, and in 1875 vertebrate neurology became an established course in the university .... (Gage, 1911, p. 360).

[10]For details of Leuret's experiment, see Marshall, L.H. and Magoun, 1998, pp. 183–185.

Fig. 1.1. The overriding interest of Burt G. Wilder, M.D., during his long tenure at Cornell was the comparative study of human brains as influenced by gender, education, and moral values. Image from Denny-Brown, 1975, p. 76.

His successor, James W. Papez, noted in a detailed report on Wilder's brain 4 years after his death:

> He was a man of strong convictions, and for whatever claimed his interest his enthusiasm knew no bounds. In his lectures on physiology and hygiene he advocated the exercise of all functions … with moderation …. In his speech and composition, he strove for clearness, consistency, correctness, conciseness, and completeness. These he called his five C's. He placed clearness first (Papez, 1929, p. 289).

Papez also had noted, in part:

> [Wilder's emphasis] was always to the actual specimen, or the dissection which showed the structures as they are in nature. So necessary a part of instruction did these studies … become that he opened a laboratory [and began] the collection of specimens [which] soon formed the nucleus of a zoological museum …. In 1910, on his retirement … the neurological division of the museum comprised some 1600 [brain] specimens preserved in alcohol (Ibid., pp. 287–288).

Forty percent of the brains were human, from embryo to adult. As had Jeffries Wyman earlier, Wilder and many of his contemporaries believed that only by studies of brains of people of known character could a "true" correlation between intellect and brain structure be discovered, and he was thus able to acquire the brains of a considerable number of distinguished individuals, such as the Cornell psychologist, Edward B. Titchener, economist Jeremiah W. Jenks, pathologist Theobald Smith, feminists Rosica Schwimmer and Alice Chenoweth Day, and others.

Wilder's influence was felt far beyond the classroom. A participant in the early meetings of the American Neurological Association (ANA), Charles K. Mills (1924/1975, pp. 34 passim) wrote: "My recollections of the work of Wilder are of special interest. At almost every meeting ... some valuable morphological and anatomical studies of the brain were presented by him .... [D]uring this first quarter of [ANA's] history ... Wilder began his propaganda for the revision of encephalic nomenclature. He was earnest and enthusiastic on this subject." In 1884, at Wilder's suggestion, the ANA's president appointed a Committee on Neuronymy (Wilder's term for "neural terms"), including H. H. Donaldson, C. K. Mills, E. C. Seguin, and E. C. Spitzka, with Wilder as chairman. Twelve years later, in 1896, the ANA unanimously adopted the committee's report, advocating some forty single-word terms that have since become enshrined in our textbooks and medical dictionaries and are still in use. Wilder's penchant for mononyms again came to the fore in his Cartwright Lectures of 1884, when he proposed "alinjected" as a single-word equivalent of "alcohol injection." In Mills's recollections of the early days of the ANA, he continued,

> We had a joke about this .... Members of the Association were sometimes accompanied [at meetings] by their wives, who naturally had a great admiration for high-sounding neurological phrases, and on these occasions the former would camouflage their invitations to take a drink by remarking, 'What about Wilder's alinjected ventrad?'

Fun aside, Wilder was highly regarded by his contemporary practicing neurologists and, in 1885, was the first of two nonprofessional, academic figures elected to presidency of the ANA; the other was Henry H. Donaldson, in 1937. As might have been anticipated, the title of Wilder's presidential address – dealing with neurological nomenclature – was, "Paronymy versus Heteronymy as Neuronymic Principles"!

Papez and His Measurements

To set an example for potential donors, Wilder had willed that his brain should be added to the Cornell Brain Association. After Wilder's demise, James W. Papez published a study of his brain (1929) that found a sclerotic kink in the left ophthalmic artery which had caused atrophy of the left olfactory and piriform lobes, together with the hippocampal region and the olfactory centers in general. Eight years later, Papez's major conceptual publication, "A Proposed Mechanism of Emotion" (1937) referred to many of the neural structures atrophied unilaterally in Wilder's own limbic system.

Two years before Papez provided his account of Wilder's brain, he had published (1927) the results of what is probably still the most comprehensive study of the brain of a normal, achieving, human female, that of Helen H. Gardener (1853–1925; Fig. 1.2), also known as Alice Chenoweth Day. After an earlier period in teaching and administration, during 1875 to 1900, this crusading woman had lived in New York City, writing magazine articles, stories, and books. About 1888, she became involved in the struggle for woman's suffrage. At the turn of the century, she moved to Washington and, as vice-president of the National Woman Suffrage Association, contributed importantly to securing the amendment to the Constitution giving women the franchise. In 1920, she was appointed by President Wilson as the first woman member of the U.S. Civil Service Commission.

One of Gardner's best-known books was *Sex in Brain*, reprinted by demand at home and translated into eight languages. A statement made by

Fig. 1.2. The detailed analysis of the brain of Helen H. Gardener fostered the early-century belief that unusual "endowment" of the cerebral associative areas was related to specific achievement. Photograph from Papez, 1927, p. 80, Figure 1.

a prominent physician, that the brain of a woman was inferior in 19 different ways, had challenged her interest.[11] She spent many months studying brain anatomy and showed that the female human brain was probably not different from that of man's. In this study she was supported by E.C. Spitzka, a prominent New York neurologist and serious student of the anatomy of the human brain. In 1897, she willed her brain to the Cornell Brain Association for subsequent study. When she died in 1925 at Walter Reed General Hospital, Washington, her brain was removed and preserved in 10 percent formaldehyde. On arrival at Ithaca, the weight of Gardener's brain, 1150 grams, was considered reasonably high for her body size, about 5 feet and weighing 106 pounds. Papez concluded that certain morphological measurements of "cerebral endowment" in the associative areas indicated potential achievement, and that Mrs. Gardener's brain showed three such areas of high development of cortex for "scholastic and literary work."

As background, Papez pointed out,

> Many ... have proved that the human brain embodies in its essential sensory and motor parts a primitive simian [ape-like] pattern .... in addition, there are certain cortical association areas small in the simia which have been so vastly enlarged in the human brain that they constitute features distinctive of man .... (Papez, 1927, p. 66).

Papez's emphasis on the associative areas as indicative of mental capacity was a wholesome steadying influence on the functional localization story. It erased the European preoccupation with cranial and cerebral signs of mental faculties and focused attention on the interconnectedness in brain function. Before moving into the 20th century, however, the interest in general comparisons of overall brain "power" continued around the socially sensitive influences of age, gender, and race. The excitement of finding some hallmark of exceptional intelligence in the human brain has persisted well into the 20th century. When suitably stained sections from a small block of Albert Einstein's preserved brain were examined microscopically, there was only a faint, statistically insignificant suggestion of an unusual finding in an association area (Diamond, Scheibel, Murphy, & Harvey, 1985). A contemporary view of human brain comparisons may serve as ballast to a stable viewpoint: "I am, somehow, less interested in the weight and convolutions of Einstein's brain than in the near certainty that people of equal talent have lived and died in cotton fields and sweatshops" (Gould, 1978–1979, p. 40).

[11] "I began ... by questioning the arguments and logic of the medical pseudo-scientists from their own basis of facts. I ended by questioning the facts themselves ... ." (Gardener, 1893, p. 100).

THE PHILADELPHIA STORY

Turning to the other American port of entry of biomedical science, Philadelphia's claims to distinction at the turn of the 20th century included being among the ten most populous cities in the world and first in the country in home ownership (Bissinger, 1997), perhaps the worst and best of labels, respectively. Such prosperity was not lost to neuroscience: among the many contributions of Philadelphia's citizens and institutions, the Turner's Lane Military Hospital (Fig. 1.3; later merged with the Hospital of the College of Physicians of Philadelphia), was created in 1862 to specialize in casualties with head wounds, under command of a Philadelphian who subsequently became the outstanding neurologist of his time, Silas Weir Mitchell. His associates at the hospital were W.W. Keen, a neurosurgeon and G.W. Morehouse. Philadelphia was also the site of the earliest American organized research unit for brain study: in 1906 the Department of Neurology of the Wistar Institute became the sole U.S. representative to the internationally organized Commission on Brain Sciences and in 1912 the Institute of Neurological Sciences was established at the University of Pennsylvania.

Fig. 1.3. Turner's Lane Military Hospital, Philadelphia, shown in 1862, was the first in the United States to specialize in diagnosis and treatment of nervous system disorders. From Burr, 1929, facing p. 105.

A Question of Race

At Philadelphia, the counterpart of Boston's Jeffries Wyman was Joseph Leidy (1823–1891). He was among a number of gentlemen attending the Congress of American Physicians and Surgeons in Washington D.C. in 1888 who organized the Association of American Anatomists, its object being the advancement of the anatomical sciences. The choice of president of the new society was Professor Leidy, "admittedly the most distinguished scientist of the country identified with anatomy" (Piersol, 1911, p. 71). An M.D. graduate of the University of Pennsylvania in 1844, Leidy was appointed Professor of Anatomy there in 1853, and held the post for almost 40 years. In 1871, with establishment of the university's "Biological School," he additionally became Professor of Zoology and Comparative Anatomy. One-third of his 599 publications were on paleontology of North American fauna and secured his international reputation. "The appearance of Darwin's *Origin of Species*," Leidy wrote, "was as though I had hitherto groped almost in darkness and that all of a sudden a meteor flashed upon the skies" (quoted in Ibid., p. 85). Leidy and Asa Gray, the Harvard botanist, were among the earliest American scientists to appreciate the significance of the theory of organic evolution and to explain its teaching. Leidy's studies in comparative anatomy included the human brain, and he requested from his friend, William Hammond, on medical duty with the Army in Kansas, a brain specimen of a full-blood Native American. After a frustrating search, Hammond invited him to visit Ft. Bragg, "to advance the interests of science" (Blustein, 1991, p. 36) but Leidy did not take up the invitation.

A contemporary Philadelphian, Weir Mitchell, wrote of Leidy with warm affection:

> He had no literary tastes ... but he was an unsurpassed observer and a joyous comrade, one of the Biological Club which arose out of the Academy of Natural Sciences .... We had science of the highest; good talk; never were gayer dinners! The club lasted thirty years with no death .... (quoted in Burr, 1929, pp. 136–137).

The founding president of the American Association of Anatomists, Leidy held office for 3 years. Though not strictly a neuroanatomist, Leidy's ranging interests extended into that field and his was the first member's brain to be acquired for study by the American Anthropometric Society mentioned earlier. As the anatomical association became a major organization advancing the development of the neural sciences in this country, through its first half-century a number of its presidents were drawn from that field: Burt G. Wilder, 1898; Charles Minot, 1903; Franklin P. Mall, 1906; Ross G. Harrison, 1912; G. Carl Huber, 1915; Henry H. Donaldson, 1916; George E. Coghill, 1934; and S. Walter Ranson, 1938. The lacuna in the post-First World War period extending through the 1920s and early 1930s may be explained in part by the contemporary rise of interest in endocrinological research.

Parker's Conclusions

As Burt Wilder was the leading protégé of Jeffries Wyman at Harvard to devote his career to study of the brain, so Anthony Jackson Parker (1855–1892) was the chief student of Leidy's at the University of Pennsylvania to spend his briefer career in investigation of simian and human brains. A contemporary wrote,

> He matriculated in the medical department ... in 1874 and while there ... Professor Leidy became greatly interested in [him] as did also Dr. Henry C. Chapman .... Under the stimulus of Leidy, [Parker] studied the protozoa ... and invertebrate forms, while he diligently dissected the great mass of vertebrate material [provided] by Professor Chapman. He was especially fortunate in having placed at his disposal a large number of brains of apes and monkeys. With the aid of the coroner he collected quite a number of negro brains (Spitzka, E.A., 1907, p. 265).

In his expanded graduation thesis on the cerebral convolutions (Parker, 1878), he noted several distinctions from Caucasian brains that led him to the conclusion that the simplicity of the Negro brain was more marked than in any normal "white" brain he had seen and that the former was more closely related to the ape type than was the latter.

Parker's conclusions were not without rebuttal, albeit extremely delayed. Thirty years later, in 1909, the Hopkins biologist, Franklin P. Mall, wrote that Parker's conclusions were "careless and Superficial" and could not be supported, any more than the claim that intelligence correlates with human brain complexity. Relatedly, in 1920, J.J. Keegan summarized his study of the brains of North American Indians, "[T]here is no discernible difference from the average brain of the white race" (Keegan, 1920, p. 60).[12]

S. Weir Mitchell's Long Arm

The influence of one individual on the early Philadelphia scene extends well into the 20th century. Like Oliver Wendell Holmes, Sr., Silas Weir Mitchell (1829–1914; Fig. 1.4) was another 19th-century figure who combined medical practice and research in the health sciences with substantial literary accomplishments. His father, John K. Mitchell, had been appointed to the chairs of chemistry at the Franklin Institute in 1833 and of medical practice at Jefferson Medical College in 1839, thereafter becoming one of the city's

[12] Attempts to "prove" a difference based on race have continued to this day, one of the most contentious being that of William Shockley (Pearson, 1992).

Fig. 1.4. The neurologic and psychiatric contributions of S. Weir Mitchell, M.D., LL.D., F.R.S., to early progress of American neuroscience were significant and extensive. Portrait painted by Frank Holl.

foremost physicians. As Weir later recalled in his unpublished autobiography, "One of my greatest joys was to go with my father to his chemical laboratory .... [It] was to me like a fairyland, and sometimes I was given chemicals to take home; the joy with which I saw the beautiful colors of the precipitates that I made I can never forget" (quoted in Burr, 1929, pp. 28–29).

Completing his undergraduate study at the University of Pennsylvania in 1847, he "decided to be a doctor, much to my [father's] disgust. He said ... you have brains enough but no industry" (Ibid., pp. 43–44). As a medical student at Jefferson, Weir worked in analytical chemistry during the spring and summer for 2 years and, for a few months, made up prescriptions in a drugstore. He gained his M.D. in 1850, presenting a thesis that related his early background in chemistry but, of all subjects, was on "The Intestinal Gases." While in medical school, he found anatomy boring but was "captured" by Dunlingson's lectures in physiology. He then pursued a year of advanced study in Europe, half of it in Paris, where he attended Claude Bernard's lectures in physiology, the subject that had attracted him in medical school. As A.J. Carlson (1938, p. 475) pointed out, "[I]n his formative years ... he came under the influence of the two ablest teachers of physiology of that period both in America [Dunlingson] and in Europe [Bernard]."

Snake Venoms and Toxins

On return to Philadelphia, for several years Mitchell lectured on physiology at an organization for extramural teaching. He wrote: "[T]his I was glad to do as I always had some leaning in that direction, and had it not been for the fact that I failed later in my efforts to become a professor of that branch, [physiology] would have been my life work" (quoted in Burr, 1929, p. 69). Weir carried home from France the concept of experiment. His first paper on uric acid crystals was the result of getting through work in the late afternoon and then remaining in the laboratory all evening, sometimes to one in the morning, a "slight meal" being brought in from a neighboring inn. Combining the two fields of his scientific interests – biochemistry and physiology – while building up a practice, he was able to gather enough data to publish "Researches Upon the Venom of the Rattlesnake," in the series *Smithsonian Contributions to Knowledge* in 1860, in which he addressed the unsolved question of the mechanism of its toxicity. Twenty years later, his attention was again called to the subject by an extraordinary incident, the sudden recall, on seeing a snake-like ravel from a rug, of the early work and the realization that the poison of serpents must be a double and not a single poison. Mitchell and a colleague, E.T. Reichert, collected serpents and set about splitting the poison into its components. In 1886, they published their classic paper, "Researches Upon the Venoms of Poisonous Serpents" (Mitchell & Reichert, 1886) again a contribution to the Smithsonian series.

An Assistant from Afar

In 1900, a young, penniless Japanese, Hideyo Noguchi (1876–1928), presented himself at Mitchell's door. He had come to Philadelphia in search of Simon Flexner, whom he had met in Japan, but Flexner, just appointed at the University of Pennsylvania, had not been able to help him. Mitchell hired him to take up the analysis of snake venoms using new biological approaches to toxicology and immunology. The following year, Noguchi was invited to demonstrate his work before the National Academy of Sciences, to which Mitchell had been elected in 1865, just 2 years after the academy's founding. Later still, in further support of his protégé, Mitchell obtained funds from the Carnegie Institution, of which he was a trustee, for publication of a handsome volume of Noguchi's entire project, *Snake Venoms* (1907). Also about then Mitchell persuaded the Carnegie Foundation trustees to revive publication of *Index Medicus*, the accumulative catalogue of current medical literature that had been suspended for lack of government funds.

When the Rockefeller Institute for Medical Research was established in New York City with Simon Flexner as director, Noguchi was brought into its program. Shifting his research to study of the spirochete of syphilis, in 1913 Noguchi identified this organism in the diseased brains of general paretics. This was the first substantial evidence of a neural substrate for a psychotic syndrome and was widely recognized as the most outstanding research accomplishment in the initial century of American psychiatry (Whitehorn, 1944).

Whatever factors prevented Weir Mitchell's appointment to the chair of physiology either at Jefferson or the University of Philadelphia, his national stature in this field was of the highest. In addition to his work on snake venoms, his physiological experiments on the cerebellum (1869) provided a theory of that organ's function as augmenting the cerebrospinal motor system. He originated the idea of forming the American Physiological Society and with Newell Martin of Hopkins and Henry P. Bowditch of Harvard issued invitations to the organizational meeting in New York City, 1887, where his influence and prestige helped promote membership in the new society.

Weir Mitchell's rising reputation as a neurologist was enhanced in part by his publication of *Injuries of Nerves and Their Consequences* (1872) based on studies undertaken at Turner's Lane Hospital at Philadelphia during the Civil War (see p. 20). The treatise included detailed descriptions and suggestions for treatment of a wealth of neurological and psychic phenomena, and he became equally renowned as an alienist.[13] At the initial meeting of the American Neurological Association in New York in 1875, Weir Mitchell was unanimously elected president, but declined. Thirty-four years later, at the 1909 meeting in Philadelphia, at the age of 80, he was again easily elected and happy to accept.

In the 19th and early 20th century, the differentiation between neurology and psychiatry was by no means pronounced and combined specialization in neuropsychiatry was commonplace. In the 1870s, therefore, it was not unusual for Mitchell to extend his earlier preoccupation with neurology to the development of a therapeutic regimen for overcoming neurasthenic states, especially in women. This culminated in his widely publicized "Rest Cure," in which the first step was to provide "the firm kindness of a well-trained hired nurse" (Earnest, 1950, p. 83). Next was rest in bed for a month, during which the well-fed patient remained motionless, but was massaged and her muscles contracted by electrical stimulation. "At the end of this period, the subject was usually eager to accept the order to get up and take some exercise. There is, however, the famous anecdote of a woman who refused to follow Dr. Mitchell's edict. After [ineffectual] persuasion, he threatened, "If you are not out of bed in five minutes – I'll get into it with you." He removed his coat, then his vest, but the patient did not move. When he started to take off his trousers, a very angry woman leaped out of bed" (Ibid., p. 83).

[13] Most recently, Richard D. Walter has published a superlative "medical biography," *S. Weir Mitchell – Neurologist* (1970) in which this material is presented in full elaboration.

Another such incident was recounted by Webb Haymaker (1970b, p. 482):

> Some of Mitchell's ideas strike knowledgeable psychiatrists today as very sound. But sometimes he resorted to strange diagnostic measures in functional illnesses .... As consultant to a lady considered sick unto death, he once sent all assistants and attendants out of the room, then soon emerged himself. Asked whether she had any chance for survival, he answered 'Yes, she will be running out of the door inside of two minutes; I set her sheet on fire. A case of hysteria.' His prediction proved correct.

Psychiatry Assailed

As a sequel to these developments, in 1894, when the psychiatrists were planning a special program to celebrate the 50th annual meeting of their American Medico-Psychological Association (later the American Psychiatric Association), it was natural for them to invite Weir Mitchell to address them. He promptly declined, however on the ground that he could speak only critically. When told they still desired him to speak, he reflected that it would be worth his while to address them and agreed. Commenting on the isolation of psychiatry from the rest of medicine, Mitchell said,

> You were the first of the specialists.... You soon began to live apart, and you still do so. Your hospitals are not our hospitals; your ways are not our ways.... Where, we ask, are your annual reports of scientific study, of the psychology and pathology of your patients? (Mitchell, 1894, pp. 414; 422).

A rebuttal was shortly provided by Dr. Walter Channing (1894, p. 171), who deftly pointed out:

> Fifty years ago, when ... the Association of Medical Superintendents [was formed], it was not a question of knee-jerk, or ankle-clonus, or reaction-time which confronted them, but how to house the then already large numbers of insane, who...were suffering the tortures of the damned in alms houses and in their own homes; and from that day to this the pressure has never relaxed for more accommodations.

It may be noted that 50 years after the fact, some of the contributions to the American Psychiatric Association centenary volume, *One Hundred Years of American Psychiatry*, were still defensive against Mitchell's critique of their profession. The association had also learned its lesson – it took no chances and invited no external critic to deflate the aura of accomplishment and celebration.[14] Significantly, John C.Whitehorn, head of the department of psychiatry

[14] For a brief discussion of the ultimate reaction to Mitchell's and others' criticism and the accompanying change in psychiatric theory of mental illness, see Grob, 1998, p. 200.

and Phipps Clinic at John Hopkins School of Medicine, in his centenary article, "A Century of Psychiatric Research in America," pointed out that Weir Mitchell's 1894 blistering chastisement did not take account of the actual state of scientific investigation in psychiatry in the eighteen-nineties. He called attention to the distinctive American contributions in psychobiological understanding and study of "personal relationships" and then sketched briefly his ideal hospital, designed and organized primarily to stimulate and encourage scientific investigation and progress. Inconsistently, Whitehorn then described the subsequent development of "Intensive laboratory research in the basic medical sciences of pathology, physiology, and biochemistry – as related to psychiatry" – at the McLean Hospital under Edward Cowles at Waverly, Massachusetts; at the Boston Psychopathic Hospital under E.E. Southard; at St. Elizabeths Hospital, Washington, under William Alison White; and at Kankakee, Illinois, the Worcester State Hospital, Massachusetts, the New York State Pathological Institute, and the Phipps Psychiatric Clinic at Johns Hopkins, several of which had been under the sequential direction of Adolf Meyer – the Johnny Appleseed who dropped research laboratories into psychiatric hospitals, as he made his way across the country and up the professional ladder, between the 1890s and the First World War.

Percival Bailey Enters the Fray

The American Psychiatric Association did not long escape a contemporary chastisement, however. At its annual meeting in 1956, held in Chicago, Percival Bailey, then director of the Illinois State Psychopathic Institute in that city, was invited to give the Academic Lecture. As a member of the U.S. National Academy of Sciences, Percy (as he was called by his friends) had earlier been selected to prepare a biographical memoir of Silas Weir Mitchell, in which, strangely, there was no mention of Mitchell's address to the association in 1894. There must have been some unconscious association, however, for in 1956, though Mitchell's body was amouldering in the grave, his soul was both marching and egging Percival Bailey on. Bailey's Academic Lecture on "The Great Psychiatric Revolution" was a bloody assault on the soft underbelly of American psychiatry of the period and its preoccupation with Sigmund Freud and psychoanalysis.

Percy's subsequent elaboration of the subject, *Sigmund the Unserene, A Tragedy in Three Acts* (Bailey, 1965), extended the mayhem to a full-scale draw and quartering of "Sir Sigi." Presented as the Norman Wait Harris Lectures at Northwestern University, they provided a critical analysis of the three stages of Freud's intellectual career: the first concerned with neural science, the second with metapsychology (speculation about the origin, structure, and function of the mind), and the third with cultural, historical, and mystical problems – for which, presumably, Freud was awarded (as Bailey pointed out) the Goethe

Prize for literature, rather than the Nobel Prize. In a foreword, Percy's friend, Roy R. Grinker, Sr., wrote of Bailey's contributions to several disciplines and longstanding interest in psychology and psychiatry, and asked: "Why should a scientist, who has ... made order out of the chaos of heterogeneous gliomas, ..., epilepsy, and [neurosurgical relief of anxiety]-be so interested in Freud and psychoanalysis?" (Bailey, 1965, p. vi).

Bailey provided an explanation in his concluding lecture. During the long period, 1918 to 1952, as a neurosurgeon, he stated,

> I was...very busy with somatic problems of the nervous system .... Then, in 1952, developing cataracts made it impossible for me to work longer with microscope and biopsy. I accepted a position in the field of psychiatry. It now became necessary for me to familiarize myself with the developments in psychiatry ... especially with psychoanalysis which was usurping an ever increasing dominance of the field .... It occurred to me to wonder ... whether there might not be embedded somewhere in the chaotic confused theory of psychoanalysis an adequate Way of Life based on science rather than faith. I never found it (Ibid., pp. 104; 105).

The parallels between Mitchell's and Bailey's attacks go beyond the similarity of the charges – lack of experimentation and innovative ideas. Both men came from a laboratory background with experience at the bench, Mitchell in chemistry and Bailey in neuroanatomy and pathology. An interesting study could perhaps trace the origins of the movement in the 1930s toward the development of the Ph.D./M.D. program in American medical schools back to those blasts against the establishment.

William Hammond, Father of American Neurology

Turning back to the early figures, in 1888, in the opinion of a colleague: "Professor Hammond ... has perhaps been among the busiest of active practitioners in America ... and [his] investigations and records have established him as the father of, and *the* authority in American neurology" (letter quoted in Blustein, 1991, p. 94). Again, fortunate contacts with Philadelphians were to lead to a distinguished career and service to fellow humans.

Naturalist

The son of a physician, William Alexander Hammond (1828–1900), was born in Annapolis, Maryland. He studied medicine at New York University and gained his M.D. at the age of 20. The following year he entered the U.S. Army as a surgeon and served at frontier stations in New Mexico and then in Kansas, at Fort Riley. Combining military service with natural inquisitiveness and ability, he was able to engage in scientific work in collecting animal specimens, mainly for the Academy of Natural Sciences, under Joseph Leidy, in Philadelphia. He virtually deluged the Academy with specimens of insects, mammals, reptiles, fossils, and other curiosities. The largest consignment included 480 examples of 40 different species of reptiles, in 1 year. Hammond himself studied metabolic processes in the living black snakes that were easily available to him. After leaving the military, by 1860 he had accepted the chairs of anatomy and physiology at the University of Maryland, briefly established a practice of medicine in Baltimore, and had spent some time with Weir Mitchell as well. With their common interest in toxins and snake venoms and similarity in age, it is not surprising that they collaborated scientifically, alternating as first author in two papers presented to the Academy of Natural Sciences of Philadelphia (1859, 1860) and that they exchanged mutual respect and friendship.

Organizer
At the outbreak of the Civil War, Hammond returned to the Army and in March, 1862 held the position of Inspector of Hospitals. By the end of April, he had been promoted by President Lincoln to head the medical department of the Army, as Surgeon General. Soon after this, General Hammond appointed Weir Mitchell as contract surgeon to the 400-bed hospital on Turner's Lane, Philadelphia, and he additionally arranged that all patients with gun-shot wounds or other neural maladies were to be first sent to Turner's Lane. As McHenry commented (1969, pp. 326–327): "American neurology was cradled and developed in the Army during the Civil War period, largely under the impetus of Hammond and Mitchell."

After the war, Hammond established himself among leaders of New York medicine and limited his practice principally to clinical neurology. He reached the summit of his career by creating an institution in which neurology could be taught, discussed, and perpetuated. He told the faculty and students of this innovative New York Post-Graduate Medical School that

> I am one of the founders of this school. I shall always regard that fact as the most honorable of all the events of my professional life .... [The school is not intended] to increase the number of Doctors but to add to the qualifications of those already graduated .... [M]edical [S]cience ... is progressive and it is difficult for a practitioner to keep pace (quoted in Blustein, 1991, pp. 100; 101).

It was to provide for the wants of such men that the school was organized and it became an instant success. By 1890, some 850 students were attending the Post-Graduate and its rival, the New York Polyclinic, and a dozen similar institutions had been organized across the United States.

Hammond's endeavor to combine neurology and psychology in medical teaching was tested at Bellevue's Medical College, where in 1872 he gave a course on diseases of the mind and nervous system, which was promptly dropped when he resigned a year later and "Diseases of the nervous system" was separated from "Psychological medicine and medical jurisprudence" (Ibid.). But in 1874, he became Professor of Diseases of the Mind and Nervous System at his Alma Mater, University Medical College, from which he had graduated in 1848. That crisis of ambiguity was repeated 25 years later in Clarence L. Herrick's failed program at President Harper's University of Chicago.

By the end of 1874, Hammond had taken steps to launch yet another organization, this one, national in scope – the American Neurological Association, which held its first annual meeting in June, 1875. Thereafter, he devoted much attention to the organizational work of the association and for a decade contributed one or more papers to each annual meeting and participated actively in its discussions. When its organ, the *Journal of Nervous and Mental Disease* moved from Chicago to New York in 1876, Hammond became an associate editor. After the founding of the American Neurological Association, the growth of neurology in America was rapid (see Volume 2).

It was rather ironic that Hammond, whose first interests and researches were carried out with difficulty and in comparative isolation at military outposts in Indian Territory, should ultimately have achieved prominence in neurology as a clinician and organizer. He "was an uncommonly tall and large man with a voice so powerful that it could be heard up-wind in a hurricane. He had a penchant for theatrical action, which he exercised as playwright, lecturer and novelist" (Haymaker, 1970a, p. 448). He carried out original physiological experiments on viper poison and wrote on the cerebellum, but perhaps his most important contribution was publication (1871) of the first American textbook of neurology, based on his own clinical experience and notes from Charcot's lectures.

The Philadelphia story, set here in the late 19th century, exerted an influence over 20th-century neuroscience in many directions. Not only did that city's scientific institutions promote inquiries into brain anatomy and function, but the caliber of its established physicians and their social milieu affected the decisions of younger, less secure practitioners. For example, in 1920, Samuel D. Ingham, approaching the end of an appointment in neurology at the hospital of the University of Pennsylvania, considered his options for a successful practice in Philadelphia. In the words of his son, "He found himself with several eminent, established, and competent neurologists who were too young .... He thought he would be a second-rate physician no matter how well he did for a good many years if he stayed" (Ingham, 1981). He headed West, and in so doing, helped establish a school of neurology in Southern California (see Chapter 15).

Chapter 2
Passage to the Twentieth Century

AN AMERICAN SCHOOL OF COMPARATIVE NEUROLOGY

In the mid-19th century, a time when scientific inquiry engaged some of the best minds in Europe, the medical field was producing increasing numbers of trained graduates who were attracted to research and teaching rather than to solely clinical pursuits. The study of nervous system morphology appealed to many of them, especially in Germany, with progressively improved microscopes, new fixative agents, sectioning and staining methods, and a relentless curiosity about how the brain functions. The dominance of Germany in those developments established specialized postdoctoral training at one or more European universities as a prerequisite to a successful career in brain or behavioral sciences among the freshly professionalized ranks in America. Exposure to the laboratories, clinics, and lectures of preeminent figures was highly influential in shaping the direction of the neurological sciences in America.

The European Precedent: Ludwig Edinger's Memoirs

The diligently kept memoirs of Ludwig Edinger (1855–1918), supplemented later by his physician-daughter, Tilly, afford an unusual opportunity to reconstruct the career and impressions of one of the most important of those figures.[1] Born into a well-to-do Jewish family in Worms am Rhein, at an early age Edinger took refuge from the formal and spiritless teaching at the local gymnasium in the gift of a small microscope; as he peered into the world of nature, his career in biology was under way.

[1]Much of this section is based on the translation by Frances Keddie O'Malley of a condensation of the 361-page Memoir, prepared to celebrate the 100th anniversary of Edinger's birth and the 50th anniversary of the Edinger Institute (Krücke et al., 1959).

Education

At the University of Heidelberg, Edinger was exposed to the ideas of Karl Gegenbaur, Ernst Haeckel, and to Willy Kühne's view of the cell as a living structure. In 1874, he moved to Strasbourg, where the pathologist, Friedrich Recklinghausen, recognized as the most gifted student of Virchow, was among the predominantly German faculty. Edinger was still pursuing his early bent for comparative histology when he met the anatomist, Wilhelm Waldeyer (1836–1921) and completed his first neurological study, on the innervation of the skin of a snail (1877). Later, he studied the spinal cord of a man with rudimentary arms, found that certain cell groups were missing, and concluded that they must be the source of the forearm nerves (Edinger, 1882). Edinger's next appointment was to a clinic in Giessen, where he and the director, Franz Riegel, attempted to reproduce an experimental model of asthma by severing the vagus nerve and stimulating the stump to contract the diaphragm. Again antisemitism jeopardized his advancement and Edinger spent a period visiting clinics in Berlin and Leipzig and meeting many of the great names in German medical and basic science.

Association with Carl Weigert

The most important and lasting influence on him was Carl Weigert, who taught Edinger the chrome-haematoxylin and carmine stain for myelin and other tissues. They were thrown together again in Frankfurt in 1883, where Edinger had moved at his mother's behest after residence in Paris. In Frankfurt, Edinger announced himself as a physician and neurologist, among the first in Germany to do this. In addition to a flourishing practice, Edinger followed Flechsig's footsteps and with the Weigert stain mapped fiber-tracts in the developing fetal human brain. With specimens from a nearby hospital, he found the central course of the sensory tracts, which on the authority of Meynert terminated in the cerebellum. In the 7-month fetus, in which the tracts were distinct and dyeable, he found that after crossing in the medulla oblongata they ended in the thalamus. The memoir vividly expresses Edinger's surprise and delight when Meynert, having found that the much younger man was right, visited him unannounced, probably in 1887, and bid for his friendship.

The Edinger Institute

Edinger's close association with Weigert commenced when the latter arrived in Frankfurt as head of anatomy and prosector for the state hospital, thereby greatly improving the scientific life of Frankfurt. Edinger secured space in a laboratory of a private institute built on state ground which he shared with the foreign students: Americans, English, French, and the Scandinavians with whom Weigert could speak their language. The space was so tight that someone had to stand up whenever any one wanted to leave the room. Edinger wrote of Weigert's almost daily morning visits to the laboratory and later for tea in his apartment, where everything was discussed and ideas were freely exchanged.

In the fall of 1883, the medical society at Frankfurt asked the newly arrived neurologist to give a series of lectures on the central nervous system. He was bold enough to comply and even to have the lectures published (1885), a successful venture that made him world famous. In 1886, Edinger changed the direction of his research and began to study the lower vertebrates, their behavior and the structure of their brains. He realized that in its basic functions the brain was not constructed any more simply in the lower vertebrates than in the higher ones, even man. The differences in the behavior of the higher vertebrates had, then, to be related to the increasing elaboration of the cortex, the beginnings of which could be found first in the lizard; the brain, perhaps, was composed of two parts, one for the simple functions and another (the neopallium) for complicated functions. The seventh edition of his lecture series was split into two volumes, on man and mammals (1904) and on the anterior and midbrain of fish, amphibians, and reptiles (1908). A later section with Wallenberg and Gordon Holmes included the brain of birds. Drawings in color by Edinger himself were beautifully reproduced. In 1902, Edinger at last had his own room in the institute, he was 47 years old, and had an assistant and a few students. It was here that the Neurological Institute originated. In 1914, the year the University of Frankfurt was founded, Edinger was appointed Professor of Neurology and his institute was made a part of the university but was supported by private funds from the Ludwig Edinger Foundation. Its annual reports listed a flood of foreign workers who came for postdoctoral training, including F. Carpenter, G.L. Streeter, and R.M. Strong from Columbia University, and many who later became prominent in Germany.

Contributions

Among Edinger's achievements was his demonstration of fibers connecting the olfactory apparatus with the cerebrum in reptiles, the first sensory projection to a higher level to be described. Through a combination of anatomical research and behavioral studies, Edinger attempted to establish an approach to the "true" psychology based on the coordination of mental faculties with the stage of development of the cerebral cortex starting with primitive species, thus reversing the usual pattern of working downward from man. He differentiated the vertebrate brain into the phylogenetically relative stable paleoencephalon and paleocerebellum from the neoencephalon and neocerebellum with layers subject to, in some species, spectacular changes in progressive position on the phylogenetic scale. Edinger's ability to interconnect structure, function, and clinical experience assured the great success of his textbook, but his pioneering contributions to comparative neurology were so predominantly anatomical that his parallel interest in their behavioral implications often has not been noted.

The American Founder: Clarence Luther Herrick

In the latter 19th century, the interests of American colleges and universities in natural history, biology, and zoology – then predominantly morphological in their orientation – began to elide with those of the schools of medicine which had begun to include a substantial representation of the health sciences in their curricula, similarly with a major anatomical emphasis. In this period, the term "neurology" had a broader, more generic reference to any aspect of the nervous system than does its current identity with a specialty of clinical medicine. Studies of the nervous system of animals other than man became designated, however, as comparative neurology; which might include studies of anatomic, physiologic, or behavioral nature, although the study of animal behavior was also called comparative psychology.

Out of this spectrum there gradually differentiated an "American School of Comparative Neurology," interdisciplinary in its range of interests but principally neuroanatomical in its initial focus. The developing interests of neurobiology – the structure and function of the primitive nervous system of invertebrates – were extended next to studies of the peripheral nervous systems, particularly the cranial nerves in lower vertebrates, chiefly fish and amphibians. As in the pattern initiated abroad by Edinger, these were followed by analyses of their central connections with the brain stem and spinal cord, as the first step in comprehending the organization and function of the elaborate brain of man. In this latter goal, neurobiologists were joined by comparative neurologists in the fields of anatomy, physiology, biochemistry, and pharmacology, in the improving American schools of medicine. The founding pioneer in those developments was an extraordinarily productive midwesterner denied the chance to exercise his full potential.

Early Years

As first son of a Baptist minister, Clarence Luther Herrick (1858–1904) grew up on the family farm on the outskirts of Minneapolis, Minnesota, then undeveloped enough to permit his interest in nature to develop freely. Entering high school at the age of 16, he and three friends formed a Young Naturalists Society which assembled collections and prepared reports of their studies in zoology, geology, and paleontology. With the gift of a simple microscope from his father, Clarence extended his interest to animalcules in pond water and to chick embryology. A year later he transferred to the preparatory department of the University of Minnesota as a "sub-freshman" for a 6-year course of study. He was soon appointed as assistant to Professor N.H. Winchell, director of the Geological and Natural History Survey of Minnesota and, on the award of his bachelor of science degree in 1880, joined the survey full-time, while completing his master's degree.

Clarence Luther (C.L.) saved much of the money earned during his first year of employment to spend 1881–1882 in Germany, mostly at Leipzig, studying zoology with Rudolf Leuckart. He also attended lectures in psychology and, on a short visit to Berlin, lectures by Helmholtz. In a letter home in February, 1882, C.L. wrote: "It is now my plan to return in early summer.... It seems entirely probable that by waiting until Fall I could get the 'Phool D,' but it would oblige me to take studies I could pursue to greater advantage at home ...." (quoted in Windle, 1979, p. 36).[2] In the same letter, Clarence added,

> I have also for a short time daily met the daughter of Sir Wm. Hamilton [a Scottish scholar of German philosophy] whose rich fund of varied learning has given me a favorable idea of the capacity of the English gentlewoman. Miss Hamilton was engaged in the translation into English of one of the works ... of Professor Lotze ....

Lotze's Influence

A simple paperback volume had appeared in Germany in 1881 and C.L. had purchased a copy and, emulating Miss Hamilton, undertook its translation. The book was published in English at his expense in 1885, with the title *Outlines of Psychology. Dictations from Lectures by Herman Lotze* with the addition of a chapter of his own on "The Anatomy of the Brain." In the preface, C.L. wrote that the volume would prove convenient for ordinary college use for its physiology of the nervous system and as an introductory book of psychology where the psychological side does not receive special attention in the Philosophical Department. Without doubt, Clarence's discovery and translation of Lotze's writings was his most significant accomplishment during his year in Germany, and with his added section on anatomy of the brain, in 1884 he became one of the first Americans to introduce an integration of neural and behavioral sciences in higher education in this country (see Windle, 1979).[3]

Returning home to a position as State Mammologist in the Minnesota Survey, in 1884, Clarence Herrick next accepted a temporary instructorship at Denison University in Granville, Ohio and a year later was appointed Professor of Geology and Natural History there. Denison, founded by Baptist

[2] This and subsequent quotations of C.L. Herrick's correspondence are from excerpts in the publication cited and are based on the extensive Charles Judson Herrick Papers at the Kenneth Spencer Library, University of Kansas, Lawrence.

[3] An instance of the prevalent educational attitudes of that period is provided by the first effort of C. Judson Herrick, as an instructor at a small midwestern college, to implement his older brother's goals, described on p. 61.

ministers as a literary and theological institute in 1831, in response to the impact of science on American institutions of learning at that time, had two full professorships in science and conferred the bachelor of science degree after a 4-year course of study. As a portent of things to come, one of C.L.'s first steps was to found and edit a scientific periodical for the publication of papers by the faculty and their students. The initial volume of this *Bulletin of the Scientific Laboratories of Denison University* appeared in 1885 and, relatedly, Herrick organized a Denison Scientific Association, with 27 charter members, most of whom were students, and the *Bulletin* became its official organ.

As Professor in Cincinnati

In 1888, when his accomplishments had become more widely known, an offer was made of a professorship at the University of Cincinnati, where science in relation to medicine was emphasized. There was also more space and equipment for the anatomical and physiological studies he promptly initiated. The latter consisted of stimulation of the cortical motor areas in the woodchuck, rabbit, racoon, and opossum, undertaken with his student and protégé, William Tight, who later succeeded him as president of the University of New Mexico. A series of comparative anatomical studies on the brains of infra mammalian species was also begun by C.L., with the alligator as a starter, and were continued on the avian brain by a graduate student.

Charles H. Turner

The son of the janitor of a Baptist church, Charles Henry Turner (1867–1923; Fig. 2.1) was described patronizingly by C.L. as the "ablest colored man I have ever seen and a gentleman… a Baptist and I think a Christian" (quoted in Windle, 1979, p. 76). Turner was the first African-American student to obtain an advanced degree at Cincinnati and, later a Ph.D. in zoology at the University of Chicago, in 1907. His paper on the avian brain (1891) in the first volume of the *Journal of Comparative Neurology* represented a surprising amount of work toward the B.S. degree. Before Herrick departed Cincinnati, he and Turner (1895) collaborated on a 500-page treatise describing small crustaceans of Minnesota. The year he received his Master's Degree, Turner was appointed Professor of Biology and head of the Science Department at Clark University in Atlanta, Georgia, where he stayed until undertaking graduate work at the University of Chicago leading to his Ph.D. (zoology) *magna cum laude* with a dissertation on ant behavior (1907). Turner became known internationally for "Turner's circling," the movement pattern of some ant species approaching their nest entrances. When Turner died at 57, he had been teaching biology at Sumner High School in St. Louis, Missouri for 13 years. Haines (1991) attributes this seeming lack of recognition of a world authority on insect behavior

Fig. 2.1. Although Charles H. Turner worked on avian brain as a graduate student, he studied ant behavior for his Ph.D. dissertation. Turner was the first African-American to receive a doctorate in zoology (1907) at the University of Chicago (Haines, 1991). Photograph from *Trans. Acad. Sci. St. Louis*, 24(9), 1923.

to a preference for a center of scholarly activity," in "the finest black high school in the country if not the best of any color" (Haines, 1991, p. 16). In any event, the teaching of neuroscience at the secondary level was practiced in America long before it became *de rigueur* in today's high schools.

In the spring of 1889, Charles Judson Herrick (C.J.) joined his professor-brother at Cincinnati as an undergraduate in biology. The younger Herrick later described the arrangements:

> "The biological laboratory was a large room where a dozen or more advanced students worked individually on their separately assigned topics. The director's office was a tiny enclosed alcove in one corner, but it was seldom occupied, for Clarence did his work at one of the laboratory tables surrounded by his students" (C.J. Herrick, 1955, p. 37).

During that period, C.L. formulated his comprehensive program of integrated research in psychobiology to include research in comparative and physiological psychology as well as neurology and the *Journal of Comparative Neurology* (1891) was to be a publishing and clearing house for all those fields.[4]

[4] See Volume 2 on the founding of the journal; see C.J. Herrick, 1941, for a detailed account.

Debacle in Chicago

Clarence Herrick's short-lived appointment to the initial faculty of the new University of Chicago, scheduled to open – with the support of John D. Rockefeller – in the Fall of 1892, was first broached in January, 1891 and was finally terminated on June 7, 1892, thus spanning a period of 17 months. President William Raney Harper wrote Herrick early in 1891, asking his possible interest in a position. Herrick replied that he would consider an offer if he could develop his research and teaching in the three allied fields of comparative neurology, physiological psychology, and comparative psychology. He emphasized that he was "primarily a biologist," and that he would expect university support in launching a journal devoted to comparative neurology.

After the president met with Herrick in Cincinnati, in June 1891, he made an offer in agreement with those terms and Herrick submitted his resignation to the University of Cincinnati, effective at the end of 1891. He spent the intervening period in Germany, preparing for his broader activities and purchasing equipment for his new laboratory. Before sailing in December, 1891, Herrick sent Harper an expanded outline of his proposed interdisciplinary courses of study – again emphasizing his responsibility for comparative neurology, and physiological and comparative psychology.

Harper replied before Herrick sailed:

> I understand that you are to work along the lines you indicated .... [W]e are negotiating with Professor Whitman of Clark University for the headship of the Department of Biology. The plans are by no means clearly laid out but you will be protected in the particular department of work in which you wish to labor (quoted in C.J. Herrick, 1955, p. 69).

Clarence Herrick left New York after Christmas, 1891, believing

> that everything would be provided to effect his dream of an integrated program in neuroscience. He had no objection to the appointment of Whitman, whom he recognized as a distinguished biologist, sixteen years his senior, recipient of a Ph.D. from Leipzig, director of the Woods Hole Laboratory, and founder of the *Journal of Morphology* (Windle, 1979, p. 77).

Not long after Herrick reached Berlin, he received a long-delayed letter from Harper, written in December, 1891, a letter replete with foreboding clues:

> [T]he title of your chair is still unsettled .... If you are willing to let it stand neurology, there can be no objection .... I do not quite see our way to guarantee $500.00 for the *Journal* until matters are in shape .... It seems altogether certain that we shall be able to authorize you to purchase a small outfit of microscopes, etc., for the University and yet here again ... we have no plans as yet in this direction (quoted in Windle, 1979, p. 78).

The equivocations were due to the fact that before Herrick had left for Berlin, Harper had opened negotiations with Charles A. Strong, son-in-law of

Mr. Rockefeller, to leave Columbia for Chicago and an appointment in psychology but omitted revealing that a laboratory in physiological psychology had already been promised to another. Herrick, in turn, learned of this development not from Harper but casually from a fellow visiting professor who was to teach philosophy at Chicago. Herrick's already aroused suspicions boiled over when he received the official letter of appointment from the Board of Trustees to become Professor not of Neurology but in the Department of Biology. He dispatched a conditional letter of resignation based on his perception: "In view of an ambiguity respecting the nature of the work and status of the chair . . . ." (quoted in Windle, 1979, p. 81). Herrick also wrote to Harper requesting an appointment, protesting "the propriety of permitting me to come to Europe at my own expense to prepare for work which had long since been allocated to another . . . . From the personal stand point my situation is disastrous . . . ." The exchange of letters continued, Strong seeking to relinquish part of the work in psychology, Harper refusing to change anything that would affect the son-in-law of the university's current or future benefactor, and Herrick graciously conceding the comparative aspect of psychology to be ample for his plans. In June of that fateful winter and spring, Herrick was in Harper's office hoping for an independent chair of neurology. Fate had taken over, however, for in Harper's anteroom was a photograph of new faculty, among which was that of "Henry H. Donaldson, Professor of Neurology." Earlier, at Clark University, Professor Whitman had demanded Donaldson's appointment to the chair of neurology, and Harper had acquiesced to Whitman's insistence that the whole corps at Clark should come, or none.

Owing to the tragic outcome of Herrick's dealings with President Harper and the evaporation of his plans for introducing an integrated program of neuroscience at the new University of Chicago, Herrick's second visit to Germany had lasted only 3 months. His main purpose had been to prepare for the elaborate courses he had expected to offer in the Autumn of 1892. He had not intended to spend all his time in Berlin, however, but expected to visit Leipzig again in the Spring, and also to confer with Edinger at Frankfurt. Herrick had corresponded with Edinger from Berlin but under the circumstances had to cancel the visit. As during his earlier visit, a decade before, Herrick spent much of his time auditing the lectures of several distinguished professors. He was unimpressed by the routine lectures, even those of famous men (but DuBois-Reymond "carries great style") and found the laboratory materials of the "highest order."

On his return from Berlin in the spring of 1892, Clarence had received a warm welcome in Granville. A new president had been appointed at Denison, who planned to build a new science building and initiate graduate study in biology and geology. Herrick accepted the offer of Professor of Biology there in June, 1892, together with a subsidy for the *Journal of Comparative Neurology*.

His protégé, William Tight, was given an assistant professorship in geology and botany, thus leaving Clarence with more time for his activities in research and development. A year later, C. Judson Herrick was attracted back to Denison, as a supported graduate fellow, to pursue doctoral study with his brother, with the option of spending 1 year of the program at a more developed university in the East (see below). Everything appeared to have turned out for the better until, in December, 1893, Clarence Luther came down with a respiratory infection, which culminated in an incapacitating pulmonary hemorrhage and the diagnosis of tuberculosis. He was advised to move to the Southwest for climatic reasons, and left Granville for Albuquerque, New Mexico, in July, 1894.

Three years later, he had recovered enough to accept an appointment as the second president of the Territorial University of New Mexico, a post he held for the next 4 years. A competent faculty was assembled and advanced study was initiated in biology and geology. Herrick's two outstanding graduate students were George E. Coghill (see p. 68) in biology and Douglas W. Johnson in geology. As Clarence's health subsequently deteriorated, he was given a year's leave of absence and resigned the presidency in 1901, to be succeeded by Professor William Tight from Denison University. At the age of 46, on September 15, 1904, Clarence died in Socorro, New Mexico, and was buried there.

Two Enduring Contributions

C. Judson Herrick concluded his tender biography of his older brother with an assessment of the outcome of Clarence's two major "Contributions to Psychobiology." The first was his projected

> program of integrated teaching and research that would cross the usual departmental boundaries [at the new University of Chicago] and would involve cooperative relations with other activities in zoology, anatomy, physiology, genetics, animal behavior, psychology, and philosophy in a search for scientifically acceptable principles of psychobiology (C.J. Herrick, 1955, p, 70).

In the eyes of the reverent younger brother, that pragmatic concept was so novel and the resistance to interdepartmental cooperation so great that the project was summarily rejected. The German model of autonomous departments that was adopted led to internal and external rivalries and after Harper's death was gradually replaced by more successful collaborative efforts.

The younger Herrick believed that

> [i]f my brother had been permitted to develop his project, even in a limited way ... the innovation might have set a pattern that would have significantly changed the history of science in North America during the subsequent half century and accelerated the movement toward the same objective that is now in full flood (Ibid.).

Citing current programs at Yale, Harvard, Michigan, Pennsylvania, Columbia, and several other universities, Herrick also called attention to the more informal collaborations in psychobiology which operate without administrative restrictions or overhead expenses and produce research of greater importance than those of organized institutes.

Clarence Herricks' second contribution implemented more directly his goal of advancing an integrated neuroscience in America: inauguration in 1891 of the *Journal of Comparative Neurology* – accepting papers in neuroanatomy, physiology, and psychology. Two years later, when Clarence's life hung in the balance for several weeks, his younger brother, though a graduate student and instructor at Denison University, was faced "literally overnight" with the problem of the journal. Knowing how much the journal meant to his brother, C.J. resolved to carry on the responsibilities in spite of his complete lack of experience. The point to note in the subsequent labyrinthine sequence is that the journal's founding great-grandfather was Clarence Luther Herrick. In 1954, the journal published its one-hundredth volume, celebrated by an article on "The History of the Journal" by C. Judson Herrick, then in his 88th year. His concluding words were:

> This chronicle is frankly a tribute to the prescience and courage of a man of vision by the one who has profited most by his guidance and teaching. It is published in the hope that this story of a pioneering adventure ... may help to perpetuate and fortify that passion for crossing new frontiers which is the vital breath of science (C.J. Herrick, 1954, pp. 755–756).

More than a decade earlier, Adolf Meyer at Johns Hopkins analyzed Clarence Herrick's work and the setting in which it took place with a less worshipful eye. Meyer had met the elder Herrick in 1892 in Berlin when the latter was preparing for his "neuro-biological" department in Chicago. In the 50th anniversary volume of the *Journal of Comparative Neurology*, Meyer's essay on the founder contained this insight: "Herrick was determined to let generalization grow close to the fundamentals of actual observation and contact, without the slightest development of pedantry, but perhaps with too definite a belief in anatomy and histology as the basic and at the same time supreme court of evidence" (Meyer, 1941, p. 21).

FUNCTIONAL COMPONENTS OF CRANIAL NERVES IN LOWER ORDERS

The chief contributions of early comparative neurology, exemplified on the Continent by Edinger and later by his protégé, Ariens Kappers, emphasized determination of the major structures of the evolving brains in the classes of vertebrates from fishes, through amphibians, reptiles, birds, and mammals.

The first of the major stages of this central neural expansion and differentiation was identified by Edinger (as already noted) as the establishment of an initial paleoencephalon, taking shape at the cephalic end of the neural tube in fish, around the central connections of the cranial nerves. The second was the subsequent elaboration of an integrative neoencephalon, overlying the rostral end of the older brain, incipient in its development in the amphibians and reptiles, expanding into a hypertrophied basal ganglion in birds, and mushrooming into the neocortex of the mammalian brain, in which it reached its greatest dimensions in the primates, and its epitome in the cognitive brain of man.

Against this background, the distinctive features of the succeeding American school of comparative neurology lay in its analysis and determination of the basic principles of the functional organization of central neural components that, when established in the lower vertebrates, provided the scaffolding on which expansion and elaboration occurred during subsequent stages of neural evolution. The studies of this American school developed, as did the paleoencephalon itself, around the central connections of the cranial nerves in lower vertebrates.

Early Investigators

Whereas the pioneering figure of this American school was Clarence Luther Herrick, its initial major accomplishments resulted from the sequence of studies by Henry F. Osborn, Oliver S. Strong, and C. Judson Herrick (Clarence's younger brother), on the central connections of the cranial nerves of fish and amphibians.

Henry Osborn and the "New Vista."

Although the front-runner, Henry Fairfield Osborn (1857–1935), was the dark horse in this troika, and his and far-ranging accomplishments in other areas doubtlessly obscured his seminal role in this achievement. His father was a founder and for many years the president of the Illinois Central Railroad and young Henry inherited many of his administrative abilities. He either founded, reorganized, or otherwise helped develop the Department of Zoology at Columbia University, the section on paleontology in the American Museum of Natural History, of which he was long the president, the New York Zoological Park, the New York Academy of Sciences, the New York Aquarium, and the Brearley School in New York City. In 1923, he was presented for an honorary doctorate at Yale by Professor William Lyon Phelps, who emphasized the range of Osborn's activities by noting that simultaneously he had been the president of the International Congress of Eugenics, the American Bison Society, and a school for girls.

On obtaining his baccalaureate at Princeton in 1877, Osborn pursued advanced study in psychology with James McCosh and Francis Galton, in comparative anatomy with Thomas Huxley, in embryology with Francis Balfour, and in neuroanatomy with Bernard von Gudden – the latter three in London, Cambridge, and Munich respectively. As Professor of Comparative Anatomy at Princeton, Osborn continued his studies of the cranial nerves and brains of amphibians, using as staining the prolonged immersion of ammonia carmine (Von Gudden's method) he had learned in Edinger's laboratory. Osborn's assembled observations were published in 1888, in the second volume of the *Journal of Morphology*. In this and other studies more generally, Osborn aspired to discover new principles from assembled facts and believed that the important principle he had found in the amphibian brain was the subdivision of the cranial nerves into many physiological components. He attributed this "new vista" to a student, Oliver S. Strong, who had traced all the components of the cranial nerves in the brain of the frog. A similar classic on the brain of the fish was prepared by C. Judson Herrick, under Strong's direction. Osborn then commented pompously on how privileged he was to have initiated a new school of neurological research in America.

Oliver Strong's Homologies

A New Jersey boy, Oliver Smith Strong (1864–1951) was educated at Princeton, taking his baccalaureate in 1886 and a master's degree in zoology in 1888, under the direction of Professor Osborn. Next, he served as an assistant in the Allis Lake Laboratory in Milwaukee, and in the fall of 1891, shortly after Osborn was appointed the Da Costa Professor of Biology at Columbia University, Strong rejoined him there as a fellow. Four years later, his doctoral dissertation, "The Cranial Nerves of Amphibians. A Contribution to the Morphology of the Vertebrate Nervous System," which was partially worked out in the Allis Laboratory, was published (Strong, 1895) in the *Journal of Morphology*. Its gorgeous Plate XII is a reconstruction of cranial nerves V, VII, IX, and X of the tadpole, showing their different components in different colors.

Strong's investigative plan, a continuation of Osborn's work, was to ascertain the differences in fine structure, internal origin, distribution, and function of the components of each cranial nerve. His dissertation then stated: "[E]ach of the main components thus determined will be further considered as representing a system, and used as a basis of homologizing the cranial nerves of the Amphibians and those of other orders" (Strong, 1895, p. 107). In the higher fishes Strong identified three kinds of cutaneous nerves, distinguishable by their fibers, distribution, and origin. The three systems were probably originally one from which the others differentiated. Placing emphasis on the interpretation of his observations, Strong pointed out that specialization probably played an important part in the changes leading to the organization of the vertebrate nervous system. As an example, the reduction of the cerebellum in amphibians possibly is correlated with the reduction in the cutaneous sense organs. As for its bearing on classification of the nerves, "[t]he foregoing comparative study of the cranial nerves shows that *the present numerical classification is unphilosophical*."

> One principal cause [is] that the ... serial numbering of the nerves, is based upon the conditions existing among the higher vertebrates [whose] cranial nerves ... have undergone considerable reduction of primary components .... It will therefore ultimately be necessary to remodel our cranial nerve terminology, but ... their exact composition has not yet been sufficiently determined in order to do this successfully (Strong, 1895, p. 208).

Strong's dissertation brought order to a confused jumble of cranial nerves by the application of fundamental physiology; it provided a rational basis for their classification and delineated their central connections in all vertebrates. Deemed to be the most important contribution to neurology in 50 years, this paper inspired a program of research that is still active and that became known throughout the world as the work of the American school (Smith & Elwyn, 1951).

C. Judson Herrick as Torch Bearer

As further indication of the contemporary impact of Oliver Strong's research, Charles Judson Herrick (1866–1960) with his brother's approval, took leave of absence from Denison University to do his graduate work, in 1896–1897, under Strong. He and C.J. worked at the same lab bench at Columbia College and were together several summers at the Marine Biological Laboratory, Woods Hole.

An Unusual Dissertation

Herrick's completed dissertation won the Cartwright Prize and was published in the *Archives of Neurology and Psychopathology*, the organ of the Pathological Institute of the New York State [Mental] Hospitals, where Herrick had part-time employment in the institute's well equipped laboratory at No. 1 Madison Avenue. This position offered a highly lucrative opportunity for the work Herrick was prepared to do at his own expense and the *Archives,* a new journal experiencing a dearth of papers, was eager to fill its pages with a 300-page article containing seven expensive lithographs. As Herrick explained,

> Our Journal of Comparative Neurology could not finance so costly a publication, but it was enabled to issue this contribution by reason of the generous terms offered by the New York State Hospitals Press [Utica]. The *Archives* bore the primary cost of production and the Journal purchased additional impressions at the cost of paper and press work. Several hundred more copies were

> printed, repaged, and distributed by the institute in monograph form. This dissertation, accordingly, was distributed in three issues which were identical in content except for page headings and numbers (C.J. Herrick, 1954, p. 734).

This was probably a unique bonanza for graduate dissertations, either in this country or the world.

In the introduction to his dissertation (1899) Herrick laid out the conceptual philosophy of his life-work. Because the nervous system relates the inner and outer worlds of the organism, the important role of the sense organs and peripheral nerves must not be overlooked. The comparative method, as exemplified by Edinger's work, is the only way to the understanding of the adult human brain. Because of their interdependence, study of the lower parts is indispensable to correct comprehension of the complex, higher sphere. Herrick's aim in his monograph was to trace the structure and function of the cranial nerves from lower vertebrates to man and hence to bridge the gap between comparative neurology and other fields such as psychiatry. No dodger from a challenge, his daunting goal was to work out some of the general laws of structure and function.

The consummate teacher, Herrick deplored the rote of learning the cranial nerves and their distributions and the deficiency of the textbooks. He cited Strong's research on frog (1895) as the only enlightened study published so far. A study of fishes, on the other hand, the simplicity and diversity of which present a "remarkably beautiful morphological series" of nature's experiments, would reveal the centers of the sense organs and provide homologies even to the human brain. Extending what Strong had accomplished with amphibians, Herrick's grand plan was to unravel in fishes the components of the cranial nerves as a morphologic unit. He found in their cranial nerves three sensory systems of components and two motor (in addition to the sympathetic system), each with physiologic and anatomic characteristics in common, "so that they may react in a common mode." Each of these components was then described in detail and illustrated in multicolored figures, culminating in Plate XV, showing "The cranial and first spinal nerves of *Meridia gracilis*, reconstructed from serial sections and projected upon the sagittal plane, X23 … ." This plate formed one of the three classical illustrations of the cranial nerves of lower vertebrates, together with that of Allis (1897) in *Amia calva*, and Strong (1895) in the tadpole.

Herrick's Academic Career

On his return to Denison, C.J. Herrick completed several follow-up studies of the cranial nerves and their central connections in fish. In 1907, he climbed the academic scale to the University of Chicago, succeeding Henry. H. Donaldson, who had moved to the Wistar Institute. Here, with more time and resources for his research, Herrick also ascended the phylogenetic scale and begun an intensive, long-term study of the amphibian brain which was continued for some 40 years. An extended series of papers reporting focal findings appeared in the *Journal of Comparative Neurology*, and two monograph-length publications – the first on the mud-puppy, Necturus (1933a) and the second on the salamander, Amblystoma (1948) – provided an integrative synthesis of his progress.

It may be noted that study of the amphibian nervous system had been familiar both to C. Judson Herrick and to Coghill from their graduate student days and its late adoption as a career program had not come as a novelty. After young Herrick obtained his baccalaureate at the University of Cincinnati in 1891, he taught a "settee" of sciences at a small Kansas college for a year, and then returned to Denison as a fellow in neurology under his brother who had been appointed there as Professor of Biology. Specimens of *Amblystoma puntatum* were abundant in that locality and in 1894, C.J. published a paper on the cranial nerves of this animal. Although brief reference was made to Osborn's and Strong's contributions, Herrick's account was a perfunctory description of the traditionally identified cranial nerves, wholly lacking in the brilliance of his later doctoral dissertation on Medinia.

New Directions

In parallel with Herrick's long preoccupation with the structure of fish and amphibian brains, he maintained a continuing interest in synthesizing the progress of other brains in the one comparative neurological development that could not be illumined by study of fish and amphibians, i.e., determination of vertebrate phylogeny leading culminatingly to the human neocortex. This was, by far, the most significant of all the Holy Grails of comparative neurological research, and Herrick did not shrink in assessing the potential of his own studies for this program but, in each recurring instance, he came up lacking. In an article on "Some reflections on the origin and significance of the cerebral cortex," he candidly admitted that in "all species below the frog, [the thalamus] is probably the organ of highest associations of which these animals are capable" (C.J. Herrick, 1913, p. 227).

In 1926, Herrick published a book-length account of *The Brains of Rats and Men, A Survey of the Origin and Biological Significance of the Cerebral Cortex.*[5] In his opening chapter, Herrick discussed the range of views concerning the role of the brain in human mental life and behavior.

[5] It was derived in part from a course of lectures he was invited to give at the University of California in the summer of 1924. In an introduction to a reprint edition in 1963, p. ix, Paul G. Roofe recalled: "At a dinner given in honor of the late Professor Herrick's twenty-fifth year of service at the University of Chicago, he made the following facetious remark … 'If you wish to become famous you should write a [companion] book entitled *The Brains of Cats and Women.*' "

> One extreme," he remarked, "is typified by the president of a college to whom, as a youthful enthusiast, I went many years ago with the suggestion that perhaps it would be helpful if I were to offer a course on the functions of the brain parallel with his own lectures on psychology. 'Young man,' he said, 'the brain has no more to do with thinking than the cabbage heads in my garden!' (C.J. Herrick, 1926, pp. 13–14).

Continuing with more general issues, Herrick wrote on the possibilities of demonstrating patterns of connections between anatomic and physiologic features among species, taking the cerebellum as an example:

> [I]n the simplest vertebrates the vestibular centers of the medulla oblongata are the cradle of the cerebellum. This is a strictly reflex mechanism of adjustment of equilibration and musculature tone, acting primarily under the influence of the semicircular canals of the internal ear. In more active animals with greater diversification of bodily movements the cerebellum enlarges and into it are discharged nervous impulses from the organs of muscle sense, the eyes, and all other receptors which may serve postural adjustments and orientation of the body ... in space. These are the proprioceptive senses, and thus the cerebellum was elaborated as the great central adjustor of proprioceptive control (Ibid., pp. 29–30).

Because of their sluggish inactivity, development of the cerebellum in the Amphibians is not only minimal but regressive compared with fish. Herrick's choice of Amphibians for his career research thus handicapped his contributions to the development of this neural organ as much or perhaps even more than to the development of the cerebral cortex.

In respect to the cerebral cortex, Herrick believed it to have emerged from reflex control centers responsive to external stimuli – the exteroapparatus – in contrast to the cerebellar cortex which controls coordinated responses to stimuli chiefly from within. In a summary of vertebrate cortical evolution, Herrick pointed out that

> the center of highest physiological dominance [and] of most effective control of behavior within the brain, has moved progressively forward from midbrain (most fishes), thalamus, and corpus striatum (reptiles and birds), to the cerebral cortex (mammals) (Ibid., pp. 35–36).

In Herrick's view, the cerebral cortex exerted a largely inhibitory influence on subcortical functions and repressed as well as activated the more elementary components of behavior.

In 1933, at the National Academy of Sciences (to which he had been elected in 1918), Herrick read a paper on "The Functions of the Olfactory Parts of the Cerebral Cortex" (1933b) in which he summarized his inferences drawn from structural arrangements. He saw a differentiation of the cerebral cortex in which the olfactory system was dominant in primitive types and reduced to a subordinate position in primates as the neopallium expanded and differentiated, and concluded that the olfactory cortex not only participated in its own specific way in cortical associations but also served as a nonspecific activator for all cortical activities.

> This is a generalized activity of primitive type acting on the neopallial cortex as a whole, lowering its threshold or increasing its sensitivity, or, in the case of noxious stimuli, exerting inhibitory influence (C.J. Herrick, 1933b, p. 14).

Final Summary

When 90 years of age, C. Judson Herrick (Fig. 2.2) capsulized the breadth and depth of his thinking as follows:

> I wanted to find out what these animals do with the organs they have and what they do it for, with the expectation that this would help us to unravel the intricate texture of the human nervous system and show us how to use it more efficiently .... [The task is] to define as exactly as possible the relations existing between the physiological processes and the colligated mental processes .... When these relationships are adequately known we shall be able to formulate the principles of the mechanics of mental processes .... [We] have reason to believe that goal is not unattainable (1956, pp. 8; 237).

Fig. 2.2. By the end of a long career of disciplined inquiry into the phylogenetic development of structure and function in lower forms, C. Judson Herrick believed that understanding mental processes was within reach. Photograph courtesy of Robert B. Livingston.

After that book-length publication with its autobiographic statement, Herrick contributed papers to two contemporary symposia. The first was a collaborative paper (with George Bishop, 1958) on spinal lemniscus systems which surveyed the phylogeny of afferent pathways to the paleothalamus involved in activating the cerebral cortex in mammals. The authors' consensus was that the earlier findings in the brain stem of the salamander, Amblystoma, had provided a substrate for current views.

In his final paper, Herrick wrote,

> It has been my privilege to see the development of comparative neurology as an organized science almost from its beginning. This experience has led to the formulation of some general principles of the mechanics of behavior.... These are: First, properly controlled observation of introspective experience is as valid and useful a scientific method as is reflexology. Second, the distinction between analytic and integrative processes ... is fundamental and its recognition clarifies much that has been obscure. Third, the neuropil is the mother tissue from which the specialized central functional systems have been derived in vertebrate phylogeny, and it plays an important role in learning (C.J. Herrick, 1961, p. 628).

In her memoir of C. Judson Herrick for the *Journal of Comparative Neurology* (1960), his protégée, Elizabeth Crosby, recalled,

> During the early years at the University of Chicago, Dr. Herrick organized his justly famous graduate class, which drew students not only from Anatomy but also from Physiology, Zoology, and, particularly, Psychology.... [H]e was a truly inspired teacher... as, tilting back in his chair, with his legs crossed, his pipe lighted, and his face aglow with interest and enthusiasm, he discussed his favorite subjects – the evolution of the nervous system... of various vertebrates, the... structural and functional activity of animals, including man, as evidenced by their behavior, and the philosophical significance of the anatomical and the physiological patterns of the nervous system. His profound scholarship, his great knowledge of the detailed structure of the nervous system in various forms, and his intense interest in the discussions, held his listeners enthralled. This was graduate teaching at its best (Crosby, 1960, p. 6).

A Personal Encounter

In a Macy Conference on *The Central Nervous System and Behavior* (1960), one of the participants remarked:

> I would like, if I may, to recall an August Saturday, many years ago, when a group of young graduate students from Ranson's Institute of Neurology in Chicago escaped the city heat and the confines of the laboratory and their studies, to the Indiana dune country on the southern shore of Lake Michigan. The rolling palisades of yellow sand, coming up from the blue lake, with scrub pine and oak at the top, made an ideal vacation setting. About noon, however, an Indiana summer thunderstorm came up, and there was a deluge.

> To protect the basket lunch, we approached a frame cottage at the top of the dune, to ask if we could put it under shelter. A canvas was stretched out from the side of the cottage and under it, in a deck chair, was a rather slight, tanned, clean-cut man in pajama bottoms. Near him were two or three volumes of philosophy which he had been reading. Putting his books aside, and with the most gracious cordiality and friendliness, he invited us to join him for lunch.
>
> As we lunched, sheltered from the rain, he began to tell us about his scientific life and how he had been motivated to begin a study of the comparative anatomy of the nervous system, although he emphasized that it was undertaken as a basis for understanding the maturation of function and behavior rather than from a strictly morphologic interest. He told us also how, through the loss of his brother, who had died very early in his career, he had derived an incentive to continue to carry out his brother's plans and goals.
>
> He went on to the associations that he had enjoyed with Coghill, who had similarly made efforts to investigate the maturation of behavior in lower vertebrates, and he described camping trips they had taken in the Grand Canyon area of the West. He explained how his interest had become increasingly aroused by the philosophical aspects of his study of the brain and behavior.
>
> I do not believe that any graduate students ever enjoyed a visit more than this group .... As you may have guessed, this was C.J. Herrick .... It seemed [sic] appropriate ... to recall that occasion, because just a few weeks ago, Dr. Herrick passed away at the age of 91 (Magoun, 1960, pp. 27–28).

In 1969, James O' Leary and George Bishop of Washington University Medical School, St. Louis, published a centennial essay, "C.J. Herrick, Scholar and Humanist" which, more than most, epitomized his major conceptual contributions to neurobehavioral sciences, interrelated with those of George Coghill. To paraphrase the essence of this essay, integrative mechanisms which concern neuropil and respective roles of total and partial patterns of behavior contain the rudiments of the Herrick-Coghill scientific philosophy. In Herrick's scheme of brain function, neuropil is the seat of integrative mechanisms, explicitly for Necturus, and implicitly for man. In lampreys, because their activities are mostly mass movements, no specialized integrative apparatus is required and much of the brain is composed of an undifferentiated relatively equipotential neuropil. In the primitive generalized amphibians – Necturus and Amblystoma – neuropil is distributed throughout the brain, but increases in amount rostrally, perhaps forecasting encephalization.

Localization of function is a universal concept in neurological thought. Conventional localization resides largely within stable analyzer mechanisms, where it occurs in spatial patterns. Embryogenesis leads to the elaboration of this stable architectural framework of nuclei and tracts which mediate the stereotyped, inherited behavior evolved in all species which have survived the ages. Phylogenesis is manifested in the individually modifiable behavior, in which the connectivities of neuropil evidently play the major role.

Another aspect of Herrick's philosophy dealt with the relation of mind to brain. This was deeply colored by concepts of total and partial patterns of behavior, the latter individuating out of the former. He saw the origins of mentation in motility, as had Sherrington, Coghill, and others. In this sphere, he conceived another kind of localization, in the image of neuropil, in terms of dynamic patterns of action. This kind of localization takes place in fields instead of cell groupings, and various recurring patterns of performance are fabricated within a frame of modifying, conditioning, or inhibiting influences. Localization of mental function is to be considered, if at all, in terms of such field patterns. Conceptions of mental patterns are usually total ones and, for Herrick as for Coghill, such total patterns have ultimate significance both for movement and the mind. Moreover, integrative agencies come to expression in total patterns and the apparatus of synthesis is nearly identical with that of integration.

In Herrick's view, partial patterns come to expression through the system of analyzers, consisting in the framework of nuclei and tracts, with extensions along sensory and motor axons to the peripheral receptors and musculature. Thus, except for the neuropil, all of the nervous system is analytic in function, with local sign – as in reflexes – the prime expression of its activity. In embryogenesis, it was the thesis of Coghill that partial pattern develops out of and remains subordinate to total pattern.

In contrast to Coghill's and Herrick's views, in higher forms such behavior did not seem to emerge gradually from simpler antecedents. Rather, as pointed out by Washington University's embryologist, Viktor Hamburger (1934), it appeared to be activated suddenly and performed, even at first, with relative precision. In the chick, neuronal and behavioral differentiation proceeds in a cephalo-caudad progression, concordant with the progress of motility, a prelude to the breaking of the shell. The analogous inception of mammalian motility (determined at length by William F. Windle; see below) became a central issue in the controversy over the primacy of total versus partial patterns. O'Leary and Bishop (1969) concluded that Coghill's thesis was not inclusive enough to encompass all the advances made in intervening years in the study of higher species.

Maturation Order of Locomotion

George Coghill and Amphibians

The indebtedness expressed repeatedly by C.J. Herrick to his mentor and friend, George Ellett Coghill (1872–1941) was predicated on the extensive intertwining of Coghill's contributions to the American school of comparative neurology with the substantive work of the Herrick brothers. Their careers overlapped first at the University of New Mexico, then at Denison University, and later still at the Wistar Institute in Philadelphia. On growing up in the rural and religious life of the Illinois farmland, Coghill entered the seminary at Brown University in Rhode Island but withdrew "in great perturbation of mind" (C.J. Herrick, 1949) and went West to recover his equanimity and replan his future. In Albuquerque, New Mexico he came under the influence of the elder Herrick, Clarence (see above), and decided that study of the nervous system was prerequisite to pursuit of his interest in psychology and philosophy. With Herrick's help and inspiration, Coghill commenced a life-long exploration of Amblystoma. Returning to Brown, he completed his Ph.D. with a dissertation on that animal's cranial nerves (1902). Among other findings, it filled in some gaps in the published work of the younger Herrick on the same cranial nerves (C.J. Herrick, 1894). Their careers overlapped again in Granville, Ohio before Herrick departed for Chicago and Coghill succeeded him at Denison.

During appointments at the University of Kansas (1913 to 1926) and the Wistar Institute (1926 to 1936), Coghill published a famous series of 12 papers with the generic title *Correlated anatomical and physiological studies of the growth of the nervous system of Amphibians*, a series still unfinished at the time of his retirement in 1935. Coghill was the first representative of the American school to implement in vertebrates the precept of the school's founder and Coghill's mentor, Clarence Herrick – to interrelate morphological studies of the nervous system with observations of behavior. Coghill had not gained experimental competence in that direction from the Herricks, neither of whom had any to provide, so he left the experiment to Nature, which had evolved a magnificent behavioral transformation in the life of every lowly Amblystoma – from fertilized egg into a fish-like, water-dwelling larva, and thence into a land-dwelling salamander. Through the course of this "metamorphosis" Coghill carefully observed the segmental steps of the animal's movements, from the initial neck flexion, through a "coil," followed by an "S" movement leading to free-swimming locomotion (see Fig. 2.3). Dividing this sequence into six stages, he killed adequate numbers of animals at each stage to provide detailed information on the concomitant maturation of the central and peripheral neuromuscular systems which undergirded their expanding spectrum of behavioral capabilities.

From this analysis, Coghill concluded

> [T]here are two processes that are operating simultaneously in the development of behaviour. The one is expansion of the total pattern as a perfectly integrated unit; the other is the individuation of partial systems which eventually acquire more or less discreteness (Coghill, 1929a, pp. 88–89).

That quotation is from the publication of three lectures delivered by invitation in 1928 at University College London. In the United States, appreciation of Coghill's serial contributions to the subject of brain and behavior was slow in coming and the invitation to London provided the unusual sequence of his

Fig. 2.3. Through careful observation of salamanders, George Ellett Coghill discerned swimming movements during development and proposed a theory of locomotor maturation. Photograph at age 54 from Oppenheim, 1978, p. 44. Diagram from Coghill, 1926, plate 7.

gaining recognition abroad before that at home. This was elaborated in a letter to Herrick (quoted in C.J. Herrick, 1949, p. 45).

> It may interest you to know that ... the late Sir Grafton Elliott Smith and Prof. H.A. Harris [in England], felt that they had a real part in discovering Professor Coghill to the larger scientific world. They felt that University College had been the midwife at the delivery of his classical 'Anatomy and the Problem of Behaviour' and that as Englishmen had shown the Alps to the Swiss they had revealed Coghill to American science.

Although belated, Coghill's achievements were recognized by his presidency of the American Association of Anatomists (1932 to 1934), his award of the

Elliott medal of the U.S. National Academy of Sciences (1934), and his election to the academy in 1935.

An Incautious Extrapolation

The latter part of Coghill's career was marred, however, by controversy over his attempt to extend the conclusions of his amphibian studies to the development of birds and mammals. In a survey of Coghill's work it was pointed out that in early publications "he eschewed theoretical generalizations" (Oppenheim, 1978, p. 54), but in 1929 he began to review the literature on vertebrate forms including man, thereby attracting more attention than the lowly salamander. Coghill's views came under serious attack in what later became known as the Coghill-Windle controversy, attributed in part to misunderstanding of Coghill's terminology, such as "total pattern" by many in the field (Hooker, 1952).

Mammalian Embryos and William Windle

Initially lacking any knowledge of Coghill's work, William Frederick Windle (1898–1985; Fig. 2.4) independently commenced his well known

Fig. 2.4. Anatomist William F. Windle studied the development of locomotion in mammals, including human embryos, and the effects of asphyxia on behavior at developmental stages in monkeys. Photograph from the William Frederich Windle Papers.

studies on the development of fetal reflexes and their structural correlates.

> If Coghill is to be admired for his lifelong devotion to Amblystoma, Windle's efforts, although representing a different approach, are equally noteworthy. In a brief ten-year period (between 1930 and 1940), Windle published detailed reports on embryonic and fetal behavior correlated with neural and physiological development in chicks, ducks, rats, cats, sheep, and humans .... He clearly had a greater familiarity with the facts of mammalian fetal behavior than anyone before or after him .... (Oppenheim, 1978, pp. 55–56).

Although they shared a common interest in determining when locomotion appears and correlating that with its anatomic substrate, differences in staining methods and animal species separated the two experimenters. Coghill worked on amphibian locomotor maturation, and Windle on maturation order in mammalian embryos. When Windle attended Denison College in 1919, he was given an assistant's job that included cleaning out some histologic slides made from windowpane glass of amphibian larvae stained with Lyons blue, a general stain, that were assumed to be Coghill's discards from many years before. Ten years later, Windle had completed his doctorate with Ranson at Northwestern University Medical School and with a graduate student, Miles Griffin, began to study behavioral development as shown by fetal movements in kittens. They sent their paper (Windle & Griffin, 1931) to *The Journal of Comparative Neurology*, and Windle received[6] a "warm letter of acceptance" from C.J. Herrick, the editor, in which "he commented on the significance of our studies in relation to George Coghill's theory of emergence of motor behavior from a total pattern." As noted, Coghill's studies at Wistar at that time constituted the only major research program directly interrelating the early school of comparative neurology with functional development. In Amblystoma, he found that initial mass-action patterns, involving both the trunk and limbs, preceded capability for more focal movements and elaborated a principle of progressive organization from the whole to the part. Coghill had published (1929b) his article in which he extrapolated his findings in amphibians "to the few recorded observations on human fetal movements. This actually was an unfortunate attempt, and he drew an unjustified assumption."

In his second paper on the subject, Windle (1931) reported the neurostructural mechanism in kittens that was "quite unlike" that in amphibians, and by 1933 he had a third paper ready based on stimulation of cat embryos and fetuses. Taking it personally to Herrick, Windle remembered vividly how Herrick's face changed while reading it. He rejected the paper, saying that if

[6]This and subsequent quotations are from the transcript of an interview (CON WIN) conducted 2 March 1979 at the Neuroscience History Archives, Brain Research Institute, University of California, Los Angeles.

the findings were correct, "then Coghill's theory was wrong and Coghill's whole life's work was 'for nought.' " The paper was instead published (Windle, O'Donnell, & Glasshagle, 1933) in *Physiological Zoology* the same year that Coghill delivered his presidential address at the annual meeting of the American Association of Anatomists, in which he "vigorously criticized" Windle's work.[7]

A year later, Windle had the opportunity to show Coghill the "elegantly stained silver preparations" that had formed the basis of his articles and believed that he had convinced him, judging from a letter to Herrick after the visit in which Coghill wrote that "the findings in cat fetuses might be correct." That was the last personal contact between the two men; the final effort on Windle's part to resolve the controversy by experimentation was at the University of Cambridge on a few pregnant ewes supplied by Sir Joseph Barcroft; unfortunately the results were inconclusive. At home, Windle arranged to examine at the operating table some human fetuses of unanesthetized patients, and confirmed from direct observation and movie film records that the neuromuscular apparatus of the eight-week human fetus is highly excitable and capable of reflex movement (Fitzgerald & Windle, 1942).

A decade later, there was general acceptance (Hooker, 1952) of the *perioral* area as showing the earliest sensory perception of external stimuli, due largely to the very detailed and expert review by Tryphena Humphrey (1951) of sections from tested human embryos. Her studies of the extent of trigeminal fiber growth demonstrated that the morphologic substrate for a probable reflex pathway was in place to coincide with the first manifestation of behavior.

Windle carried his interest in behavioral development to experimental work on the effects of asphyxia on nervous systems at different stages of maturation. During his tenure at the National Institutes of Health, after establishing the Laboratory of Neuroanatomical Sciences in the National Institute of Neurological Diseases and Blindness, he initiated a series of experiments using rhesus monkeys at the free-ranging colony on Caye Santiago, Puerto Rico (see Chapter 10).

FOREBRAINS OF REPTILES, BIRDS, AND EVENTUALLY MAMMALS

Functional Organization

Not all early American investigators of the nervous system were preoccupied with the cranial nerves of lower orders. Some pursued higher inquiries, both

[7] The published version of the address (1933) contains merely a statement of differences, suggesting that the offensive comments were expurgated.

anatomically and in more advanced forms than the amphibians – the forebrains of reptiles and birds and, later, of mammalian brains. That progression was facilitated by "a fundamental plan of the organization of the nervous system on a functional basis" (Rasmussen, Herrick, & Larsell, 1940, p. 21) devised by John Black Johnston (1868–1939). The "extensive literature in comparative neurology was chaotic, with few accepted guiding principles" (Ibid., p. 20), when Johnston applied his intimate knowledge of vertebrate brains to sort and arrange them and to generalize about primitive structures and their relation to higher orders.

J.B. Johnston

A contemporary of Herrick and Coghill, Johnston was born in a small Ohio community, enrolled in zoology at the University of Michigan, and gained his baccalaureate in 1893. Continuing in graduate study in zoology, he obtained his Ph.D. in 1899, with a monograph-length dissertation on "The Brain of Acipenser. A Contribution to the Morphology of the Vertebrate Brain." With help on the use of the Golgi stain from Oliver Strong at the Marine Biological Laboratory, with whom he could discuss points from Herrick's dissertation (1894) on the cranial nerves in Meridea, Johnston's paper was published in 1902.

Each of these remarkable publications, like Oliver Strong's in 1895, was illustrated with a greatly enlarged reconstruction of the midsagittal plane of the head and brain of the fish or frog, showing in four colors the functional components of the peripheral cranial nerves (Strong; Herrick), and the major central neural stations and fiber bundles of the brain. Johnston's rendering was easily the largest, being fully two feet long when unfolded. All three dissertations were promptly regarded as classic contributions to the nascent American school of comparative neurology.

From Michigan, Johnston went to West Virginia University where he constituted the "head, body, and tail" of the Department of Zoology. In 1904–1905, he spent a leave abroad at the Naples Zoological Station and at Freiburg. Two years later, he was attracted to the University of Minnesota where his later career was occupied with university administration, together with national issues in higher education, the latter through his involvement in the American Council of Education.

Organizing Biology

While still at West Virginia University, Johnston published his widely known textbook, *The Nervous System of Vertebrates* (1906), which presented a pattern of biological organization extensively followed by subsequent neuroscientists. The initial chapters explored the four main functional divisions of the nervous system corresponding to their neural activities: 1) reception of somatic stimuli, 2) dissection of somatic movements, 3) reception of visceral stimuli, and 4) direction of visceral activities. Corresponding to these four types of activities, there are four anatomically distinct divisions of the nervous system: somatic afferent and efferent, visceral afferent and efferent divisions.

After explaining the representations of each division in each segment of the body and therefore their fundamental importance, Johnston noted a point of contrast between the somatic and visceral divisions:

> Although somatic afferent impulses may produce somatic reflexes directly without sensation, very commonly sensations are produced. When present the sensations are definitely localized and the responses may be consciously directed. Visceral afferent impulses, on the other hand, usually produce reflexes without sensations. When present, the sensations are vague, general, poorly localized, and consciously directed visceral activities are very exceptional if not abnormal. The somatic activities [contrastingly] are par excellence related to the conscious life (Johnston, 1906, p. 101).

Most of the latter portion of Johnston's book was devoted to "Centers of Correlation." The functional divisions throughout the brain and spinal cord are "quite distinct" from each other; their specialization has resulted in unspecialized material (neurons) which connect the divisions and segments. Such neurons are scattered widely throughout the spinal cord but in the brain, many remain near the ventricle wall. Dendrites of both types of neurons mingle with the fiber tracts and are called *substantia reticularis*. In the brain this "indifferent material" is proportionately greater and the reticular neurons are closely related to the functional divisions in which they are embedded. The phylogenetic history, morphology, and functional relations of these special centers of correlation constitute the bulk of the volume.[8]

John B. Johnston was one of the few early comparative neurologists, together with George Coghill, who attempted to test the functional application of their morphologic observations and conclusions by actual neurophysiological experimentation. His four-page paper (1916) on "Evidence of a motor pallium in the forebrain of reptiles" showed that under chloroform anesthesia, electrical stimulations of a comma-shape area involving the anterior and lateral portion of the general pallium of turtles evoked movements of the eyes, jaw, neck, legs, and tail. A decade later, at Johns Hopkins, Charles Bagley, with Curt P. Richter (1924) and Orthello R. Langworthy (1926), elicited motor responses by electrical stimulation of the dorsomedial surface of the alligator's cerebral hemispheres: The head turned to the opposite side, the tail to the side stimulated, and the contralateral legs were flexed. No focal

[8] Johnston's idea of neurons "left over" from formation of fiber tracts serving as connections between functional divisions and segmental centers was reflected in the thinking of his student William F. Allen (1932) in his comments about nucleus reticularis.

responses were obtained, all movements were of this mass variety. At about the same time, T. Koppanyi and J.F. Pearcy (1925) at the University of Chicago were unable to evoke responses from the turtle's cortex, but mild stimulation of the deeper portion of the forebrain – the corpus striatum – produced slow rhythmical swimming movements of all four legs.

In 1939, F. Bremer at the University of Brussels, with visiting fellows R.S. Dow from this country and G. Moruzzi from Italy, provoked movements of the neck, jaw, and legs by *faradic* stimulation of the forebrain of the turtle. These motor effects were certainly not cortical in origin, however, for they were not abolished by prolonged cocainization of the cortex, and therefore were attributed to a diffusion of current to subcortical centers. An elegant British review (Goldby & Gamble, 1957) of the work tentatively concluded that the reptilian cortex may have slight functional and structural resemblances to the cortex of mammals; there is, however, nothing corresponding to the mammalian motor area or pyramidal system and decortication leads to no obvious loss of function under laboratory conditions.

Johnston's Role

Although Johnston may not have been fully correct in the interpretations given to his pioneering observations, the pattern of combining morphological studies with experimentation provided an exceedingly fruitful paradigm that expanded dramatically and overran all subdisciplines of neuroscience by the middle of this century. As he wrote in 1906 (p. 95), "[T]he nervous system can be understood only by studying it from the standpoint of the work which it does.... For this reason, the term mechanism is preferable... or mechanism implies... structure in action."

Extension to Man

Gotthelf Carl Huber (1865–1934)

Born in India of Swiss missionary parents, when he was 6, Huber moved with his family to the United States, where his father occupied parishes in small communities of Indiana, Ohio, and New York. On graduating from the Medical School of the University of Michigan, in 1887, Carl was appointed assistant demonstrator in anatomy and continued to serve that university for the next 47 years with limited leaves for advanced study and research at Berlin (1891–1892), Prague (1895), and Philadelphia at the Wistar Institute of Anatomy and Biology (1911–1912).

In 1903, he was appointed professor of histology and embryology; from 1914 he rose subsequently to professor of anatomy, director of the anatomical laboratories, and dean of the graduate school at Michigan. Nationally, he was a highly regarded member of the American Association of Anatomists, serving as its secretary-treasurer from 1902 to 1913, and president in 1914–1915. For almost 30 years he was a member of the editorial board of the *American Journal of Anatomy*, and managing editor of the *Anatomical Record* from 1909 to 1920. He remained on the advisory board of the Wistar Institute from its establishment in 1905 until his death in 1934. He served on the National Research Council's Medical Fellowship Board for 14 years and, in the last 8 of them, was the board's chairman.

Beginning Collaboration

Huber's initial research in neurology commenced in 1889 with W.H. Howell, who recalled that when he was appointed professor of physiology and histology at Ann Arbor, Michigan, his assistant in histology was Huber. Together, Howell and Huber competed for a prize of $250 offered by Weir Mitchell to the American Physiological Society for the best piece of experimental work on nerve degeneration and regeneration, and they won the prize. As noted earlier (Chapter 1), Weir Mitchell's interest in this field had developed from his earlier care of Union Army soldiers with peripheral nerve injuries from gunshot wounds in the Civil War. In a tenuous scientific bridge between those two wars, a quarter-century after this research with Howell, Huber returned to the problem of surgical repair of injured peripheral nerves during the First World War, and at the same time he taught the fundamentals of nerve degeneration and repair to groups of army surgeons, thus exerting a great influence on peripheral nerve surgery.

Preceding this resumption of research on nerve regeneration, however, in the earlier portion of his career, Huber – like Picasso – had a blue period, but one marked by his mastery of the capricious methylene-blue stain for study of the finer structures of the peripheral nervous system. He published observations on the motor nerve-endings on all types of muscles, as well as sensory endings in tendons, the meninges, and the tooth-pulp. His most extensive studies with this technique explored the comparative anatomy of the sympathetic nervous system of vertebrates and won him international recognition while still in his early 30s.

In addition to these varied contributions to the peripheral nervous system, in midcareer Huber undertook major studies of the embryonic development and adult structure of the glomerular vascular rete and uriniferous tubules of the mammalian kidney, research that corrected the long-respected but erroneous concepts of Carl Ludwig and Rudolf Virchow, dating back to the mid-19th century.

Elizabeth Caroline Crosby 1888–1983

The concluding portion of Huber's career consisted of intensive study of the comparative morphology of the central nervous system of vertebrates, undertaken in closest collaboration with one of America's great neuro-anatomists, Elizabeth Crosby. Her move to the University of Michigan in 1920, bringing her rich background of doctoral study and dissertation research with C. Judson Herrick, was clearly a major factor in the inception and extended accomplishment of this joint Huber-Crosby program. Because of the importance of her contributions, the balance of this program's activities may be presented in relation to Dr. Crosby.[9]

In his unpublished biography of C. Judson Herrick, P.G. Roofe wrote,

> The Story of Dr. Crosby's coming to study under Herrick is now almost legendary .... The early Crosby-Herrick association began shortly after graduation from Adrian College in 1910, when she appeared before Professor Herrick in his office at the University of Chicago, insisting that she came to study biology with him .... She had taken honors in Latin and mathematics at Adrian, but only a single elementary course in biology .... Herrick tried to convince her that she should spend at least a year of preparatory study. 'NO!' she said, she had come to study with Professor Herrick! Later, he wrote, 'She did what I told her was impossible and she did it with distinction.'[10]

On completing a master's degree in 1912, Crosby was granted a fellowship in the anatomy department and, in 1915, was awarded her Ph.D. *magna cum laude*. Her dissertation, on the forebrain of the American alligator, was published and promptly became a standard reference on that subject (Crosby, 1917). Some years later, Dr. Crosby went to Europe to visit several institutions renowned as centers of neurological research; she ventured to call, without an appointment, upon the most eminent comparative neurologist of Italy on the opening day of the semester, with the professor burdened by academic duties. When her card was sent in to him, however, he immediately came out and greeted her with, "Oh! Doctor Crosby, of Alligator mississippiensis," and then devoted several hours to a conference with her (C.J. Herrick, 1959).

After her doctoral studies, Crosby devoted the next 5 years to the care of her aging parents in Petersburg, Michigan, where she had grown up and attended grade and high schools. In 1920, when her return to research seemed feasible, she asked Professor Huber if he would allow her to continue the study of comparative neurology in his laboratory at the University of Michigan. She also wrote Herrick of her plans, and his letter to Huber on March 17, 1920 is included in Paul Roofe's remarkable biography. Herrick's elegant letter

[9] We rely on an unpublished work of Paul Gibbons Roofe (1899–1988) who turned from a theological calling to study anatomy under C.J. Herrick, gaining the Ph.D. in 1934. After a career in teaching and research (he was chair of anatomy at the University of Kansas at Lawrence 1945 to 1962), he secured the personal papers from Herrick's estate for the Kenneth Spencer Research Library at Lawrence. We are greatly indebted to Professor Roofe's daughter, Mrs. Nancy R. Dunn, for permission to quote from the manuscript.

[10] Roofe's quotation is from C.J. Herrick, 1959, pp. 15–16.

praised Crosby's work for thoroughness, accuracy, and grasp of the morphological problems involved, and as an excellent laboratory teacher. Although shy, she "delivers the goods," and is fully qualified for a good university position. To remain home to care for her parents, Crosby had taught at the local village schools and "they desire her to remain as superintendent ... but it is crime for a woman of her abilities to be immersed in a village school system" (Roofe, p. 302).

Accepting an appointment in anatomy at the medical school of the University of Michigan in 1920, Crosby initiated a program of collaborative research in comparative neurology with Carl Huber which continued until his death in 1934. She stated that he inspired most of her research and he, in turn, regarded her as one of the outstanding investigators in this field, as well as a superb teacher. Beginning with the study of the thalamus of the alligator, their collaboration extended to the avian diencephalon and, thence, to the reptilian optic tectum.

Amsterdam and a Unique Legacy

In 1926, Crosby spent a period of research at the Institute of Brain Research at Amsterdam, and returned for further work there in 1931. Following her second visit, the institute director, C.W. Ariens Kappers, author of the early compendium, *Die Vergleichende Anatomie des Nervensystems Wirbeltiere und des Menschen* (1921), requested that Crosby and Huber undertake its English translation and revision with him. "With characteristic energy, Doctor Crosby threw herself into the task of revising and enlarging the material in the original text, and adding more recent data" (Woodburne, 1959, p. 22). Two years after Huber's death, the resulting two-volume *Comparative Anatomy of the Nervous System of Vertebrates, Including Man*, some 2,000 pages, was published. During the 5 years involved, Elizabeth Crosby devoted a great amount of work to completion of this project.

After Huber's death, Crosby assumed direction of teaching and research in neuroanatomy in the department and was promoted to professor of anatomy in 1936. Her interest in and encouragement of graduate and postdoctoral students was remarkable. Over a 30-year period, some 40 Ph.D.s completed their graduate study and initial research at Michigan under her supervision. An additional 30 advanced students, who didn't take doctorates, spent various periods in work with her. A large number of clinical neurologists and neurosurgeons attended her postgraduate courses to prepare for specialty board examinations and broaden their understanding of the nervous system. In appreciation of her educational and research contributions to these professional fields, she was made an honorary member of the American Neurological Association in 1956 and was designated an associate of the Harvey Cushing (neurosurgical) Society in 1953.

Fig. 2.5. Elizabeth C. Crosby's golden reputation as neuroanatomist and teacher was international. She excelled in teamwork, and applied her extensive knowledge to clinical conditions. Photograph from *J. Comp. Neurol.*, 112 (8): frontispiece, 1959.

Crosby's Mark

When this great lady (Fig. 2.5) at age 70 retired from active teaching at the University of Michigan, her colleagues, former students, and other friends paid tribute to her in an entire volume of the *Journal of Comparative Neurology*, "HONORING ELIZABETH CROSBY." In the volume's opening contribution, C. Judson Herrick (1959) again showed his great admiration for this gifted neuroanatomist. After eulogizing those few who have contributed in two domains – growth of knowledge and enthusiastic teaching – he wrote that she belonged in that small elite group in one of the most critical and difficult areas of current science – the study of the brain and its functions. He pointed out that because the human brain has developed from simpler structures and lower organization, the comparative approach, where Crosby's work excelled, was most promising. Paul Roofe's inexhaustible biography contains an additional, warmer, and more personal tribute by Herrick to Elizabeth Crosby:

> During the many years of our intimate association and collaboration I have profited much from your work, and still more from your example of selfless

> devotion to high ideals of science, personal integrity, and service to others (Roofe, ms. p. 761).

Roofe concluded:

> It is imperative that to this account of Elizabeth Crosby should be added that she is now directing a complete and wholly revised edition of the famous Kappers-Huber-Crosby encyclopedia of *The Comparative Anatomy of the Nervous System.* Around her, at the University of Alabama at Birmingham, has gathered devoted scholars who are diligently and rigorously attacking basic comparative neuroanatomical problems with sophistication, both in knowledge and newer experimental techniques available in anatomy and physiology .... Dr. Crosby also maintains continued contact with neuroscience at the University of Michigan, to which she commutes by air every few weeks (Roofe, pp. 161–162).

The extraordinary regard which Herrick and others held for Elizabeth Crosby was widely shared for in 1980 she was the second neuroscientist[11] to be awarded the National Medal of Science.

[11] Hallowell Davis at Washington University, was the first, in 1976.

Chapter 3
Consolidation and Discovery

As the program of the American school of comparative neurology moved into the second quarter of this century, its general investigative focus had continued a gradual climb up the evolutionary scale from an initial preoccupation with fish, to an emphasis on amphibians, and thence, to a broad concern with reptiles and birds. From this latter springboard, during a critical 5-year period, 1927 to 1932, three talented young men broke through a figurative taboo and carried the school with them into a mushrooming range of interests in research on the mammalian brain.

In the last ring of the old bell, in 1929, G. Carl Huber and Elizabeth C. Crosby, in their monograph-length paper on the avian diencephalon, expressed essentially the same goals and ideals that had motivated Clarence Luther Herrick and resulted in both the figurative school and his very real *Journal of Comparative Neurology* almost 30 years earlier. Huber and Crosby wrote:

> It may not be beside the point to emphasize that the present study was not undertaken with the idea of contributing to knowledge of the bird brain as such .... Its purpose is rather to provide an anatomic basis for work on behavior in birds, since such work should enrich our knowledge of the general processes of learning and reasoning and of psychological activities in general; to add some small part of the story of the phylogenetic development of the nervous system; to provide an adequate anatomic basis for experimental work on the diencephalon and tectum in a form easily available for experimental purposes; and, above all, to build another step in the stairway to a more complete and adequate knowledge of these higher centers in man (Huber & Crosby, 1929, pp. 6–7).

Experimental studies using birds were already underway in America. Robert Yerkes had used the crow in studies of animal learning at his Fairfield Station.[1] At about the same time, Charles O. Whitman, in the later years of his career at the University of Chicago, had maintained a large colony of pigeons at his home and spent much of his time observing their behavior. In the 20s,

[1] Yerkes's field station was established on acreage in Fairfield, Connecticut, where several experimentalists (W.B. Cannon was another) interested in animal studies spent their summers (see Chapter 10).

with the advice of C. Judson Herrick, Fred T. Rogers (1922) had begun studies of stimulation and ablation of the cerebral hemispheres of the pigeon at the University of Chicago and later (1924) extended them to the opossum at Baylor University, Houston, Texas.

FILLING GAPS BETWEEN ANIMAL CLASSES

Long before midcentury, most research both in physiological psychology and neurophysiology in this country had begun to utilize mammalian forms – the more ubiquitous of which were the albino rat, the domestic cat, and the macaque or other monkeys. In the 30s, both Yerkes and John Fulton had already begun experimenting with chimpanzees at Yale. The detailed microscopic studies of the brains of fish, salamanders, turtles, and birds – which had formed the *métier* of the American school of comparative neurology – had not yet, however, provided an adequate base for research on the larger and more developed brains of these mammals. This was particularly the case in studies of the cephalic brain stem and its overlying cerebral hemispheres.

This gap between the submammalian concerns of the comparative neurobiologists and the needs of neurophysiological and behavioral scientists for substantial knowledge of the anatomy of mammalian brains was first bridged, during 1927 to 1932, by the accomplishments of E.S. Gurdjian, D.McK. Rioch, and W.R. Ingram – the first two as research fellows at the newly identified Laboratory of Comparative Neurology at the University of Michigan, and the third as an assistant professor at a newly established Institute of Neurology at Northwestern University Medical School in Chicago. The two impresarios responsible for guiding and directing the research of these young investigators were Elizabeth Crosby at Michigan and Walter Ranson at Northwestern.

The Mammalian Diencephalon

Elisha Gurdjian

Doctor Crosby's responsibility for advanced instruction at Michigan had begun long before her promotion to professor of anatomy in 1936, as noted previously. The first student to conduct research directly under her supervision was Elisha S. Gurdjian (1900–1972; Fig. 3.1), who was born in Smyrna, Asia Minor and came to this country in 1920 to complete his scientific and medical education at the University of Michigan. His master's thesis and doctoral dissertation were recast for publication in a series of three papers: "The olfactory connections" (1925), "The diencephalon" (1927), and "The corpus striatum of the albino rat" (1928). In their aggregate 180 pages, Gurdjian finally initiated a collaborative association between comparative neurologists and the growing number of behavioral and neurophysiological scientists in this country who

Fig. 3.1. Studies on the rat by Elisha S. Gurdjian during the 1930s bridged the work of neurologists and that of behavioral and neurophysiological scientists who were using mammalian models. Photograph 1956 from the National Library of Medicine.

were using mammalian animals in their research. Gurdjian's selection of the albino rat was an inspired choice for this important initial step and long overdue, for Adolf Meyer was reputed to have imported white rats from Switzerland (see Chapter 9) for Donaldson's neurological program at the University of Chicago in the 1890s. In brief, at the turn of the century, John B. Watson had introduced their use in experimental psychology at Chicago, as Linus Kline and Williard Small had independently at Clark University. Later, Watson had conveyed white rats to John Hopkins, where the careers of both Karl Lashley and Curt Richter began with their use. On Donaldson's move to the Wistar Institute at Philadelphia, a major breeding colony of white rats had been established to supply the growing research needs for these animals throughout the country.

Gurdjian's initial paper, in 1925, opened with an explicit salvo, "Our goal is a study of the mammalian thalamus" (p. 128). It described the nuclear groups of the olfactory forebrain and their interconnections with the diencephalon. His second paper provided a detailed description of the nuclei and connections of the rat's thalamus and hypothalamus, as his third paper did for the corpus striatum. Gurdjian noted that E. Horne Craigie (1925), in his atlas of the rat's brain, published 2 years earlier, treated the thalamus very briefly and the striatum and amygdaloid complex only sketchily. It may be noted here that

Fig. 3.2. With a series of anatomical studies on the dog brain, David McK. Rioch demonstrated the significance of species differences and the need for accurate descriptions of the diencephalon. Photograph from Brady and Nauta, 1972, frontispiece.

J.B. Johnston's earlier comparative study (1913) had been limited to the morphology of the septum, hippocampus, and pallial commissures.

David Rioch and Species Differences

Shortly after Gurdjian had departed for Rochester, New York, to gain his surgical training, a postdoctoral fellow of the National Research Council, David McKenzie Rioch (1900–1985; Fig. 3.2), spent 1928 to 1929 in the Laboratory of Comparative Neurology at Michigan, studing the diencephalon of carnivora, with the "help, instruction, and advice" of Elizabeth Crosby. In the pattern of Professor Huber, David had also been born in India. Gaining his M.D. at Hopkins in 1924, he completed 2 years as house officer at Peter Bent Brigham Hospital, Boston, and another 2 years at Strong Memorial Hospital, Rochester, New York, before going for postdoctoral training successively at Michigan, Amsterdam, and, in 1929, Oxford.

Like Gurdjian, Rioch published his *Studies of the diencephalon of carnivora* in a sequence of three papers. As the dog's brain is considerably larger than the rat's, so the aggregate 320 pages of Rioch's three papers were almost double that of Gurdjian's publications. In his introduction, Rioch pointed out that the recent interest in the diencephalon had accumulated much information on its anatomy, especially in lower forms and that the data demonstrated the need for more accurate studies to form the basis of experimental work on that region's functions. Also needed was a revision of the nuclear masses composing the diencephalon of the dog, which "is a commonly used laboratory animal and is an average representative of the carnivora." Serial sections of the brains of cats were used for comparison, and "marked differences" between the two were noted (Rioch, 1929a, p. 3). Of the aggregate 35 figures in Rioch's three papers, 30 were from sections of the dog's brain, 4 were from the cat's, and 1 was from the brain of avesia, a small carnivore of West Africa.

Rioch's unusual choice of the dog may have harked back to that animal's early use by Harvey Cushing in his hypophysectomy studies, first at the Hunterian Laboratory at Hopkins and later with Percival Bailey and Frédéric Bremer at Peter Bent Bent Brigham Hospital in Boston. During the '20s and '30s, the dog was also used widely at medical schools throughout the country for teaching and research in visceral physiology, as well as in experimental surgery. The only neurophysiologists who come to mind as using the dog in that period are Allan Keller and associates (1936) and R.K.S. Lim and collaborators (Chen, Lim, Wang, & Yi, 1937) at Peking Union Medical College, in China.

There was, however, another early neurosurgeon, preceding Cushing, who preferred to use cat, rather than the dog, in experimental studies of the brain. This was Victor Horsley in London, who in 1905 persuaded his physiologist associate, Robert H. Clarke, to design and oversee the construction of a stereotaxic metal frame that could be attached firmly to the head of a cat – at the ears, orbits, and jaw. With the aid of a movable electrode carrier mounted on the frame, fine electrodes could then be oriented through small openings in the calvarium, to deep structures of the brain, either to stimulate or to make focal lesions at selected targets. To assure the replication of precise electrode positions in a series of brains, the experimental animals had to possess essentially identical brain-cranial relationships. Such species constancy is far greater in cats than in dogs, in which the range of different breeds displays a wide variety of cranial shapes and sizes, with corresponding variations in the relations of these crania to their enclosed brains.

Walter "Rex" Ingram at Ranson's Institute

In 1930, Stephen Walter Ranson revived the use of the Horsley-Clarke instrument, which had lain essentially fallow since its development a quarter century before. Ranson conceived its revival as a major research resource in the development of his new Institute of Neurology at Northwestern University,[2] and his young associate Walter R. ("Rex") Ingram (1905–1978; Fig. 3.3), was assigned the task of preparing an atlas of the cat's brain from serial sections stained alternately for myelinated nerve fibers and cell groups. All precautions were taken

[2] See p. 72 for a detailed account of this instrument and its revival.

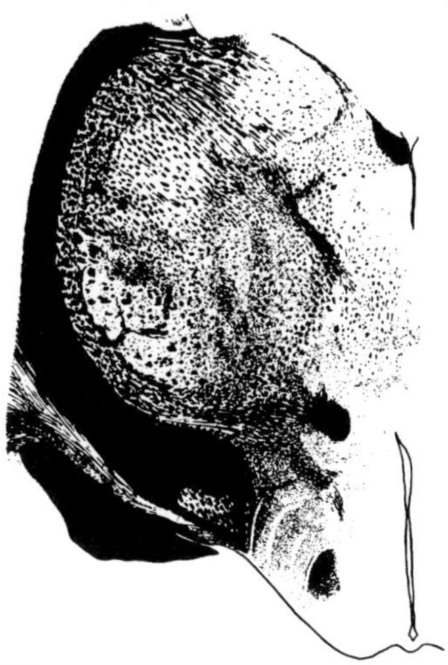

10 Drawing of frontal section no. 148 of cat's diencephalon stained by Weil's method. × 10.

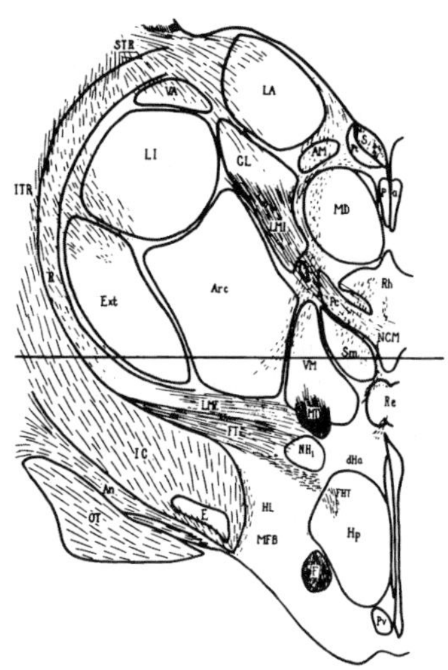

11 Outline drawing of frontal section no. 149 of cat's diencephalon, stained with cresyl violet, showing nuclear outlines for comparison with figure 10. The zero horizontal plane is indicated by a horizontal line. × 10.

Fig. 3.3. An atlas of the cat's brain, prepared under the supervision of Walter R. Ingram (above), was the first map of subcortical nuclei based on electrical exploration of the region. From Ingram, Hannett, & Ranson, 1932, pp. 380, 381; Figures 10, 11.

so that, from these sections or an atlas prepared from them, Horsley-Clarke coordinates could readily be determined for any neural structure inside the brain.

Ingram's previous research experience had been in amphibian endocrinology pursued for his dissertation for the Ph.D., under W.W. Swingle in the Department of Zoology at the University of Iowa. In an autobiographical essay, Ingram (1975, pp. 181–182) commented on some of those developments:

> In retrospect I am amazed that the work progressed so rapidly .... I can see why Dr. Ranson was content to let the youngsters shed the blood – and

tears .... He stood by and observed, making occasional suggestions or criticisms and held me to the line. He had no advantage of experience in this expanding situation but was able to stabilize our goals of high precision and honest observation. As a matter of fact, I never saw him personally use the instrument of which he was so proud, although he showed and explained it reverently to distinguished visitors ....

....

The rapidity of progress is indicated by the publication in 1931 of four communications concerned with the developing methods and some results. Those were hard, intense, and sometimes bitter months – a clear dispassionate backward look is not easy. However, the inevitable frustrations and balks were transitory and the total abiding impression is one of reward and reinforcement. Certainly Dr. Ranson was rewarded to the point that some exuberance leaked from cracks in his normal shell.

The study that laid the basis for these developments, and provided the model for ranging further expansion of the stereotaxy field, was published by W.R. Ingram, F.I. Hannett,[3] and S.W. Ranson in 1932 as "The topography of the nuclei of the diencephalon of the cat." It contained 24 paired figures, arranged on facing pages, with the left showing fiber connections and the right delineating cell groups of thalamus, subthalamus, and hypothalamus. Its stated purpose was "to fix the nuclear boundaries as definitely as possible and [provide] a series of illustrations ... of practical value to the experimenter." This project was obviously aided greatly by Rioch's just-published papers on the diencephalon of the dog and, at Northwestern, as at Michigan, "The writers wish to extend their especial thanks to Dr. Elizabeth Crosby for her kindness in giving aid and suggestions at a crucial point in the devlopment of this study" (Ibid., p. 335). This novel paper differed substantially from its precursors on rat and dog, however, in the direction earlier emphasized by Huber and Crosby. Its purpose was to provide the anatomic basis for study of deep brain structure using multiple neurophysiologic techniques: electrical stimulation, focal lesions, coupled with recording of evoked potentials and the electroencephalogram.

Infrahuman Primates

With the upward movement of American comparative neurology into the subprimate mammalia, principally at the University of Michigan during 1927 to 1932, its culminating ascendancy into the primates followed rapidly through the balance of the 1930s. In 1934, James W. Papez and L.R. Aronson at Cornell

[3] Frances I. Hannett had just completed her medical studies and before beginning her internship, she obtained a master's degree for her contributions.

University published two papers on the thalamic nuclei of *Macacus rhesus*. As they pointed out:

> If investigations are to be carried into the human thalamus it is necesary to have a complete knowledge of the thalamus of some higher primates to complete the gap in the series leading up from the lower forms. [The macaque] offers particularly favorable material owing to its availability, the convenient small size of its brain and the intermediate position which this animal occupies in the order of primates .... Many features which are small and indistinct in the brains of lower mammals stand out as large and distinct structures in this more highly developed brain (Aronson & Papez, 1934, pp. 27–28).

Implementing those remarks, the authors presented 17 figures of serial transverse sections of the diencephalon in which the labelled nuclei were elegantly depicted. In the same year, Richard L. Crouch (1934) at the University of Missouri presented a similarly carefully worked out account of the nuclear configurations of the thalamus, extending the work to subthalamus and hypothalamus of the macaque.

Stereotaxis Revived

With these valuable papers available, Donald Atlas and Rex Ingram (1937) next published an article that matched the earlier one on the cat: "Topography of the brain stem of the rhesus monkey with special reference to the diencephalon" which provided full coordinates and extended the use of the Horsley-Clarke instrument to this animal. Shortly thereafter, George Clark (1939), also at Northwestern, devised an adapter that enabled use of this instrument on the rat. By the end of the decade, therefore, resources had been developed and were being applied at Ranson's Institute in stereotaxic studies of the brains of rat, cat, and monkey – the three species with which this instrument is still busiest today.

Ranson's revival of the Horsley-Clarke instrument truly opened a neurophysiologic Pandora's box, first at his institute, then more widely in this country, and soon internationally. It also contributed substantially to the expansion of traditional comparative neuroanatomy, if in no other way than by the increasing number of atlases prepared for a steadily widening range of animals. By 1975, 40 years after the initial stereotaxic atlases by Ingram and his associates appeared in 1932 and 1937, more than one hundred (106) had been published, and from the two orignal species, cat and monkey, atlases for the brains of 30 species were available.[4] By that time at least 15 descriptions of stereotaxic instruments for use on the human brain had been published, but only four human stereotaxic atlases (Krieg, 1975, pp. 132–137).

[4] These data are from an excellent brochure on *Stereotaxic Atlases* (1991) and a supplement (1994), prepared for David Kopf Instruments, 7324 Elmo St., Tujunga, Ca 91042. See Volume 2 for the distribution of species for which atlases have been published.

PIONEERS IN PRIMATE STUDIES OF THALAMOCORTICAL RELATIONS

The movement up the phylogenetic scale of American comparative studies of the brain, which reached the subprimate mammalia during 1927 to 1932, gained its ultimate attainment in primate studies of the thalamocortical relations throughout the balance of the 1930s. This work had the added distinction of utilizing experimental methods, introduced in 19th-century Europe by Marchi, Nissl, and Von Gudden, which differentially stained injured neurons after lesions of the brain. Following destruction of the nerve cell body, the Marchi method colored the degenerating myelin sheath of its amputated axon. Reciprocally, after destruction of distal axonal terminals, the Nissl/Von Gudden methods detected the retrograde degenerative changes in their parent cell bodies. These manipulative techniques were applied to studies of the architecture and connections of the most cephalic levels of the brain – best developed in the primates – the nuclear groups of the thalamus and the sensory areas of the cerebral cortex, to which their fibers are projected. The animal used was routinely the macaque monkey, imported in increasing numbers from India in the 1930s, which had been introduced in America to several laboratories by the end of the decade, as noted.

Stephen Polyak

A towering figure in those primate studies in comparative neurology was Stephen Lucian Polyak (1889–1955; Fig. 3.4) who had emigrated to America from Yugoslavia in 1929. After a year at the University of California, Berkeley, he spent the balance of his career at the University of Chicago, first in neurology and then in anatomy. In a memorial for Polyak in 1955, his long-time friend, Heinrich Klüver (who had himself emigrated from Germany in 1923), wrote:

> The tumults of Europe in recent decades have induced many of their most able scientists to come to our shores, to our great advantage. Among these was an adventurous Croatian … who declined a professorship … at Zagreb because he saw in this country better promise of freedom to carry on a cherished program of research (Klüver, 1955, p. 3).

Early Life

Young Polyak began the study of medicine at the University of Graz, Austria. With the onset of the First World War in 1914, he served in the medical corps of the Austro-Hungarian Army. Captured on the Russian front, he enrolled in the medical school of the New-Russia University in Odessa. After the Revolution, he joined a Serbian group that was able to leave Russia by a circuitous route and for the balance of the war, he served in the Serbian army in Macedonia and Albania. On returning to civilian life in Yugoslavia in 1920, the University of

Fig. 3.4. The portrait of Stephen Polyak, painted by Christian Abrahamsen in 1953, was the frontispiece of *The Vertebrate Visual System* (1957), a large monograph embellished by insightful historical vignettes and published posthumously.

Zagreb granted him a medical degree and an appointment as assistant in the Department of Neurology and Psychiatry.

For the next 8 years, Polyak was active in neurological research at Zagreb and with support from the government and the Rockefeller Foundation, he served as a visiting research fellow for various periods at Obersteiner's Institute at the University of Vienna, with G. Elliot Smith at the University of London, with Ramón y Cajal at Madrid, and in 1926–1927, with C. Judson Herrick and Karl S. Lashley at the University of Chicago. On returning to Zagreb in 1928, Polyak found his position had been given to another. At this juncture, he was invited to join the University of California as assistant professor of neuroanatomy and began his career in this country in January, 1929.

Three years later, his monograph, *The Main Afferent Fiber Systems of the Cerebral Cortex in Primates,* was published as Volume 2 of the series "University of California Publications in Anatomy." In a lengthy foreword, Polyak ambitiously proclaimed that

> The investigations of the three chief afferent systems [somatosensory, auditory, and visual] in primates, published herein represent only part of a more extensive

experimental study of the fiber systems of the cerebral cortex.... (Polyak, 1932, p. vii).

Polyak acknowledged the locales of the work (with a more extensive study promised) as Chicago, Zagreb, and Berkeley, and his indebtedness for "kind mediatorship" in turn to Herrick, Lashley, and especially to H.M. Evans, for help in publication. As regard methods, he attempted to interrupt the afferent paths of these chief systems close to their diencephalic origin and trace the subsequent degeneration to their primary projection in the cortex. In spite of "carefully laid plans" the lesions involved extensive neural tissue and the only clear result from his animals was to push the boundaries of the somatosenory cortex to the entire precentral agranular region, the entire postcentral gyrus, and a portion of the parietal agranular region.

The Retina

Although Polyak had intially envisioned extending these studies to a more ranging survey of cortical systems, during the balance of his career, the focus of his research was directed first to *The Retina*, a 600-page monograph published in 1941: and then to *The Vertebrate Visual System*, on which he was still working at his death in 1955. Heinrich Klüver, who had promised to complete this *vade mecum* and get it published, spent the next 2 years at the task. Polyak's 1390-page monograph, which appeared in 1957, within its field somewhat resembled Kappers, Huber, and Crosby's *Evolution of the Nervous System... Vertebrates and Man* (1936), but it rested much more extensively on the author's personal contributions to the subject. In addition, one of its most unique and appealing features was Polyak's scholarly presentation of the historical aspects of the development of the knowledge of the eye and visual system from classical antiquity to the present. In his foreword, Klüver called attention to Polyak's "joy of discovery in tracing the origins of our present-day knowledge", as well as "his deeply felt desire to... do justice to the men on whose shoulders we stand...." (Polyak, 1957, p. vi).

A. Earl Walker, Neurosurgeon

A protégé of Stephen Polyak, who easily matched if not exceeded his mentor, Arthur Earl Walker (1907–1995; Fig. 3.5) was born in Winnipeg, Canada, gained his M.D. at the University of Alberta in 1930, and the next year completed his internship at Toronto. Emigrating to the United States for a residency in the division of neurology and neurosurgery at the University of Chicago, he promptly began a study of thalamocortical relationships in the macaque monkey to elucidate the role of the thalamus in sensation. Although he spent 1935 to 1937 as a Rockefeller fellow with John Fulton at Yale and Bernandus Brouwer at the University of Amsterdam, Walker was explicit in attributing his involvement in study of the thalamus to Polyak. In 1938, in the preface to *The*

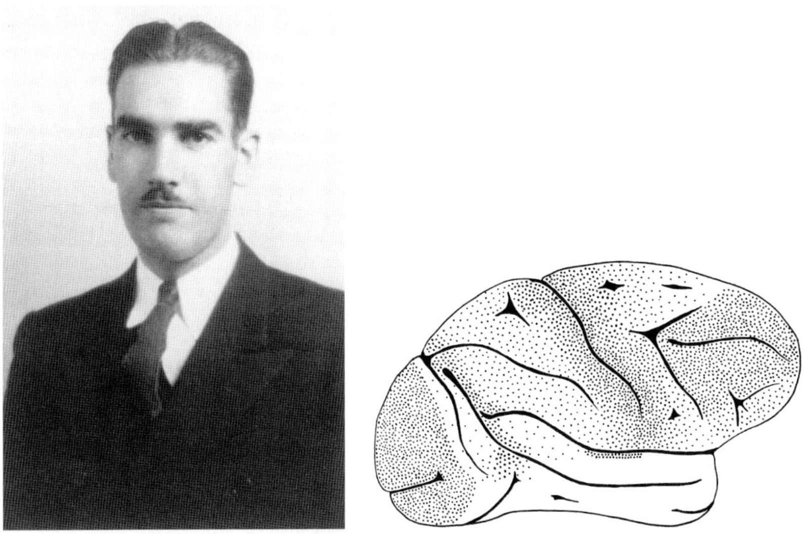

Fig. 3.5. The work and writings of neurosurgeon A. Earl Walker (photographed ca. 1940) are noted for elegant and clear-cut studies of thalamocortical interrelations. Stippling of the stylized drawing of primate right cerebral hemisphere shows densities of projections to cortex from thalamus. From Walker, 1983, p. 191, Figure 68.

Primate Thalamus, he elaborated:

> At the conclusion of his monograph on the somatosensory, visual, and auditory systems, Dr. Stephen Polyak pointed out the paths for future research in this field and suggested the means by which new trails might be followed. The present study, begun at his instigation, has since been guided by his wise counsel (Walker, 1938, p. ix).

It should be noted, however, that Walker's investigative approach to the study of thalamocortical relations was the inverse of Polyak's, for he first extirpated areas of the cortex and later identified the distribution of the degeneration in related nuclei of the thalamus. With this method, it was more feasible for Walker to make focal and well defined lesions in the exposed cortex on the surface of the brains than had been possible for Polyak, who had to plunge knives into the hidden, deep-lying, and much more compactly distributed thalamic structures. As a result, Walker's results possessed a clean-cut quality, enhanced by a characteristic clarity and elegance of presentation.

Continuing with the study of the somatosensory system emphasized by Polyak, Walker pointed out the distinct, well localized terminal projections of the sensory systems in the thalamus, without diffuse distribution. In contrast to Polyak's conclusions, however, Walker's results clearly delineated a differential distribution of the cortical projections from two major divisions of the ventral thalamic nucleus. The lateral division's fibers he found projected to the motor and premotor areas of the precentral convolutions. The posterior division projected to cortex entirely within the central sulcus and postcentral convolution. In this projection, the horizontal somatotopic organization of the nucleus becomes arranged vertically, with fibers from the face area of the nucleus terminating in the lower portion of the postcentral gyrus, those from the arm area maintaining an intermediate position, and those from the leg area of the nucleus passing to the upper end of the gyrus. It will be noted that this arrangement of somatosensory projections in the postcentral gyrus precisely matches the dorsoventral organization of the cortical motor functions in the immediately rostral precentral gyrus. The maintenance throughout the entire system of this spatial association signifies a close relation to the organization of fibers from the bodily segments.

Clinical Implications

As a neurosurgeon, Walker was well equippped to discuss the clinical relationships of the thalamus and cerebral cortex to somatic sensation in man:

> Whereas there is difference of opinion regarding the representation of sensation in the precentral convolutions, there is complete unanimity regarding the sensory function of the postcentral convolution. Stimulation of this gyrus [in man] produces [subjective sensory] paresthesias of a rather complex nature – tickling, itching, vibrations, coolness, etc. rarely pain .... Ablations of this cortex of the postcentral convolution produce severe sensory disturbances .... [A]ll modalities of sensation are absent intially, but ... the rate and degree of restitution of the different categories of sensibility vary. Pain is the first to return and light touch the most persistently disturbed. Some, notably [Henry] Head, believe that pain and temperature sensibility are relatively unimpaired following lesions of the postcentral convolution (Walker, 1938, pp. 256–257).

Walker discussed this further in connection with classical thalamic syndromes following occlusion of blood vessels supplying the thalamus, the most commonly involved being the thalamogeniculate artery. Immediately after such a vascular accident, a complete hemianesthesia is the rule. Later varyious degrees of sensibility return. The disturbances are most pronounced in the extremities, frequently sparing the face, for the softening usually does not involve the medial portion of the thalamus. The discussion continued with descriptions of timing and types of recovered sensation and heightened sensitivity to stimuli on the affected side, capable of producing extreme pain.[5]

[5] It is of interest to compare these clinical symptoms with the reports of intractable pain in S. Weir Mitchell's causalgic patients with gun-shot wounds of the peripheral nerves (see Chapter 1). If the pain were due to epileptic discharge in the thalamus, might that in Mitchell's patients, with irritating scars in peripheral nerves, have been attributable to "kindling"?

In concluding, he wrote:

> This discussion of its pathology has emphasized the complex functions of the thalamus. It is the mediator to which all stimuli from the outside world congregate and become modified and distributed to subcortical or cortical centers, so that the individual may make adequate adjustments to the constantly changing enviroment. The thalamus thus holds the secret of much that goes on within the cerebral cortex (Walker, 1938, p. 277).

To the remarkable number of American neuroscientists who have made important historical contributions during their busy careers – Herrick, Cushing, Fulton, Larsell, Haymaker, Penfield, and others – A. Earl Walker should be added for his editorship of and personal contributions to the comprehensive *History of Neurological Surgery* (1951).[6]

Fred Mettler's Contributions

The related research of Frederick Albert Mettler (1907–1984) at Cornell described the thalamocortical connections to the four lobes of the brain. Born in New York City, Mettler pursued his undergraduate studies at Clark University and completed his Ph.D. under Papez at Cornell in 1933. His dissertation research on "Corticofugal connections of the cortex of macaca mulata," was published in four papers on the frontal, parietal, temporal, and occipital lobes (Mettler, 1935–1936). Like Walker, he made surgical lesions of the cortex but then, like Polyak, used the Marchi method to trace the distribution of degenerating corticothalamic fibers. In the occipital paper, Mettler acknowledged his indebtedness to Papez for his time and suggestions.

To begin with the frontal lobe, following Mettler's lesions in the precentral gyrus, there was widespread degeneration in the lateral nucleus, especially in its anterior part. Lesions in the prefrontal cortex were also followed by degeneration in the anterior portion of the lateral thalamic nucleus. At about the same time, Paul M. Levin, at the division of neurology of the University of Chicago, using the same method and animal, reported (1936) that the corticothalamic system of the frontal lobe arose predominantly in the prefrontal region and terminated chiefly in the rostral portion of the lateral and ventral thalamic nuclei and, in smaller degree, in the medial nucleus. At least three-quarters of this extensive projection arose from Brodmann's area 9. Smaller projections from areas 4 and 6 passed to the lateral and ventral thalamic nucleus. Levin wrote that Ramón y Cajal had suggested that these fibers may

[6] A historical manuscript by Walker, *The Genesis of Neuroscience*, was titled and published posthumously in 1998 by the American Association of Neurological Surgeons in which the origins of the discipline are claimed exclusively for the neurological sciences.

be inhibitory over corticopetal sensory impulses and thus modulate attention. Again, the author's indebtedness to Polyak for suggesting and guiding the work was expressed.

Of the parietal lobe Mettler wrote verbosely that its chief corticofugal connections were to the lateral thalamic nucleus and that there appeared to be an anteroposterior correlation between them. He noted that the pulvinar is definitely within the visuopsychic receptive circuit. After lesions of the occipital lobe, corticogeniculate fibers follow the course of the geniculocortical fibers, but no corticotectal fibers were degenerated. Additionally, the posterior two-thirds of the lateral nucleus received fibers, as did all of the pulvinar. With respect to lesions of the temporal lobe, Mettler found that the temporal cortex sends fibers to the anterior part of the lateral thalamic nucleus, and posterior portions send fibers to the posterior part of that nucleus and to the pulvinar. The superior temporal gyrus sends fibers to the medial geniculate body.

Walker's Conclusions

With his talented capacity for synthesis, Walker stated that evidence from different animals clearly showed both the existence of a corticothalamic system and the importance of its well organized arrangement in the normal functioning of the thalamus. Based on Mettler's studies and Levin's work, it was evident that the cerebral cortex receives a spatially well organized system of fibers from the thalamus and sends to it a system that is organized but not as precisely. As for the significance of this system, some writers held that it was inhibitory, others including Brouwer, with whom Walker worked for a period in Amsterdam, believed it served to permit the cerebral cortex to modulate the sensitivity of the primary receptive centers. As a mechanism of "sensory attention," the complex anatomical organization of the corticothalamic system would not seem an "anatomical waste," but would increase the efficiency of a sensory modality, both by an absolute increase in its sensitivity, and relatively by inhibiting the sensitivity of other sensory mechanisms.

Brouwer's Visit

By a curious coincidence, early in January 1937, Professor Bernardus Brouwer of Amsterdam, visiting this country to participate in the program of the Association for Research in Nervous and Mental Disorders on "Localization of Function in the Cerebral Cortex," was invited to deliver a lecture at the Boston Society for Psychiatry and Neurology. He selected the title "Centrifugal Influence on Centripetal Systems in the Brain" and much of his presentation dealt with the role of the corticothalamic projections. Pointing out the great significance they must have for physiology, he added that they are also valuable for

> the psychological process which we call 'attention.' [In this state] some nerve cells have to be activated, other cells have to be inhibited, because undesirable stimuli have to be prevented entering consciousness .... Hence we accept that there is also a centrifugal side in the process of [somatic] sensation, of vision, of hearing .... I believe that a further analysis of these descending tracts to pure sensory centers will also help physiologists and psychologists to understand some of their experiences (Brouwer, 1933, p. 627).

The suggestion of Brouwer and of Walker that corticothalamic connections provide a substrate for "sensory attention," is certainly pertinent for those fibers that project to the sensory relay nuclei of the thalamus – the nucleus ventralis posterior and the medial and lateral geniculate bodies. The observations presented above, however, leave the impression that the numbers of corticothalamic fibers focally directed to these relay nuclei form a relatively small proportion of the total cortical projection. The great bulk of such corticofugal fibers appears both to arise from associational areas of the cortex and to terminate in the associational nuclei of the thalamus – in the lateral nucleus, the pulvinar, and the medial nucleus. If true, this would not necessarily preclude their involvement in "sensory attention," but it might complicate, rather than simplify, an understanding of the mechanism involved. More recent electrophysiological and EEG studies of the neural substrates of alerting or orienting attention have emphasized the role of the phylogenetically old, intralaminar nuclei of the thalamus, so yet a third category of the thalamic nuclei may be implicated.

A British Critic

During almost 30 years, between 1935 and 1964, studies of functional localization in the cerebral cortex in this country were enlivened by a series of denigrating critiques. The publications of each and all of the investigators mentioned here were the recurrent targets of contentious, critical, transAtlantic barbs from the facile pen of F.M.R. Walshe (e.g., 1957). Writing from Queens Square Hospital, London, reputedly his "only sport was intellectual fencing," in support and defense of Jacksonian principles and the tradition-worn concepts of the British clinical neurological school. Appropriately, all of Walshe's articles were published in the journal of that school, *Brain,* of which he was soon appointed editor, a post he held from 1938 to 1953. Walshe was succeeded by Lord Brain (formerly Dr. Russell Brain), the first occasion since *Pflüger's Archiv für die Gesamte Physiologie,* that a respectable journal included the name of the editor in its title!

Chapter 4
Electrophysiology Overtakes Morphology

During the early decades of the 20th century, physiological studies of the brains of lower vertebrates were scarce compared with the deluge of research by neuroanatomists, whose interest in the functional significance of the structures they so meticulously described was based largely on inference. There were, of course, some notable exceptions. William Loeser (1905) at the University of Kansas repeated on the frog the 19th-century decortication studies in higher animals of Goltz (1881). The later work of Johnston, Rogers, Bagley, and others on the effects of cortical stimulation and decerebration in mammals, amphibians, and birds has been mentioned. A review (Aronson & Noble, 1945) of those investigations, however, pointed out that there is no cerebral cortex in the avian brain, and claims of an excitable motor pallium in reptiles probably rested on a spread of current to the subcortical striatum. Although most of the results of stimulating the more developed mammalian cortex did little more than confirm and extend the 19th-century observations of Eduard Hitzig and David Ferrier, the intensity and scope of the work accomplished in the general field of physiology, the publication of experimental details, and the luster of the sophisticated laboratory settings had created a strong European attraction. It was inevitable, then, that the major American figures should seek to strengthen their knowledge of nervous system function from that source and initiate a steady intercontinental exchange which continues today.

TRANSATLANTIC CROSSINGS OF STUDENTS AND LECTURERS

The spread of experimentation in neurophysiology from Continental and British centers was fostered by the two-way traffic of students and lecturers across the Atlantic during an era when the New World lured senior scientists to visit and lecture while the Old World drew younger men seeking postdoctoral training in established laboratories. This counter-current effectively enlightened and instructed both young and old scientists and accelerated the diffusion of functional concepts of the nervous system into the traditional morphological studies. With new research ideas and teaching programs, the transfer of apparatus and methodologies became a major element in the growth of neuroscience in the United States. As noted in relation to neurology:

> Investigative neuroscience began in earnest during the 1870s, when a few key individuals spent periods of training abroad and brought back to the United States the idea of carrying out research on the nervous system then prevailing in England and on the Continent, especially in Germany. Michael Foster's laboratory at Cambridge, Carl Ludwig's Physiological Institute in Leipzig, and other centers provided models leading to the introduction and development of investigative neuroscience in the United States (Frank, Marshall, & Magoun, 1976, p. 557).

Eastward Passage

Ludwig's American Student

The first of many American students to visit Ludwig's institute was Henry Pickering Bowditch (1840–1911), Harvard M.D., who spent 1869 to 1871 at the institute and "discovered everything that was then lacking in American science and medicine" – rooms for chemistry, microscopes, experimental physiology, vivarium, and library. "Most of all it was animated by the research spirit of Ludwig, who daily and personally supervised his students, each of whom pursued original investigation of a problem of his choice." (Ibid., p. 557). On return to his Alma Mater, Bowditch had administrative backing to establish the first experimental physiological laboratory in America. Among the early topics investigated were the innervation of the heart and the all-or-nothing nature of its contraction, vasomotor nerves, spinal cord tracts, and the physiology of the cerebral cortex. Bowditch's successor in 1906, Walter Bradford Cannon (1871–1945), although he did not go abroad until well after his student years, built on the foundation set in place by his predecessor. He extended the prestige of Harvard to eminence in neuroscientific research through his own work on the autonomic nervous system as well as by assembling in his department an illustrious group of neurophysiologists, Alexander Forbes, Pauline and Hallowell Davis, Robert Morison, and Edward Dempsey among them.

At Sherrington's Laboratory

Many of the next generation of American postdoctoral students were attracted to the congenial laboratory of the renowned Charles Scott Sherrington (1857–1952) at Liverpool and, subsequently, Oxford. Sherrington's imprint is

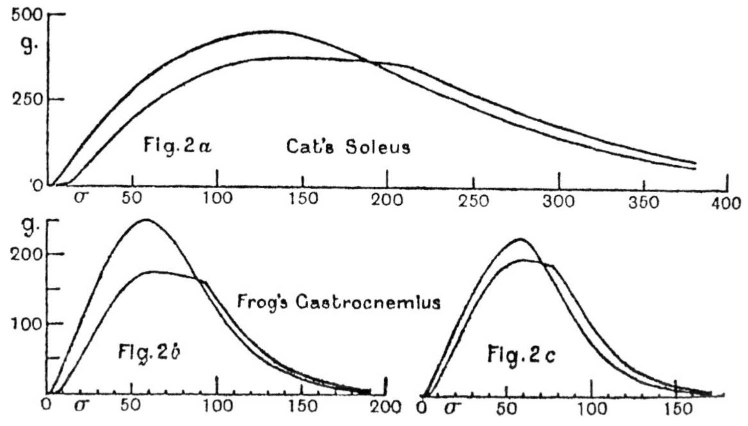

Fig. 4.1. Superimposed successive myograms of isometric muscle twitches in cat (2a) and frog (2b, c) showing the "angle" (lower traces) recorded by a "friction" myograph and the smooth curve (upper traces) obtained by a redesigned "frictionless" myograph. (From Cooper & Eccles, 1929, p. iv.)

seen on the distinguished later careers of those who spent time there: The neurosurgeons Cushing and Penfield and neurologist Denny-Brown continued to carry out experimental laboratory work in parallel with their clinical pursuits, combining the basic and applied facets of neuroscience to the benefit of both. John Fulton likened Oxford to his "second home" and spent an accumulated 6 years there from 1921 to 1930. Toward the end of that period, his career was considerably shaken by the discovery that the inflection in the myogram of muscle contraction (Fig. 4.1) was due to an artefact, an event best described in his own words:

> A ... serious calamity has now befallen us and everyone in the [Oxford] lab feels as though we had been shaken by an earthquake .... In making a new model of a frictionless myograph, we found to our astonishment – Eccles was the first to see it – that the 'angle' was not present ... and there is no question but that it was due to an over-damped recording system. It seems almost incredible that friction and damping could have so marked an effect .... [1]

Fulton's Angle

In his monograph, *Muscular Contraction and the Reflex Control of Movement* (1926), Fulton devoted some 20 pages to conditions for observing the "angle" and factors influencing its configurations. Additionally, his index identified

[1] Letter J.F. Fulton to Alexander Forbes, 30 November 1929. Forbes passed the letter to "HD." (Hallowell Davis) with the notation "Read this and shudder." Quoted with permission, the Countway Library, Harvard University.

another 20 pages on which it had been mentioned. The significance he attached to it is indicated by his statements,

> one may safely infer that the 'angle' characterized the myogram of each individual fibre, and is, therefore, a feature of fundamental importance in muscle physiology .... The question of the ultimate causation of the 'angle' must of course be left open, but that it represents the fundamental discontinuity in the contractile process must be recognized (Fulton, 1926, pp. 115, 259).

John Eccles provided a more substantive account of the angle, and how it was straightened out. After discussing the dependence on the optical myograph for all quantitative investigations in the 1920s and Fulton's systematic studies of muscle twitches, Eccles described the redesign of the bearing of the torsion rod and the smooth rise and decline of the record. The discrepancy had resulted from a frictional resistance. Many years later there was a piquant aftermath. Lord Adrian wrote to Eccles in 1952 saying how much he appreciated

> the way in which I had renounced my hypothesis of electrical transmission at ... synapses. He went on to remark that he was reminded of the charming manner in which, many years before, Sherrington had admitted errors arising from the faulty myograph design. 'But, Jack,' he wrote, 'This was long before your time!' (Eccles & Gibson, 1979, pp. 52–53).

Westward Bound

Travels of Brown-Séquard and Horsley

Among the foreign visitors during the 19th century whose impact lingered, two were notable. The lecture tour in 1832 of Spurzheim espousing phrenology and its consequences were described on pp. 4–5. No less influential were the Toner Lectures of Charles Édouard Brown-Séquard (1877) illuminating the concept of hemispheric dominance and the "educability" of the "lesser" half of the brain. This peripatetic British native (he was born in Mauritius), after failing to secure advancement in France due to lack of citizenship, attempted a career in America. His trail, through Boston, New York, Philadelphia, and Richmond brought no fulfillment to his desire to teach and carry out research in physiology, and in 1878 he finally returned to Paris, became a French citizen, and was appointed to succeed Claude Bernard at the Collège de France, where he extended Bernard's discovery of the vasomotor nerves.

Of Sherrington Himself

One of the most prestigious visitors arrived in 1904 to deliver the Silliman Lectures at Yale in the person of Charles Sherrington. His ideas on patterns of spinal reflexes and their interaction were extremely well received and were published as *The Integrative Action of the Nervous System* (1906), a classic in neurophysiology. The next year, the first public demonstration of the stereotaxic apparatus, designed by physiologist Robert Clarke, was described by Sir Victor Horsley to the British Medical Association meeting in Toronto (Anon., 1906, p. 633). That was not Horsley's first trip to North America; in 1888 he and David Ferrier were among several foreign delegates to the Congress of American Physicians and Surgeons in Washington. Both had been invited to present papers at the day-long symposium on cerebral localization; Horsley's was on "Observations on negative variation in the spinal cord and on the relation of clonic spasms to cortical stimulation." Eager American participants included the young neurosurgeons William Keen and Roswell Park. It is clear that even a century ago, the nervous system was an attractive area of research – all five papers at the initial meeting of the American Physiology Society the same year were on that subject (Brobeck, Reynolds, & Appel, 1987, p. 25).

A Few Permanent Visitors

When Johns Hopkins sought to engage men who would bring prestige to his university during its formative years, he recruited Henry Newell Martin (1848–1896), born in Ireland and with degrees from the Universities of Cambridge and of London. Attracted in 1876 as Professor of Biology at John Hopkins University, Martin's interest in the nervous system was focused on the sense organs. His major influence on American neuroscience was through his students. One of them, W.H. Howell, worked with G. Carl Huber (see p. 60) at Michigan before returning as professor of physiology when Johns Hopkins School of Medicine opened in 1893. A second influential student of Martin's was H.H. Donaldson. After study abroad, Donaldson joined the psychologist G. Stanley Hall in a move to Clark University in 1892, spent some time at the University of Chicago, and finally settled down at the Wistar Institute of Anatomy and Biology in Philadelphia for the major part of his career. His research on the growth of the brain from individual and species perspectives set the classical standard for the developmental characteristics of the experimental rat (1895, 1908).

Two additional European neuroscientists, who came to stay and thereby multiplied their impact on the formative maturation of an American tradition in neuroscience, were Jacques Loeb (Chapter 13) and Adolf Meyer (Chapter 9). Both men were from Middle Europe and on arriving in America were exposed to East Coast and Midwestern institutions before settling into their berths in Berkeley and Baltimore, respectively. This slim influx pales in comparison with the numbers of scientist-émigrés who arrived during the 4th decade in time to participate in the just-opened era of exciting research in electrophysiology of the nervous system.

Contributions to Embryology

At the turn of the 20th century, the field of major research interest in university departments of biology and anatomy departments of medical schools was embryology – principally human embryology. From the 1880s, the outstanding embryologist was Wilhelm His, Professor of Anatomy at the University of Leipzig, an institution notable also for Carl Ludwig's Institute of Physiology (see Chapter 3).

Franklin P. Mall

Fortunately for the United States, in 1883 Franklin Paine Mall (1862–1917), with an M.D. from the University of Michigan, went to Germany for 3 years of advanced study, the first at Heidelberg for acclimatization, and the other 2 at Leipzig, with His and then with Ludwig. At a meeting of the International Association of Scientific Academies (see Volume 2), in 1901 at Paris, Professor His advocated the establishment of national institutes in the two fields of embryology and neurology. The appointed commission designated a group of already established institutes of neurology, however, and the death of His in 1904 and the First World War prevented any subsequent analogous programs on national institutes of embryology. As professor His's most outstanding earlier student and then-current status as professor of anatomy at Johns Hopkins, Franklin Mall had been chosen the U.S. member of the International Neurological Commission. When Charles S. Minot, Professor of Anatomy at Harvard, was later invited to join, Mall wrote him: "Don't for a moment hesitate to join the *Hirnforschungs Institute* .... It will be easier for you to get human embryos ... ." (quoted in Sabin, 1934, p. 289).

Mall's interest in a comparable institute for his specialty persisted and in 1913 he composed an erudite "Plea for an Institute of Human Embryology" (Mall, 1913), which was presented to the Carnegie Institution for consideration. Awarded the same year, a grant for embryological research enabled Mall to secure the appointments of George L. Streeter from Michigan and Adolph Schultz from Zurich as research associates. For its first 2 years, the program occupied the anatomical laboratory of the medical school and then the Embryological Institute was moved to the adjacent new Hunterian Laboratory at Hopkins.

After a quarter-century, Mall succeeded in capturing the internationally known collection of embryologic material that His had been assembling. At Mall's death in 1917, in his honor Professor Wilhelm His, Jr. of Berlin presented to the Carnegie Institution of Embryology in Baltimore a priceless collection of manuscripts, microscopic preparations, and instruments used by his father. Thus the institution Mall founded carried to fruition the work and goal of Mall's mentor, Wilhelm His, Sr. In 1973, the entire collection was transferred on "permanent loan" to the University of California at Davis, where it is still called the Carnegie Institution of Embryology, with Dr. Ronan O'Rahilly as director.

Herbert S. Jennings

Many of the earlier geneticists, both British and American, were reluctantly caught up in the eugenics movement but as the basic field of genetics advanced in the first third of the century, they began to separate themselves from the eugenicists. Together with the embryologists and endocrinologists, the geneticists became advocates of a new mode in biology – experimentalism. Among them were Herbert Spencer Jennings (1868–1947), Professor of Zoology at Johns Hopkins, with paramecia and Ross G. Harrison in the Osborn Laboratories at Yale, with tissue culture from amphibian embryos. Embryology's continuity in the United States, carried on in part by the work of Coghill and of Windle, was described earlier (Chapter 3).

SOMATIC LOCALIZATION IN PRIMATES

Origins Abroad

After Gall's cranioscopy had laid the foundation of the doctrine of localization, the excitability of the motor cortex in carnivores, monkeys, and apes became a dominant investigative theme with Fritsch and Hitzig's report (1870) of their relatively crude experiments carried out in the latter's makeshift laboratory in Berlin. Their findings on dogs were quickly confirmed by David Ferrier in London and, recognizing the neurosurgical possibilities inherent in using species more closely related to humans, he reported seeing movements produced by faradic stimulations of sites on the ape's cortex which he boldly extrapolated (1876) on a diagram of the human brain. Added confirmation came from Horsley and Schäfer (1888) using monkeys and Beevor and Horsley (1890) on the cerebrum of an orang-outang. In Sherrington's Liverpool laboratory the visiting Harvey Cushing noticed the relatively clumsy local procedures and offered to "open" the cranium in several infrahuman primates. In addition to finding "silent areas," fading of sites of effective stimulation without sharp boundaries was reported (Grünbaum & Sherrington, 1903).

John Fulton at Yale

The preeminent pioneer American center for stimulation experiments and its corollary – ablation – was Yale University's Laboratory of Physiology developed by John Fulton when he was appointed there in 1929. From his lengthy training (see p. 83) with Sherrington, Fulton's role in disseminating Sherringtonian concepts in American soil cannot be overestimated (Hoff, 1962). One of the great nodal points of neurophysiology, Sherrington "was master of the phenomenological study of neurophysiology .... [W]ith the light [optical] myograph and the string galvanometer, he was analyzing the phenomena he had earlier described in external outline ... [including] the competing central states of excitation and inhibition" (Ibid., p. 19). Subsequently those methodologies imported from England promoted the emerging independence of neurological surgery from its attachment to neurology (O'Leary, 1974).

Education and Training

Born in St. Paul, Minnesota, John Farquhar Fulton (1899–1960; Fig. 4.2) was named for his father, a physician influential in the founding of a medical school at the University of Minnesota. Young John matriculated at that university in 1917 and, influenced by a professor in English history, extended his historical readings while working part-time in the public library. Transferring to Harvard, he graduated in 1921 and was named a Rhodes Scholar at Oxford. Receiving a B.A. from Oxford with honors in physiology, he was appointed Christopher Welch Scholar and university demonstrator in physiology, which permitted him to work in Sherrington's laboratory. Like most of the young

Fig. 4.2. John F. Fulton in 1933 and one of his "patients," photographed by R.V. Light.

men there during that period, his research was directed to the physiology of skeletal muscle and its reflex control. Completing his D.Phil. in 1925, with a remarkable thesis on muscular control (see p. 83), he returned to Harvard as a medical student, graduated in 1927 and, attracted by the new specialty of brain surgery, spent the following year on the neurosurgical service of Harvey Cushing. On his appointment to professor at Yale, Fulton developed a laboratory of legendary distinction. Its golden era was the early 1930s when the future leaders in the neurological sciences, not only from the United States but from European and South American countries, China and Japan spent time there. Additionally, Fulton organized the *Journal of Neurophysiology* in 1938[2] and with Robert Yerkes established an experimental primate colony which was the precursor of the colony at Orange Park, Florida.

During the Second World War, Fulton was committed to military duties but at its end he completed the manuscript of a biography of Harvey Cushing (1946). By then his health began to suffer from his frenetic pace and his activities were directed increasingly to the history of medicine, culminating in the organization of a Department of History of Science and Medicine at Yale. His international distinction as historian and collector of rare editions in those fields was added to his scientific career which had yielded an orderly progression of studies that promoted the relation between experimental and clinical neurology.

The Potential of Neurophysiology

The first independent phase of Fulton's career was inspired by Cushing and the latter's interest in cerebral physiology, specifically the spasticities and rigidities observed in humans. Fulton hoped to experimentally reproduce and study in the highest apes the movement disorders encountered in the clinic (Walker, 1960, p. 347). The experiments by British investigators during the last part of the 19th century were repeated under more ideal conditions with anatomical controls and resulted in more meaningful data.[3] Those stimulation and ablation experiments, combined with the behavioral observations that derived from Fulton's collaboration with psychologist Carlyle Jacobsen, had an enormous impact on what came to be known as psychosurgery (see Chapter 10). Fulton's first interest was linked to the motoneuron as "spokesman for the nervous system" (Hoff, 1962) and association with Cushing turned his attention to the effects of cortical lesions in producing paralysis often accompanied by enhanced segmental reflexes. Fulton (1926, p. 528) expressed his thinking that "the outstanding feature of the central nervous system is the progressive

[2]The term "neurophysiology" is often attributed to Fulton, but according to Richard Jung (1975, p. 481), was first coined by Oskar Vogt in 1902.

[3]The influence of Cushing was paramount here, according to Walker (1960) and Hoff, (1962).

phylogenetic dominance of the cerebrum … ." Further, that dominance, he believed, was possible only because the cerebrum can inhibit or excite lower centers; its removal released the lower centers as manifested in spastic paralysis. Evidence was accumulating that the pyramidal tract was not the chief culprit in the spasticity seen after cortical lesions in primates, for lesions sharply restricted to the motor areas produced a flaccid paralysis (Hoff, 1962). When followed by bilateral lesions in the premotor area, permanent forced grasping and spasticity resulted (in chimpanzee). Subsequent work at Yale by Paul Bucy (1933), on leave from Chicago, brought the distinction between the effects of motor and premotor areal lesions into sharp focus.

Two Fulton Collaborators

Margaret Kennard

At a 1932 meeting of the Association for Research in Nervous and Mental Disease, Fulton presented a paper with Margaret A. Kennard on "A study of flaccid and spastic paralyzes produced by the lesions of the cerebral cortex in primates" (Fulton & Kennard, 1934). First pointing out the distinction between the flaccid paresis caused by experimental lesions of motor cortex and the motor deficits seen in cases of capsular hemiplegia in man, Fulton explained their wish to find whether or not human spastic states could be induced by lesions of the cortex. A second goal was to explore the gradual return of power after a lesion in the motor area. As Leyton and Sherrington (1917) had already studied the effect of motor area lesions, they confined their work to the premotor area, the cortical surface designated by Brodmann "area 6" as distinct from precentral "area 4."

Margaret Alice Kennard (1899–1972; Fig. 4.3), a New Englander, graduated from Bryn Mawr College and obtained the M.D. at Cornell. Her work on recovery of function after selective ablation in nonhuman primates at various ages has been called "some of the most influential investigations" (Finger & Stein, 1982, p. 136) in showing the relationship of experience and behavior. After removal of discrete areas of motor cortex (Brodmann 4 and 6), she tested the animals' continued development and found that age at the time of operation was a prominent factor in the amount of recovery of function in the contralateral limb – the younger the animal the greater the recovery, which was never complete, however (Kennard, 1938). Her series of experiments brought attention to the developmental aspects of behavior and the capacity, or lack of it, for brain reorganization in young or old respectively, an approach that has been of sustained interest especially to neuropsychologists.

"Pilgrimage to Yale"

A member of the Academica Sinica, Hsiang-Tung Chang, recalled (1984) his strenuous search for training in neurophysiology. In 1942, while displaced

Fig. 4.3. In association with Fulton's program at Yale, Margaret A. Kennard demonstrated the relationship between the extent of recovery from a motor cortex lesion and the age at operation – the younger the better. Photograph from Finger, 1999, p. 270, Figure 1, with permission.

from his assistantship by the invading Japanese, he was teaching physiology at an Army medical school in the interior of China. Starved for news of science, he journeyed 40 miles to the Red Cross Library, "the only place in China where one could find current journals and books published in the Western world" (Ibid., p. 101), came across Fulton's *Physiology of the Nervous System* (1938), and wrote to the author asking to study in his laboratory. Three months later, a cable came: "Yes letter follows" and in another 15 months, after securing the essential papers and funds and arduous travel by way of India, New Zealand, and California, he arrived in New Haven.

During the period 1943 to 1956, Chang gained his Ph.D., was a postdoc at Johns Hopkins with Woolsey and Bard, returned to Yale, then moved to the Rockefeller Institute for Medical Research as an associate. He was called to the Institute of Physiology, Academia Sinica in 1956 and after a Fogarty fellowship-in-residence at the National Institutes of Health, Bethesda, in 1981 he organized and directed the Brain Research Institute, Shanghai. In his research, Chang applied the ingenuity and perseverance that had sustained him through the Japanese invasion and political crises such as the Cultural Revolution. His pent-up research persona was released in his coauthorship of five publications in 1947 alone. His work has touched most of the active topics in neurophysiology in the mid-20th century: reverberating circuits, thalamo-cortical relations, spinal cord afferent projections, and finally the neurology of acupuncture. Chang is but one early example of the internationalization of U.S. neuroscience that took place in midcentury, in large part due to the magnetism of a few early exemplars such as the laboratories of S. Walter Ranson and John Fulton.

Dusser de Barenne and Neuronography

The ascension of American research to the mammalian level of the phylogenetic scale during the 1920s and early 1930s, together with its concentration on the functional roles of cortex, thalamus, and basal ganglia in the primate brain, promptly converted the comparative school into a multidisciplinary *Anlage* of contemporary experimental neuroscience. Such unmatched progress was due principally to utilization of new experimental methods during the decade-and-a-half before midcentury. One of the important advances was Dusser de Barenne's innovation of physiological "neuronography," in which local application of strychnine to the brain reveals interconnections that can be traced by amplified electronic recording. With the involvement of Joannes Gregorius Dusser de Barenne (1885–1940; Fig. 4.4) in the study of corticothalamic mechanisms underlying sensation, this field expanded its earlier preoccupation with localization and connectivity – hallmarks of the American school – into a paradigm of current neuroscience which merges physiology, pharmacology, behavior, and other fields with neuroanatomy, to the benefit of each perspective.

Education

Born in the Netherlands in 1885, as son of a municipal official in Amsterdam, young Dusser de Barenne entered the city's university and completed his medical education in 1909. He spent the next 2 years there as an assistant and became imprinted with the study of the physiological action of strychnine on the central nervous system, a subject that occupied most of his subsequent career. He managed to continue his research between 1914 and 1918, while serving as a medical officer in the Dutch Army. The most significant study of that period was on sensory localization in the cerebral cortex with some 30 cats (Dusser de Barenne, 1916), from which he concluded that application of strychnine to an "active zone," but not elsewhere, resulted in constant and typical signs of excitation which affected both cutaneous and deep sensitivity.[4]

[4] The equipment at that time was not sensitive enough to record any but the primary volleys and thus the localizations of monkey cortical connectivities were clear and reliable (Lettvin, 1989a).

Fig. 4.4. Through strychnine-induced neuronography, Joannes G. Dusser de Barenne identified many corticothalamic projections involved in sensation. Photograph from *J. Neurophysiol.*, 3: 282, 1940.

Cutaneous effects were manifested by spontaneous excitation, paraesthesia, hyperaesthesia, and hyperalgesia and deep effects by abnormal hypersensitivity to pinching and pressure of the bones, tendons, and muscles. In cutaneous sensory functions, he showed that both sides of the body were represented in the cortex of one hemisphere, but the opposite side most strongly, whereas deep sensibility was represented only contralaterally.

In their obituary of Dusser de Barenne, in the *Journal of Neurophysiology*, of which he had been a founding co-editor, John Fulton and Ralph Gerard (1940) pointed out that during 1918–1930, his most important contribution was the result of a visit in 1924 to Sherrington's laboratory at Oxford, where he extended the findings in cats to rhesus monkeys (Dusser de Barenne, 1924). That classic paper demonstrated the major functional subdivisions of the sensory cortex – the areas for the leg, arm, and face – and was the first[5] of an important series of functional localizations in the cerebral cortex. The obituary continued:

> By this time, Dusser de Barenne had, by common consent, became the foremost of the younger generation of Dutch physiologists, [when] the three most

[5] The co-editors apparently did not know of the 1916 paper with an almost identical title.

> important chairs of Physiology and Pharmacology became vacant in Holland almost simultaneously. Any one of them [he] would have filled with distinction ... had it not been for religious restrictions in the Dutch universities, which were intolerable to a free-thinker of Dusser de Barenne's outspoken tendencies. And so it came to pass that Holland allowed the United States to claim one of the most distinguished physiologists the continent of Europe has ever produced (Fulton & Gerard, 1940, pp. 284–285).

Move to America

In the summer of 1929, Dusser de Barenne attended the two international congresses held in this country – at Boston in physiology and a week later, at New Haven in psychology. After a year he returned to New Haven as a Sterling Professor, to establish and direct the Laboratory of Neurophysiology,[6] at Yale. In the remaining, most productive decade of his career, with a succession of young associates, he pursued a variety of investigative programs in which strychninization, thermocoagulation, and aspects of electrophysiology and neuroanatomy applied to sensorimotor functions of the cortex, basal ganglia, and thalamus were a recurring theme.

In presentation of his studies at the Association for Research in Nervous and Mental Diseases, Dusser de Barenne (1935) pointed out that from monkeys upward the cerebral cortex is characterized by a well developed central sulcus, bordered by pre- and postcentral gyri. In 1874, Hitzig had shown in macaque that excitable foci for motor responses lie in the precentral frontal cortex, an observation later extended to the brain of higher apes by Sherrington and associates. From extirpation experiments, it was next found that the somatic sensory area lay exclusively in the postcentral gyrus and adjacent parietal cortex. This was first confirmed in man by Harvey Cushing (1909) who found that electrical stimulation of the postcentral gyrus in conscious patients evoked cutaneous sensations in the opposite half of the body, but neither he nor subsequent neurosurgeons could elicit such sensory events by stimulating the precentral cortex.

Strychnine as Facilitator

When Dusser de Barenne began his study of monkey cortex, however, distinct and typical sensory excitation had been induced by minimal quantities of strychnine applied to foci in wide areas both anterior and posterior to the central sulcus and the areal subdivisions had been determined, as noted. Within each subarea, focal strychninization "sets on fire" the cortex for each segment of the body, but irradiation across boundaries to adjacent subareas for other portions of the body was never observed. When typical sensory excitation

[6] Not to be confused with Fulton's Laboratory of Physiology at the same institution.

was so induced focally the entire remainder of the cortex of that subarea could be anesthetized with procaine or extirpated, in both hemispheres, without terminating hypersensitivity to peripheral stimuli. This led Dusser to propose that, in addition to exciting the entire regional sensory area of the cortex, local cortical strychninization also set on fire those portions of the thalamus in which sensibility of that portion of the body was represented. In support of this assumption, he pointed to the concept of a close functional correlation between thalamus and cortex proposed by Henry Head and Gordon Holmes (1911), as well as the more contemporary studies of Polyak and others, which again emphasized close relations between thalamus and cortex. Culminatingly, in 1938, in collaboration with McCulloch, by electronic recording of "strychnine spikes" Dusser de Barenne verified both of his earlier hypotheses: that local cortical strychninization hypersensitized the entire subareas of the sensorimotor cortex and that under those conditions the corresponding portions of the thalamic projection nuclei also "fired."

Another Novel Technique

In 1934, Dusser had entered into an extraordinarily fruitful collaboration with Warren S. McCulloch, a devoted and congenial younger colleague in research, who wrote of him:

> Characteristically, even then, depressed so much [by the death of his wife] ... he lay awake night after night thinking only that there was no way forward, no method to determine which layers of the cortex were requisite for sensation, until, in the middle of one long lonely night he began to imagine breakfast. 'Then', to quote him, 'I saw an egg cooking slowly. I jumped from bed and rushed to the laboratory, heated a brass rod in boiling water and applied it to the cortex. In twenty minutes there was the method of laminar thermocoagulation!' (McCulloch, 1940, p. 744).

Warren continued:

> We had been working for some time on the effects of various procedures – stimulation, thermocoagulation, etc., on the electrical activity of the cortex, when he finally gave permission to put on strychnine and watch the oscillograph. His face when he saw the first record is as unforgettable as the strychnine spike itself (Ibid., p. 746).

From then, neuronography became the method of choice for investigations of the functional organization of the sensory cortex and finally its relation to subcortical regions, first in monkey, then in finer detail in chimpanzee. Experiments on the chimpanzee were well underway when Dusser de Barenne died in 1940.

An unanticipated feature of the studies of strychnine-evoked excitation of the somatic sensory system was the seemingly equipotential involvement of the ventrolateral nucleus of the thalamus, as well as the precentral gyrus and premotor areas of the cortex which, in aggregate, constitute the somatic motor system. The ventrolateral nucleus of the thalamus relays to the motor and premotor cortex the input of the superior peduncle arising from the dentate nucleus of the cerebellum, an organ designated by Sherrington (1906, p. 346–347) as the "head ganglion of the proprioceptive system.... Its size from animal species to animal species strikingly accords with the range and complexity of the habitual movements of the species." It is difficult to conceive that excitation of these neural substrates for motor integration cannot intrinsically provoke the same changes in sensory function elicited by direct strychninization of the ventralis posterior nucleus of the thalamus or the postcentral gyrus of the cortex. Possibly, these discharging foci in motor regions may bombard and "set on fire" adjacent sensory areas via intrathalamic or intracortical connections. At the cortical level, however, it is not clear why such postulated levels of excitation would not simultaneously discharge corticospinal motor neurons and elicit overt motor responses. Dusser had suggested (1924) that the absence of motor effects in his early experiments on cats might have been due to the fact that his cortical application of strychnine was "slight" and probably confined to the superficial layers of cortex, where it irradiated over the part of cortex that was in close functional (sensory) connection with it.

Tower and Hines at Johns Hopkins

In the experimental analysis of the cortical motor area, two unanticipated but exceedingly cogent contributions were made by Sarah Sheldon Tower (1901–1984; Fig. 4.5) and her mentor, Marion Hines (1889–1982; Fig. 4.6), at Johns Hopkins Medical School. In a brief initial report they noted that in the cat, transection of one pyramidal tract below the pons produced in contralateral limbs a deficit in motor capability but "[s]pasticity, or other evidence of release is absent" (Tower & Hines, 1935, p. 376). Subsequent stimulation of the motor cortex produced inhibitory effects on tonic or clonic states in the affected limbs – abolition of the usual motor responses to stimulus with evocation of an extrapyramidal type of response. In rhesus monkeys a similar response to the same lesions was found, and the authors were able to demonstrate that the extrapyramidal impulses were mediated exclusively by the corticospinal tract.

A year later, in an abstract for the annual meeting of the American Physiological Society, Marion Hines filled in the last missing cipher of the picture. A strip[7] of cortical tissue about 3 millimeters wide and following the

[7] Jocularly called the "strip of Marion Hines," she gallantly persisted in using the term with quotation marks, whereas her close colleague, Dusser de Barenne, designated it (letter to Marion Hines, October 11, 1937) as 4s "(s standing for 'strip' in your honor!)."

Fig. 4.5. In close collaboration with Marion Hines (see Fig. 4.6), Sarah S. Tower differentiated the lesion causing muscular spasticity from that resulting in paralysis. Photograph from Johns Hopkins University? with kind permission.

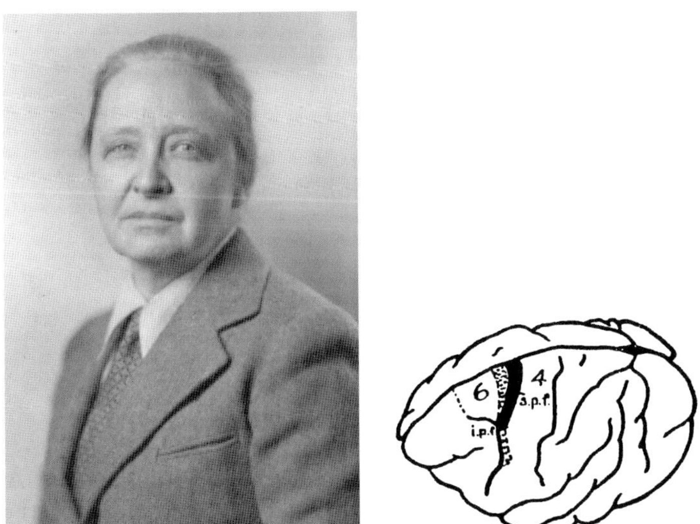

Fig. 4.6. Marion Hines extended her work with Sarah Tower (see Fig. 4.5) on muscular spasticity to localization of an inhibitory strip of cortical tissue. Photograph from Johns Hopkins University with kind permission; insert from Hines, 1937, p. 321. For an illuminating account of this team of achieving women, see p. 5 of the oral history of Clinton N. Woolsey in the archives at the University of Wisconsin, Madison.

curvature of the central fissure from the lateral inferior precentral fissure to the medial sulcus callosus marginalis was removed in seven macaque monkeys. Examination of stained sections proved the ablated cortex was taken from Brodmann's area 4. The surviving animals showed unmistakable contralateral release phenomena – clinical signs of spasticity – without forced grasping. Those results clearly were evidence of the existence of two anatomically discrete regions, the removal of which produce grasping and spasticity respectively. In a summary of her work on this complicated question, Hines (1937, p. 334) stated: "We have been able then to separate the phenomena of spasticity from its companion paralysis and to demonstrate … a new localization for the inhibition of that substrate on movement, tone."

A MIDWESTERN CENTER OF EXCELLENCE

The chance siting in Chicago of three vigorous laboratories in as many institutions during the 1930s and 1940s produced a transient powerhouse that energized the westward movement of American neuroscience. Each of the city's three major universities became home to a young and productive scientist/physician dedicated to research on all aspects of the nervous system and behavior: Percival Bailey, at the University of Illinois and Illinois Neuropsychiatric Institute (INI) on the western side of the city; Ralph W. Gerard at the University of Chicago on the south; and S. Walter Ranson at Northwestern University Medical School on the north side. Two additional centers were already active during this era: the large group at Washington University, St. Louis under Joseph Erlanger in the Department of Physiology and the more modest program in psychology with Lee Travis as its chief at the University of Iowa at Iowa City. Those centers will be described in Chapter 5 in connection with the specific contribution of each to the rapid build-up of steam that was pushing American neuroscience ahead.

Percival Bailey's Vision

When Dusser de Barenne died in 1940, Bailey, who had spent some time with him at Yale, invited McCulloch and Gerhardt von Bonin to join him in a basic neurophysiology laboratory that would complement the clinical program of the newly opened INI. One of America's most colorful and influential neuroscientists during the mid-20th century, Percival Bailey (1892–1973; Fig. 4.7) described in his collection of delightful autobiographical sketches (1969) his upward mobility from a backward region near the confluence of the Ohio and Mississippi Rivers to earning a Ph.D. in neuroanatomy from the University of Chicago and an M.D. from Northwestern University Medical School. Attracted to neurosurgery, in 1919 he made the first of several on-and-off

Fig. 4.7. Posing in the vestments of his honorary degree from the Sorbonne, 1949, Percival Bailey's energy and clear vision are evident.

associations with Harvey Cushing in Boston which lasted 9 years. Assigned to extend his mentor's earlier studies of hypophysectomy in dogs, Bailey used a novel subtemporal operative approach and found accidentally that he had produced polyuria without exposing the gland. He and Frédéric Bremer, visiting from Brussels, subsequently showed (Bailey & Bremer, 1921) that "puncture of the infundibular hypothalamus provoked adiposo-genital dystrophy with diabetes insipidus and attributed the result to disturbance of an hypothalamic innervation of the pituitary" (Magoun & Clemente, 1976, p. 236). This finding was contrary to Cushing's belief that those phenomena were directly controlled by the pituitary and was a pioneering contribution to neuroendocrinology. Two years later, Bailey's second major contribution was to neuropathology – a histological study and classification of glial tumors accumulated from Cushing's service (Bailey & Cushing, 1926).

Opportunity Denied

In 1928 Bailey was recruited by the University of Chicago because of his outstanding training and accomplishments as well as for his leadership qualities

(Blustein, 1992) with the expectation that the "vast new field" of "psychoneurology" (Ibid., p. 99) would be served by an organized center at the university. A start was made – cooperative activities of the Departments of Medicine (psychiatry) and Surgery (neurology, neurosurgery) were arranged and a "Neurology Club" was organized that drew senior figures from the Departments of Anatomy, Physiology, Psychology, Medicine, and Surgery to discuss informally their research problems and the future directions of the new field. The club embraced enthusiastically the efforts to secure funds (from the Rockefeller Foundation) for an interdisciplinary enterprise[8] which made no headway. After a frustrating decade, during which he felt his potential contribution to research had been constrained, Bailey accepted a professorship offered by the University of Illinois that gave him at the INI the room to carry out the full sweep of his training and interests; his assistant, Paul Bucy, went with him. Thus, for the second time Chicago let slip an opportunity to capitalize on the forward thinking of one of its most illustrious faculty.[9]

With the transfer of Dusser de Barenne's remaining group to the University of Illinois, the planned strychnine research program was brought to a seamless conclusion and abandoned, mainly because of the advantages of the new evoked potential technique, but it also had serious drawbacks.

> The only reservation that sometimes developed around strychnine neuronography was: Is this normal neurophysiology or is it pathological? Because strychnine alters the function of nerve cells to the point where you develop epileptic seizures, and in a sense ... you were recording discharges as you describe them in clinical neurology. But they were seizure discharges of single neurons rather than some massive area of the temporal lobe seizure or something of that sort .... I'm not trying to imply that strychnine neuronography is just a kind of miniature seizure, but there seems to be the prospect that these people could find that everything was connected with everything else, so to speak .... It seemed as though many of those neuronography records showed a far broader discharge than you get by [evoked potentials] or by making a lesion and then studying the degenerated nerve fibers anatomically to find the connections .... I think people felt that they had better control – the problem with neuronography was that it was excessive, so to speak (Magoun, 1979, pp. 120, 121, 122).[10]

Warren McCulloch, "Rebel Genius"

With his unfashionable beard and intense demeanor, Warren Sturgis McCulloch (1898–1969; Fig. 4.8) made a lasting imprint on those who heard and saw him

[8]The slow, unbelievably complex developments that resisted Bailey's vision are vividly detailed by Blustein (1992).

[9]The first episode involving Clarence Luther Herrick is described in Chapter 2.

[10]For a more detailed evaluation of strychnine neuronography, see Lettvin, 1989.

Fig. 4.8. A fleeting moment in the intense life of Warren S. McCulloch, shown with his mother at the Memphis, Tennessee airport in 1952. Photograph by A.B. Scheibel and published with kind permission.

as well as on neuroscience. Few doubted that he was a genius, and Gerard (1970) perceived the rebel.

Education and Early Work

Born and raised in New Jersey with summers at Nantucket sailing and absorbing the sea, McCulloch entered Haverford College intending to prepare for the ministry, but received his B.A. from Yale. He then decided to pursue the study of the underlying mechanisms of mind. To that end, he progressed through graduate work in psychology and medical school (M.D., Columbia University, 1927). A succession of posts – internship in neurology, experimental work in epilepsy and head trauma, with a year studying mathematical physics (New York University), then Rockland State Hospital – brought him back to Yale in Dusser de Barenne's Laboratory of Neurophysiology. This circuitous pathway was motivated partially by the philosophical questions his rearing and Quaker contacts had instilled and McCulloch finally turned to neurophysiology in search of the "significant variables" (Heims, 1991, p. 33) inherent in any explanation of mental processes. His work with Dusser, 1934 to 1941, produced a mapping of cerebral functional localizations that was essential for further studies in the higher apes. The merit of his review, "Functional organization of the cerebral cortex" (1944) attested to his achievement of solid standing in neurophysiology.

Nerve Nets

After moving to Chicago, McCulloch became increasingly preoccupied with more fundamental aspects of cerebral function – the mathematical modeling of nerve nets. With a brilliant young student from the University of Chicago, Walter. H. Pitts, Jr., McCulloch developed and tested experimentally a concept of neuronal firing and feedback loops (McCulloch & Pitts, 1943) that emerged as a precursor of cybernetics.[11]

From the experience gained during this period, in cross-town collaboration with McCulloch at the INI, where the ambience was livelier and more adventuresome than at Northwestern's Institute of Neurology under Ranson, Magoun reported that "the elite feature of Warren was his genius-like brilliance [with an] incredible range of intellectual interests .... He was always the warmest, most sociable sort of fellow with an entire spectrum of types of people. My admiration for him is as much human and personal as it is for what he did for neuroscience" (Magoun, 1979, p. 130). In evaluation of strychnine neuronography,

> You begin to see that they broke out of the confining mold of the individual discipline; they were using anatomy and electronics and neurosurgery to get at electrophysiology, for example .... The program of interrelations of cortical areas was a major contribution in the beginnings of the use of electronic amplification in the study of organizational and functional interrelationships in long pathways in cerebral cortical areas of the brain. I would say that that was one of the prime contributions made in that area. And in addition, the discovery through these means of some of these areas that had not previously been distinguished or identified like the suppressor areas. Warren was a key figure in the development of the field of biological psychiatry in this country. His broad-based concepts and vision of the notion of feedback gave a broader point of view to the theorists who were interested in it (Magoun, 1979, pp. 119, 129).

Warren McCulloch and Walter Pitts moved on to the Massachusetts Institute of Technology about 1952 where they continued the experimental work on perception and McCulloch became one of the "cyberneticians" (Heims, 1991,

[11] See Heims, 1991 for a fuller study of this aspect of McCulloch's contributions; also summarized in Marshall and Magoun, 1998, pp. 261–263.

p. 11). In his attempt to explain the human mind in mechanical terms, "[h]is romanticizing the machine was a touch of mysticism in his make-up, but his liking things mechanical was squarely in the American tradition in which he grew up" (Ibid., pp. 39–40).

Synergy of Ideas

Warren McCulloch's sociable and generous personality had a great deal to do with the influence he projected. Those who heard him talk, formally or informally, found him unforgettable and inspiring. On a more personal level, his home sheltered at least two young men who became collaborators: Walter H. Pitts, Jr. (1923–1969) and Jerome Y. Lettvin. Pitts was born in Detroit and at the age of 15, he was holding his own at the University of Chicago among the biomathematicians around Nicholas Rashevsky and studying symbolic logic with the émigré philosopher, Rudolf Carnap (Heims, 1991). McCulloch met Pitts a few years later when the former presented his ideas on the logic of the brain at Rashevsky's seminar. Still later, the important paper of 1943 was published in Rashevsky's *Bulletin of Mathematical Physics.* Pitts worked on the Manhattan Project, held a Guggenheim scholarship (1944–1945), and then followed his mentor to MIT, where he was in the Research Laboratory of Electronics as mathematician and physiologist.[12]

Jerome Y. Lettvin is another neuroscientist whose scientific roots were nourished by Warren McCulloch. Born and reared in Chicago, he attended the University of Illinois, gaining a B.S. and M.D. almost simultaneously (1942 and 1943, respectively). His neurophysiological interests were carried out in the psychology department of the University of Rochester, New York, to 1948. Moving eastward, he passed through Harvard Medical School on his way to join, in 1951, the stimulating neurocommunications group in the Research Laboratory of Electronics at MIT, Walter A. Rosenblith among them. Lettvin's experimental work (1959) extended the theoretical concepts of Pitts and McCulloch and defined the link between behavior in response to perception and the underlying anatomical elements.[13]

As for Percival Bailey, deteriorating eyesight compelled a shift in his gaze from experimental microscopic studies to psychiatry, an area that had intrigued him since his student days. He became an energetic champion of research in public mental hospitals which bore fruit in the "showplace" at Galesville, Illinois as well as at other state hospitals (see p. 163). Another manifestation of his career's refocus and its revolutionary consequences was

[12]I am indebted to J.R. Zweizig for the brief biography of Pitts in *Inst. Radio Eng. Proc.,* 47 pt. 4: 2030 (1959).
[13]For a discussion of the relation of this work on frog to that of Hubel and Wiesel on cat, see Heims, 1991, p. 243.

described on page 27. Bailey's departure from the University of Chicago diminished but did not eliminate the luster of the group dedicated to brain research that remained. The presence of leading figures such as C. Judson Herrick, Heinrich Klüver, Stephen Polyak, Nathaniel Kleitman, Ralph Gerard, and others assured the university's place among the shakers and makers in neuroscience. However, their research was individualized with no attempt among them at coordination in a planned program. Gerard's laboratory was especially productive in the evoked potential technique and inaugurated its application to neuronal activity in the central nervous system. With gifted graduate students the method was carried from peripheral nerve into the higher regions. The third and northernmost center of neuroscience activity in Chicago was the Institute of Neurology headed by S. Walter Ranson at Northwestern University Medical School. In this, the largest and most closely directed of the three groups in Chicago, all research efforts were concentrated on the lower brain stem. These programs are described in Chapter 5.

The historical development of the "fine structure" of nervous tissue that culminated in Ramón y Cajal's work was beautifully described by neurosurgeon A. Earl Walker in *The Primate Thalamus* (1938). Walker ended his introduction thus:

> By the end of 1936 the normal cytoarchitecture and myeloarchitecture of the primate thalamus had been studied, a great many contributions on the thalamocortical connections had made their way into the literature, and finally a few observations on the afferent thalamic systems had been reported (p. 12).

That summarizing sentence seems to say that the anatomical study of the central nervous system had no place to go, that progress lay only in the direction of function. Indeed, such had already taken place in the neurophysiologists' pursuit of the nerve impulse and its characteristics and causes.

Stimulation of the Human Brain

First Attempts

The first published record of direct stimulation of a living human brain in America described an event in Cincinnati, Ohio at one of the more bustling medical schools in the Midwest. The experiment was the work of Roberts Bartholow, M.D., whose patients' ulcerating cranium provided access to the intact dura mater over the parietal cortex. Using faradic current between insulated electrodes, his systematic observations (Bartholow, 1874) indicated that the dura and brain tissue were insensitive to needle penetration, and that weak current produced movement of the contralateral limbs and movement of the head and eyes away from the stimulated cortical hemisphere. Unhappily, the patient's deterioration and death a few days later and the severe criticism of Bartholow by his peers and the press marred the significance of the experiment.

Claim for Priority

Slightly more than a decade later, surgical exposure of the cerebral cortex afforded a more acceptable opportunity to stimulate the brain. Taking a cue from laboratory procedures, "osteoplastic flaps" substituted for nature's "experiments." In 1886, the daring procedure was undertaken by two neurosurgeons, Horsley in London and Roswell Park (1852–1914) at the Buffalo, New York General Hospital. Ambitious and energetic,[14] Park made the most of the new findings in cerebral localizations, assigning the responsibility for diagnosis to the neurologist and for finding the lesion to the neurosurgeon. From the literature and personal contacts he assembled a list of 67 operated cases (17 by Americans) in which

> the extirpation ... was practiced on account of lesions previously diagnosed, according to the principles of cerebral localization .... In this table are included three of my own cases, never before reported, the first of which antedates Mr. Horsley's first case, according to his published table of personal cases, by three months (Park, 1889, p. 314).

Park's table was part of a featured presentation at the first triennial session of the Congress of American Physicians and Surgeons in Washington, D.C., in 1888. His enthusiasm led him to declare:

> Indeed ... so astonishing have been the advances of the past 20 years that one finds ample justification in maintaining that, with the sole exception of the science of electricity, nowhere in the whole domain of theoretical or applied science has progress been so rapid or visible results so remarkable as in surgery. So far as operative surgery is concerned, it can now rank as an exact science (Park, 1889, pp. 301–302).

Four years later, Park again claimed his priority: "The first case attempted in this country in accordance with the principles of cerebral localization was in a patient on whom I operated Nov. 16, 1886" (Park, 1892, p. 621). His table, however, noted an even earlier case, a 14-year-old girl who had a surgery February 24 that same year, relegating Horsley even further to the rear.

Toward the end of his career in neurosurgery, Park reclaimed the first accurate localization and successful operation "in this country, where the whole

[14] The State University of New York at Buffalo mounted an exhibit on the Pan-American Exposition held in Buffalo for which Roswell Park was medical director, overseeing daily medical care and disease prevention. See http://ublib.buffalo.edu/hsl.

procedure was based upon the, then, new principles of cerebral localization" (Park, 1913, p. 304). He asked if the high hopes aroused then had been realized.

Competition

As Park noted in his table, he had several competitors in the use of localization techniques. Among them was William W. Keen, Jr. (1837–1932), a young associate of S. Weir Mitchell in several of the latter's research studies: snake venom, care and interpretation of the effects of peripheral nerve injuries (see Chapter 1), and reflex paralysis. Park's table showed that a year after Park's first case, Keen had operated using "principles of cerebral localization." Keen (1891) was the author of a report of five cerebral surgeries, in three of which he noted that he had stimulated the cortex "by faradism." He found either no response or the expected movement of contralateral muscles. When the electrode was applied to diseased tissue, a typical epileptic fit developed, leading him to ask whether or not the effect could reveal normal-appearing tissue to be, in fact, diseased. With few data, however, he wisely "submit[ted the question] for further observation and judgment" (Keen, 1891, p. 229). His case histories indicate that by the last decade of the 19th century there was wide acceptance of the reality of cerebral representation of specific muscles.

Refinement

The procedure was next reported by Harvey Cushing (1909) in Baltimore and was brought to full realization by Wilder Penfield (Penfield & Rasmussen, 1950) at the Montreal Neurological Institute[15] in carefully controlled and recorded experimental stimulations on awake patients prior to extirpation of epileptogenic cerebral tissue (see Chapter 13). As early as 1928 and as a consequence of his 6 months with Foerster, Penfield had been struck with the possibilities of physiological study of brain mechanisms offered by stimulation of conscious patients. Many years later, he recalled his thoughts:

> The electrical stimulation that must be used to guide the surgeon in his removal of the [epileptogenic] cause would perhaps tell the thoughtful surgeon many secrets about the living, functioning brain. He could learn what the conscious patient might tell him. This would help the neurosurgeon to understand the interrelationship of the mind to localized functional mechanisms in the brain (Penfield, 1977, p. 168).

[15] Penfield had visited Otfrid Foerster's clinic in Breslau in 1928 where human cerebral resections for relief of seizures were carried out. See Penfield, 1977 for his recollection of the neuropathological study of the resected tissue.

Chapter 5
New Trails on Fresh Terrain

VISCERAL AND SOMATIC PHYSIOLOGY OF THE LOWER BRAIN STEM

In tracing the history of neuroscience in the United States thus far, foreign roots have been examined by looking eastward to events and the traffic of scientists across the Atlantic Ocean. Redirecting our focus toward the opposite longitudes, as did a far-sighted American philanthropy – the Rockefeller Foundation – it is rewarding to examine China's place in the furtherance of early 20th-century U.S. neural sciences. The vivid experience of H.-T. Chang described in Chapter 4 is an example. Additional, often overlooked instances of transpacific relationships in biomedicine are found in the book, *Western Medicine in a Chinese Palace*, by John Z. Bowers (1972), the source of much of the following details.

Neuroscience at Peking Union Medical College

"PUMC" was opened with a gala, week-long, dedication program in 1921, after some 7 years of exploratory and developmental activity on the part of the Rockefeller Foundation and its China Medical Board. The name and setting of the school related to the foundation's purchase, from the London Missionary Society, of the site and buildings of an earlier Union Medical College at Peking, the only one of its kind to have been recognized by the Chinese government. For possible expansion, the Rockefeller Foundation also purchased an adjoining palace of Prince Yu. From then on, PUMC was commonly known as "Yu Wang Fu" – in English translation, "Prince Yu's Palace." To many Chinese, however, the name Rockefeller was synonymous with the Standard Oil Company which filled their lamps and, since the Chinese word for "oil" sounded like the name of the Prince, an easy pun – the Oil Prince's Palace – delighted the Chinese sense of humor.

The high goals of the new medical college were stated with clarity and conviction by its first director, Franklin C. McLean:

> 1. Primarily to give a medical education comparable to that provided by the best medical schools of the United States and Europe .... 2. To afford

opportunities for research, especially with reference to problems peculiar to the Far East.... (Bowers, 1972, p. 68).

A remarkable faculty was recruited to implement those goals. The first professorial appointment was in anatomy and was filled by Edmund P. Cowdrey, a Canadian with a Ph.D. from the University of Chicago. Cowdrey was teaching histology under Franklin P. Mall at Hopkins when he was attracted to Peking. His colleague there in neurology and embryology was a Toronto classmate, Davidson Black, who had studied with the English anatomist/anthropologist, G. Elliot Smith, and with the comparative neuroanatomist, C.U. Ariens Kappers, director of the Central Dutch Institute of Brain Research. When Cowdrey resigned in 1921, Black was designated head of the department, Kappers became a visiting professor 1923–1924, and a year after his return to Holland, his student, A.B. Drooglever-Fortuyn, joined the department to teach neuroanatomy. Samuel H. Detweiler came to PUMC in 1920 from Yale, but in 1923 left Peking to help form the College of Physicians and Surgeons at Columbia University, "where his research in neuroembryology ... was outstanding" (Bowers, 1972, p. 96).

Among the ranging investigative interests of this group, that of Davidson Black especially exemplified the second of PUMC's goals –

...research with reference to problems peculiar to the Far East." As John Bowers (Ibid., p. 96) pointed out: "Calcified structures had for many centuries been one of the most popular medicines in the Chinese formulary.... 'Dragon's teeth' and 'dragon's bones' were ... held in high esteem. A major source was the fossil remains in a limestone cliff, Chouk'outien, thirty miles west of Peking....

There, in 1927, a human molar tooth was found, which Black showed to be from a separate genus later known as "Peking man." In 1929, an almost complete skull with a jaw of *Sinanthropus pekinensis* was discovered and in all 40 specimens were finally identified. Davidson Black's studies of the skulls indicated with little doubt that the Peking man's brain was much like that of primitive man.[1]

In 1924, Wu Hsien became the first Chinese departmental head at PUMC, in physiological chemistry. He had been one of 160 students selected under the Boxer Indemnity Program for study in the United States, where he majored in chemistry and biology at Massachusetts Institute of Technology. In 1917, Wu began graduate study in biochemistry at Harvard Medical School with Otto Folin, and together they developed the classic Folin-Wu method of protein determination. In 1922, Wu was joined by Donald D. van Slyke, of the

[1] The sad story of the disappearance of the fossils has been related by H.L. Shapiro (1974).

Rockefeller Institute, as a visiting professor of biochemistry. Van Slyke later recalled: "In three months I completed experiments which would have required a year in any other laboratory, including the Rockefeller Institute .... Everything was ready to go and every experiment worked" (quoted in Bowers, 1972, p. 100). Later, in 1930, A. Baird Hastings, professor of biochemistry at the University of Chicago, was a visiting professor at PUMC. In the Department of Pharmacology at PUMC were Chen Ko-kuei and Carl S. Schmidt, who jointly introduced ephedrine to Western medicine.[2] The precipitating episode was described as follows:

> ... Chen purchased *ma huang* at an herbal shop near PUMC and injected it into a dog that Schmidt had prepared for the medical students' laboratory exercises. The kymograph [tracing] showed a sharp increase in the animal's heart rate and a rise in blood pressure .... Chen and Schmidt then undertook their classic studies which demonstrated the cardiovascular effects of ephedrine.... Soon ephedrine became one of the leading drugs in the therapeutic armamentarium of physicians in America and Europe (Ibid., p. 104).

On Schmidt's return to the University of Pennsylvania in 1924, Chen followed, but to Johns Hopkins Medical School to earn an M.D. in 1927. Two years later, he was appointed director of pharmacological research for the Eli Lilly Company and professor of pharmacology at the University of Indiana Medical Center.

Vasomotor Studies of Robert K.S. Lim

In physiology at PUMC, the first appointee was Ernest W.H. Cruickshank, from a long line of medical forebears in Scotland, who had pursued physiological studies on the vagus nerve at University College, London. "A rather dour man, [he] stayed apart from his colleagues on the faculty [and] left PUMC to itself" (Ibid., p. 101). Attention meanwhile was directed to a brilliant young Chinese, Robert K.S. Lim (1897–1969). Born in Singapore, the son of a physician, Lim went to Scotland where he graduated in medicine at the University of Edinburgh and spent 4 years working with its distinguished physiologist, Edward Sharpey-Schäfer. In 1923, at the age of 26, Lim was considered too inexperienced for a senior appointment at PUMC, and the China Medical Board arranged for him to spend a period with A.J. Carlson and A.C. Ivy in gastrointestinal physiology at the University of Chicago. Before the year was completed, Carlson reported that Robert Lim was eminently qualified for a senior appointment at any medical school, whether in the West or in China.

[2]Ephedrine is the principal alkaloid of a species of *Ephedra*, which, under the name of *ma huang*, has been used as a medicine in China for thousands of years.

The role that Lim played locally, nationally, and internationally was significant.

> Thus, the program in Physiology at PUMC received a great fillip with the arrival in 1924 of 'Bobby' Lim. He was short, charismatic and loaded with energy and enthusiasm. He came to Peking with ... far more facility in English, which he spoke with a broad Scottish burr, than in Chinese. 'If you closed your eyes you thought you were talking to a Scot, not a Chinese,' was the comment of one of his colleagues .... He was a major force in the general development of the school, throwing the weight of his intelligence and his personality into a wide range of programs (Ibid., p. 102).

In 1926, Lim led the formation of the Chinese Physiological Society, served as its first chairman, and relatedly, established the *Chinese Journal of Physiology*. Lim's success in building a department was attested by Walter Cannon's (1945, pp. 182–183) enthusiastic recollections of his year there in 1935. During the Sino-Japanese War (1937 to 1948), Lim was occupied with leadership positions in China's military medical programs. Resigning his post as Surgeon General, he came to the United States and returned to research in physiology, first at the University of Illinois and then as director of the Miles Medical Research Laboratory at Elkhart, Indiana, until his death in 1969.

S.-C. Wang's Myelencephalon Sympathetic Center

It was at this remarkable institution, with Robert Lim and under the guidance of its distinguished faculty, that Shih-Chun Wang (1910–1993) was able to complete his medical education and obtain his M.D. in 1935 at 25 years of age. During the next 2 years, he was research associate in physiology and participated in a series of studies that began with a review of the development of the concept of an autonomic center in the medulla oblongata (myelencephalon) of the brain. After mention of 19th-century European work,[3] the review called attention to studies in vasomotor reflex arcs undertaken in 1915 by S. Walter Ranson and his associates at Northwestern University Medical School, Chicago, which, inter alia, differentiated bulbar areas for pressor and depressor responses. Twenty years after Ranson's observations, Lim and his colleagues at PUMC undertook a series of studies on the "myelencephalic sympathetic center" in which S.-C. Wang was an outstanding associate. As a fellow of the China Medical Board, Wang spent the next 3 years (1937–1940) with Ranson, who by then was director of the Institute of Neurology at Northwestern's medical school, where they collaborated in brain-stem studies. Until the beginning of Wang's independent career, therefore, three pioneering

[3]Important early studies of a bulbar vasomotor mechanism by Karl Ludwig and his associates commenced at the University of Vienna in the 1860s and continued at his Physiological Laboratory at Leipzig in the 1870s.

investigators, Ludwig, Ranson, and Lim, had each contributed significantly to aspects of the brain stem's regulation of visceral function. Shih-Chun Wang was the only privileged young scientist to have been associated with two of those legendary figures, first with Lim, then with Ranson, during the 5-year period, 1935–1940. His subsequent work at Columbia's Physicians and Surgeons (1941–1978) extended knowledge of cardiovascular, respiratory, and related visceral and somatic functions. His research advanced the understanding of motion sickness and drugs to prevent vomiting and other adverse reactions. Wang reported (1981, p. 167) that his research career had been "profoundly influenced by the keen and exacting leadership of ... Stephen Walter Ranson."

Brain Stem and Autonomic Functions at Ranson's Institute

Because of the advanced level of sophistication in research which Wang brought to the institute from PUMC, much of his work was undertaken directly with Ranson without the collaborative participation of other associates. Both he and Ranson were eager to extend the vasomotor studies that earlier had depended on stimulation of the floor of the fourth ventricle of the brain. Ranson proposed that Wang use the stereotaxic instrument (see below) to systematically stimulate the interior of the brain stem and determine the actual components of the neural structures yielding vasomotor responses and other visceral effects. Wang quickly mastered the technique and during his 3 years at the institute, was author or coauthor of the remarkable number of nine papers, all concerned with visceral functions of the brain stem or spinal cord. In his major paper with Ranson (1939), Wang demonstrated the greater magnitude of the pressor responses evoked by stimulation of the lateral lenticular formation of the medulla in contrast to stimulation of the pontile tegmentum, and attributed the increased response to excitation of the reflex sympathetic center. In 1946, Robert S. Alexander, acting on the suggestion of Robert F. Pitts, then at Cornell University Medical College, New York, essentially confirmed those results with combined electrophysiological studies and serial bulbar transections which "left little doubt that the centers as identified by exploratory stimulation had a real functional significance" (Alexander, 1946, p. 214).

The Respiration Center and Robert Pitts

Pitts had earlier spent 1938–1939 at Ranson's institute in study of the localization and interrelations of the medullary respiratory centers, together with the mechanisms of respiratory rhythmicity.[4] In Pitts's experiments, maximal

[4] When Pitts departed the institute, he abandoned work on the central nervous system because his new laboratory could not afford the purchase of a Horsley-Clarke instrument, and he is better known for his subsequent work in renal physiology.

Fig. 5.1. Photographed ca. 1942, Ruth Rhines, Ph.D., M.D. (1910–1982) was a product of Northwestern University, receiving her undergraduate and both graduate degrees (1942 and 1949, respectively) in neuroanatomy from the medical school. She was one of the earliest neuroscientists and rarest women to hold both the Ph.D. and M.D. degrees and produced significant work in nervous system embryology, based on electron microscopy and tissue culture.

inspiration was evoked by stimulation of the reticular formation of the caudal medulla. Dorsal to this inspiratory area, stimulation of the reticular formation elicited expiration. These reciprocal responses were localized within the bulbar region which previous investigators had defined as the respiratory center. Comparison of the distribution of responsive points indicated that Pitts's expiratory area overlapped the distribution of Wang's vasopressor area, and analogously, Pitts's inspiratory area corresponded closely in its distribution to Wang's depressor area. Later, during the mid-40s, Magoun, with the able collaboration of Ruth Rhines (Fig. 5.1), explored the influence of stimulating the brain-stem reticular formation on ongoing somatic motor activity initiated reflexly, either in decerebrate rigidity, or from the motor cortex. Throughout the brain stem, the distribution of facilitatory sites corresponded closely to Wang's vasopressor region and Pitts's expiratory center. Reciprocally, inhibitory centromedial bulbar sites corresponded closely to the vasodepressor and inspiratory sites.

Functional interrelationships between these mechanisms were later explored by Hoff, Breckenridge, and Spencer (1952, p. 178), who stated:

> It is now clear that the focal point of suprasegmental integration in the somatic nervous system is the balanced operation of facilitatory and suppressor

> divisions of the reticulum of the brain stem .... When either of these is selectively damaged, the other becomes thereby disproportionately active.

The authors pointed out the coincidence of brain-stem sites whose stimulation produced maximal decerebrate rigidity and those that resulted in apneusis after vagotomy, and stated: "apneusis is a form of decerebrate rigidity of the respiratory system" (Ibid., p. 179). They noted that signs of parasympathetic dominance (e.g., bradycardia) were akin to signs in the "appropriately" decerebrated animal. Although it was established that the lowest medullary level was the site of a "cardiovascular center," that did not mean that was where incoming afferents were coordinated into larger functional patterns. The evidence indicated that suprasegmental integration occurred in the pontine region including parts of the reticular facilitatory system and activators of the bulbar suppressor system. Pontine regulation of the autonomic system constituted a link between hypothalamus and medulla that called for re-examination of hypophyseal relations with more caudal regions.

In summary, the structure of central regulation had been shown to derive from the association of somatic and autonomic regulations. This ability of the nervous system to integrate behavior depended on the simultaneous action of those systems, which "undoubtedly finds one of its fundamental reasons in the all-inclusive representation of autonomic, respiratory, and somatic subdivisions in the reticular facilitatory and suppressor mechanisms" (Hoff, Breckenridge, & Spencer, 1952, p. 187). Relatedly, the work of Cannon and his associates at Harvard was proceeding toward the concept of homeostasis (see pp. 129–130). This vital principle of the regulation of the mammalian interior environment came to full recognition in a span of about one hundred years with help from three continents – Europe, Asia, and North America – and along the way generated many scientific careers and questions.

Posture and Locomotion[5]

The first significant research project in the United States to utilize the Horsley-Clarke stereotaxic instrument was not to study the autonomic functions related above, nor to extend the instrument's inventors' original cerebellar study, but to rediscover decerebrate rigidity, first described by Sherrington in 1896. That master neurophysiologist had already noted the reciprocal innervation of antagonistic muscle groups "in such distribution and sequence as to couple diagonal limbs in harmonious movements of similar direction" (Sherrington, 1898, p. 332). Ranson and his graduate student, Joseph Hinsey, while at Washington University, St. Louis (1924–1928) had

[5] This section is compiled largely from Magoun's chapter in W.F. Windle's tribute, *Stephen Walter Ranson* (1981), which describes in detail the early systematic search for the neural substrates of posture and locomotion.

been studying the mechanisms involved in the maintenance of postural contraction. On their return to Chicago and during those initial years of the institute, two of their research accomplishments were to show by various levels of decerebration that a "mid-brain" cat, with only the subthalamus and caudal hypothalamus remaining attached, was able to walk; and, second, to identify the neural substrate involved by electrically stimulating the rostral cut surface of the truncated brain stem. As Hinsey recalled much later, "We could elicit a tegmental response which was quite characteristic, involving flexor contraction of one forelimb, extension of the other, and curvature of the spinal column ... ." (Hinsey, 1961, p. 15). Some rough aspects of the definitive study (Ranson & Hinsey, 1930) that placed the response in the reticular core of the brain stem were provided by Hinsey[6] and illustrate Ranson's obstinacy as well as the growing pains of science when in the throes of discarding habitual methods and concepts.

Decerebration by Anemia

During Ranson's period in St. Louis, he had routinely prepared decerebrate animals by the so-called anemic method. This technique had been developed during his earlier years at Northwestern by his colleagues, Lewis J. Pollock in neurology and Loyal Davis in neurosurgery. Under ether anesthesia, with the animal on its back, the carotid arteries were ligated in the neck, the mouth was spread open and, with a dental drill, an aperture was made through the roof of the palate to expose the brain stem, so that the basilar artery could be ligated as it traversed the base of the pons. The entire brain ahead of this ligature was thus deprived of its blood supply and immediately became functionless.

In this anemic method, the cranium was never opened beyond the small hole at the base and the brain was never exposed. Not a drop of blood was spilled, no further anesthesia was required, and the animals displayed a prominent decerebrate rigidity. Dr. Ranson had contributed to the development of this technique – he had loaned the dental drill to work out the method – and ever afterward maintained something of a proprietary attachment. After all, it was the only substantial advance in the operative aspects of decerebration since Sherrington's early development of the guillotine!

Transitional Pains

On return to Chicago, Hinsey proposed that they adopt the traditional method of decerebration in order to gain greater access to, and opportunity for analysis of, the brain-stem mechanism involved. As Hinsey (1980, p. 9) recalled:

[6]Hinsey recorded his recollections on audiotapes which he forwarded to the Brain Research Institute for transcribing in reply to twin requests from W.F. Windle, who was gathering material for *Stephen Walter Ranson* (1981) and H.W. Magoun, already working on his projected history of U.S. neuroscience.

> Ranson's attitude toward transection decerebration really took a lot of very careful maneuvering around to bring about the permission for its use. Of all the things I did while there [with Ranson], the most significant was this substitution of the transection method, which made it possible to center attention on the midbrain, thalamus, subthalamus, and hypothalamus. If we had stayed with the anemic method, the whole subsequent, wonderful development at the institute might not have taken place. It was the visceral representation in that part of the nervous system, and the studies that were so beautifully done in this field, that were responsible for Dr. Ranson's recognition ....

And again:

> I think about how long I argued with him regarding the chains of neurons, the reverberating circuit. He wasn't about to buy that, either. I carried a pad with me; people on the Elevated thought we were nuts, because I'd diagram and we'd argue back and forth .... Eventually he would come around to a different point of view ....

The concept of reverberating circuits as the basis for afterdischarge formed the high-water mark of Ranson's and Hinsey's excursion into the field of spinal reflexes, muscle tonus, and posture.

In Search of Precision

In the early studies by Ranson and his collaborators, to mark the points of stimulation that produced the tegmental response in cats and monkeys, a cat-hair was routinely placed on the cut surface to orient the stained sections later, a crude and uncertain method at best. It was then that a Horsley-Clarke stereotaxic instrument, constructed by a clever machinist from the original drawings of 1908,[7] was brought in and soon dominated the studies of posture and locomotion. It is highly probable that Ranson had seen the original instrument much earlier on a visit to Horsley's laboratory at University College, London, and later, with Hinsey, at Washington University, St. Louis, where another unit (Fig. 5.2) was used by Ernest Sachs, professor of neurosurgery. In Chicago, 1935–1936, a replica was applied to stimulation of the cerebellar nuclei (Magoun, Hare, & Ranson, 1935) and a biphasic postural pattern of the limbs and trunk as in the tegmental response was found: at cessation of the stimulus there was a tonic rebound to a posture the reciprocal of that during stimulation which often lasted several minutes.

By 1940, suggestions had appeared that the tegmental response was "evidence ... that the subthalamus contains a specific mechanism which governs the alternation of the legs in walking ... ." (Waller, 1940, p. 305). Earlier, Ingram and Ranson (1932) had decided that the response should not be

[7] Reintroduction of the instrument was described in Magoun and Fisher, 1980; Magoun, 1981a, 1985. It made possible the insertion of fine electrodes into the interior of the brain for stimulation, recording, or focal damage to a precise region of brain tissue.

Fig. 5.2 The second Horsley-Clarke stereotaxic apparatus, constructed in London in 1908, was used in St. Louis and in 1957 was presented to Horace W. Magoun (above), "the new torchbearer".

considered a discarded, phylogenetically old pattern no longer useful in mammalian behavior, but might have an important role in the act of walking.[8] In the meantime, a Scotsman, Thomas Graham Brown (1882–1965), who had worked in Sherrington's laboratory, developed his views (T.G Brown, 1913–1914) on walking or running "as the product of a potentially rhythmic nervous apparatus ... capable of beating rhythmically under balanced excitatory and inhibitory stimulation ... ." (Adrian, 1966, p. 26). Brown introduced the treadmill to test the decerebrate cat's ability to execute walking movements at various speeds and produced a movie film but never published his findings (Lundberg & Phillips, 1973). By the 1970s, the search for the neural mechanisms had swelled to the point that at least nine reviews of recent advances in the neural control of posture, movement, and locomotion appeared in rapid succession.[9] In the next decade this broad field

> had become one of the most active investigative areas in current ... neuroscience. The 'John the Baptist' of this impressive development, crying out in that neuroscientific wilderness just before and after [the First] World War, was obviously Thomas Graham Brown. Succeeding pioneers in the 1930s ... [who] provided the initial substantial foundation of brain-stem and cerebellar studies, were Joe Hinsey, Rex Ingram, Kendrick Hare,[10] and their mentor, Stephen Ranson, who

[8] That was, of course, the understatement of the decade.
[9] See Magoun, 1981a, p. 121 for these citations.
[10] Editor's note: With his usual self-effacement, the author did not mention himself.

had established and directed an exceedingly productive Institute of Neurology in that period and, at its inception, had revived the Horsley-Clarke instrument, the use of which made it all possible (Magoun, 1981a, p. 122).

EMERGENCE OF NEUROENDOCRINOLOGY

The ancient and tenacious belief that waste products accumulating within the cerebral ventricles trickled down through the funnel of the infundibulum to the pituitary gland and passed out to the nostrils was completely dispelled only after experimental work demonstrated the importance of the pituitary as a gland of internal secretion. Among those who took the first steps, early in the 20th century, in attacking an endocrine problem – observing the effects of ablation of the cells in question – were three American scientists who independently studied the influence of thyroidectomy or hypophysectomy on maturation in tadpoles.

Problems of Growth

The Thyroid Gland and Bennet Allen

In 1903, Indiana-born Bennet Mills Allen (1877–1963; Fig. 5.3) gained his Ph.D. at the University of Chicago under Charles O. Whitman with a thesis on the embryonic development of the gonads that initiated his career-long preoccupation with the role of endocrine glands in growth and development. After appointments in zoology at Wisconsin and Kansas, in 1922 he joined the faculty of the "Southern Branch"[11] of the University of California.

Improving on the earlier method of tadpole hypophysectomy by cautery, Allen (1918) surgically removed in frog larvae the cell-cluster destined to become the thyroid or pituitary gland and found that in each case the tadpoles never changed into adult frogs (Allen, 1924). Similar experiments were being carried out independently at the University of California, Berkeley, by Philip Smith. Curiously, Smith from the West Coast and Allen from the Midwest presented their initial results in separate papers to the Western Society of Naturalists in San Diego, in August, 1916 and both were published in *Science*. Compounding this coincidence, at the meeting of the American Association of Anatomists at New York a few months later, E.R. Hoskins and Margaret Morris, from the Department of Zoology at Yale presented their own findings following thyroidectomy in amphibia (Hoskins & Morris, 1916–1917). They had reasoned that extirpation of the gland before it became functional would reveal its relation to growth. The concept was timely, for by

[11]So dubbed by the pretentious faculty at Berkeley.

Fig. 5.3. Zoologist Bennet Mills Allen was among the pioneers of neuroendocrinology in demonstrating the dependence of maturation on thyroid and pituitary gland integrity.

then the swelling wave of experimentation had begun to enliven the static anatomical features of embryology and endocrinology at three productive outposts across the United States: in the East at Yale, where Ross Harrison had made major contributions to experimental embryology; in the West at Berkeley, where Herbert M. Evans was assembling a remarkably active group, and between them, at the University of Kansas where Bennet Allen's work interrelating embryology, endocrinology, and neurology was leading the parade.

Each group had found that removal of the thyroid anlage produced in the adult an immature brain, smaller and undifferentiated compared with the normal controls.[12] Allen dug deeply into the problem and noted that the brains of amphibia without thyroid or pituitary tissue were also hydrocephalic with thin cerebral walls (Allen, 1924).

[12] Bennet Allen's research constituted literally the first on the brain reported from the anlage of the Brain Research Institute of the University of California, Los Angeles.

Fig. 5.4. Philip. E. Smith's skillful operating technique in very small animals was essential in demonstrating the multiple effects of anterior lobe hypophysectomy. Photographed in 1930.

Anterior Pituitary and Philip Smith

Philip Edward Smith (1884–1970; Fig. 5.4) extended the work on tadpoles to the Mammalia, with strikingly similar, clear results. He was a modest, shy, but kindly man who obtained his Ph.D. in anatomy at Cornell in 1912, and then joined the anatomy department at the University of California, Berkeley. He became adept at removal of the anlage of the anterior pituitary in tadpoles and after a sabbatical leave spent with Walter B. Cannon at Harvard he began a program of ablation and transplantation in newborn rats. In 1924, Smith, Walker, and Graeser reported that early ablation of the anterior pituitary resulted in atrophy of thyroidal, adrenal cortical, and ovarian tissues, abolishing oestrus and stunting growth. The obvious corollary, substitution, (Smith & Engle, 1927) with pituitary implants in rats and mice clearly demonstrated that the gland produced a substance that stimulated the gonads by way of the blood stream, and that hypophysectomy removed the source of the stimulus. Moving to Columbia University, they established a monkey colony and continued work on the pituitary-thyroid and pituitary-gonad relationships that constituted a major contribution to "the historic hey-day of basic research on the hormones of the anterior pituitary" (Greep, 1974, p. 4). A few years later, Collip, Anderson, and Thomsom (1933) showed that the anterior lobe extract contained separate thyrotropic and adenotropic hormones.

Multiple Hormones and Herbert Evans

Again at Berkeley, when Herbert McLean Evans (1882–1971; Fig. 5.5) arrived in 1915 as head of the department of anatomy, he recognized the importance of Philip Smith's work and commenced to assemble a group of chemists and physiologists in a broad attack on what little was known about the anterior lobe. Ten years later in his Harvey Lecture (1925), Evans could cite evidence of the effects of too little (stunting, among others) or too much (gigantism) of the crude anterior pituitary extract. Although the growth hormone continued to be his favorite, "he was the indispensable motivation for a broad attack on the ... isolation and chemistry of the pituitary hormones and their biological effects ... ." (Bennett, 1975, p. 270). Relatedly, Evans realized that success in all biological studies depends on a reliable supply of healthy animals, and perhaps his most lasting contribution was the co-development of a stable, vigorous standardized strain of rat, the Long-Evans cross between the wild Norwegian gray and the Wistar albino (ibid.). His insistence on rigid supervision and proximity of the animal quarters ensured a strain of experimental subjects as significant for endocrinology as the Osborne-Mendell and Sherman strains were for nutrition. The University of California in 1931

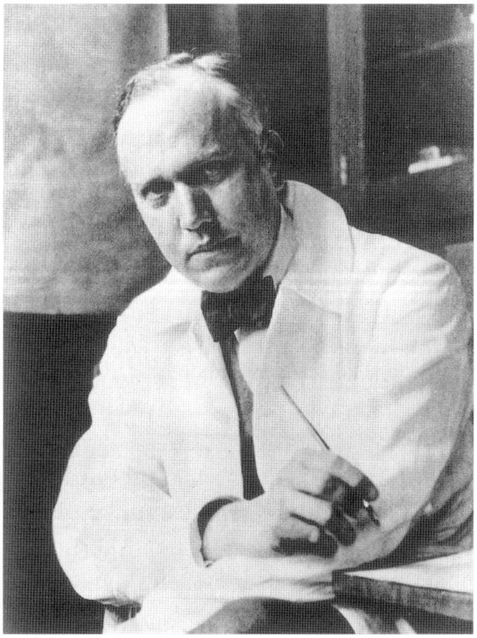

Fig. 5.5. Herbert McL. Evans made multiple advances in endocrinology, including demonstration of the growth effects (in dogs) of too little anterior pituitary extract (stunting) or too much (gigantism).

expressed its appreciation of Evans's enthusiasm and accomplishments by making him director of a new Institute of Experimental Biology. Evans reciprocated by taking a substantive role in identification and chemical and functional characterization of seven hormones of the anterior lobe of the pituitary gland. The published proceedings of a conference on links between brain and gonadal function were dedicated to Evans (Gorski & Whalen, 1966).

Neurohumoral Control: The Neurohypophysis

The early experiments gave little hint of the complex problems they were brushing against: clinical reports at home and abroad had already signaled the widespread effects of disease at the base of the brain. Patients of Marie (1886), Fröhlich (1901), and Cushing (1906) were examples, and in Cushing's case, led to a full-scale experimental attack.

Cushing and Diabetes Insipidus

In his biography of Harvey Cushing, John Fulton (1946) noted that in 1908 Cushing had been profoundly interested in the four Herter Lectures given at Johns Hopkins by the visiting Professor Schäfer, who had worked for years on the posterior pituitary, and discovered in 1898 its active principle. Cushing's experimental work on the gland commenced soon afterward, and his 16 years in Baltimore culminated in the publication of a large monograph, *The Pituitary and Its Disorders* (1912), which had originally been presented as a Harvey Lecture in New York. Cushing's friends joked about the length of the lecture and how long its delivery took. Cushing sent a copy to Schäfer who had catalyzed his experimental studies, and was likewise teased for holding the record for amplification of a single lecture.

Cushing's monograph made him the undisputed American authority on pituitary functions (Anderson, 1969) until work in his own laboratory challenged his concept of the primacy of the pituitary. In 1920, Cushing suggested to Percival Bailey, a Fellow in the surgical laboratory at Harvard, that he study the pituitary fragments remaining after partial hypophysectomy in dogs. On attempting a subtemporal approach to the gland, Bailey inadvertently cut an infundibular artery. Because of excessive bleeding, he closed the wound and returned the dog to its cage. The next morning the laboratory floor was covered with urine. Amazed that he had produced a full-blown polyuria without having touched the pituitary, Bailey spoke of it to Frédéric Bremer, in the laboratory from Belgium, who recalled a short paper on a similar finding by a French physiologist, Jean Camus.

A Major Correction

Collaboratively, the two Fellows prepared a series of dogs with punctures ("piqûre") of the infundibular region of the hypothalamus without exposing the pituitary gland and produced both diabetes insipidus and the adenosogenital syndrome. The publication of those findings from Cushing's own laboratory (Bailey & Bremer, 1921) constituted a manifesto, replacing the accepted dominant role of the pituitary with that of the hypothalamus. It is interesting that Cushing was at first opposed to publication of those results, because he was still of the view that the pituitary was primarily concerned. It was not until 1930, in his Lister Memorial Lecture at the Royal College of Surgeons[13] that Cushing finally acknowledged that, with Camus and Roussy's (1920) and Bailey and Bremer's "disclosures," his accepted views were "badly shaken" and were "a challenge which produced a veritable bouleversement of our cherished preconceptions" (Cushing, 1930, p. 124). In her authoritative history of the hypothalamus, Evelyn Anderson (1969, p. 11) wrote that these developments "brought down the curtain on Act I of the hypothalamic pituitary drama. Act II might be considered to open with the Ranson story, which was the beginning of the modern period . . . ."

Polyuria at Ranson's Institute

In keeping with his sweeping curiosity about physiological events (see Magoun, 1942), when a massive polyuria appeared after a surgical accident (as in Bailey's experience), Ranson opened a new investigative development. The classic research of Charles Fisher, "Rex" Ingram, and S. Walter Ranson at Chicago's Northwestern University Medical School[14] was presented in a 200-page monograph published with Ranson's private funds in 1938 and verbosely titled: *Diabetes Insipidus and the Neuro-humoral Control of Water Balance: A Contribution to the Structure and Function of the Hypothalamic-hypophyseal System.* Some 50 pages of the volume were devoted to the history of previous contributions to and conceptions of functions of the neurohypophysis. The story began in 1894 when Ramón y Cajal described in rat a heavy bundle of unmyelinated fibers coursing into the neural division of the hypophysis by way of the infundibular stem. More focally, the outcome of the sojourn in 1901 of Rudolph Magnus of Heidelberg at E.A. Schäfer's laboratory in Edinburgh was their joint discovery of the diuretic action of pituitary extracts. From that time, numerous functions were ascribed to this system, including regulation of water balance and implicating a role in

[13] Replete with illustrations, the lecture displayed some of the attributes of his Harvey Lecture in New York, some 18 years earlier. Although not a 350-page monograph, it appeared in two numbers of *The Lancet* (July 19 and 26, 1930) occupying a total of 19 double-column pages set in minuscule type; converted to contemporary standards of scientific journals, it would probably have occupied 40 to 50 pages.

[14] A detailed account of this work and the instrument that made it possible is found in Magoun and Fisher, 1980.

diabetes insipidus, but at no time was the evidence completely convincing. Research in Buenos Aires (Houssay, 1918) and confirmed in Paris (Camus & Roussy, 1920) revealed that injury to the hypothalamus without removal of the pituitary could cause polyuria, gonadal atrophy, obesity, and transient glycosuria. None of the early work had reported permanent polyuria and polydypsia, however. The conclusion in the 1938 monograph reads: "We believe that the function of checking the flow of water through the kidneys can be assigned to the neural division of the hypophysis with a reasonable amount of certainty on the basis of the recent work on diabetes insipidus … ." (Fisher, Ingram, & Ranson, 1938, p. 117).

When Ranson sent Harvey Cushing a copy of the monograph on diabetes insipidus, Cushing replied (26 May, 1938) with a congratulatory letter in which he gracefully disclaimed "the half-baked opinions" of his early Hopkins monograph, and added: "I suspect that you have made a discovery therein that will be much followed and imitated."

The Modern Period Opens

The monograph on diabetes insipidus was only one of the landmarks in the advance of neuroendocrinology, but in Ranson's case it also brought national attention to his institute's use of stereotaxis to provide precise and reproducible data on brain localizations. The systematic exploration of the hypothalamic and brain-stem regions that ensued was a prime example of the convergence of neural and behavioral streams to become the mainstream of neuroscience.

Neurosecretion Heresy: The Scharrers

The interrelatedness of hypophysis and hypothalamus was heavily underscored after a quarter-century of work by the Scharrers, Berta and Ernst (Fig. 5.6), first in Germany at the Edinger Institute of Neurology and finally at Albert Einstein College of Medicine, They emigrated to the United States in 1937 on a Rockefeller fellowship to work with C. Judson Herrick at the University of Chicago. In the publication, *Neuroendocrinology* (1963), they summarized research that showed two types of integrative foci in the central nervous system. The first is concerned with functions in which afferent stimuli require direct access to their foci via synapsing neurons to swiftly set off appropriate responses to evocative signals. The second type is built up gradually over a much more prolonged time than the first, such as occurs in reproductive cycles for which nerve-conducted, millisecond impulses are unsuited. To this end, in appropriate settings the nervous system has developed its own endocrine apparatus.

The presence in all animals, from worms to mammals including humans, of groups of cells in the central nervous system that exhibit certain

Fig. 5.6. The research interests of Berta and Ernst Scharrer were complementary and resulted in neurocytologic proof that neurons could act as glands. Photographs from Schmitt, Worden, and Adelman, 1975, p. 230; p. 234; left from 1964.

cytological features of gland cells had been known for some time. Two relevant discoveries the same year came from Johns Hopkins Medical School: physiologist William Howell (1898) localized the active principle of pituitary extract in the posterior lobe and microanatomist Henry Berkley (1894) reported seeing colloid-containing vesicles in that lobe. Colloidal granules in the cytoplasm of secretory neurons passing along their axons to terminal endings in blood spaces were first described in 1928 (E. Scharrer). These neurohumoral systems were extensively studied for their neurohormones may act directly on effector organs, exciting or inhibiting gonadal maturation, or stimulating smooth musculatures, or promoting water resorption by the kidney. They may activate other endocrine organs, the majority of which are not under direct neural control, and so constitute links between the central nervous system and organs of internal secretion. In an autobiographical memoir, Berta Scharrer (1975, p. 235) wrote:

> The key to the correct interpretation of such systems was the realization that the terminals of the neurosecretory neurons … fail to establish synaptic contact with other neurons or nonneural effector cells. The redundancy of this nerve supply to the posterior lobe … and the close affiliation of the axon terminals with the vascular bed, suddenly made sense.

A Subdiscipline Comes of Age

The increasing interest in all aspects of neuroendocrine research during and shortly after midcentury was marked by a flurry of interdisciplinary conferences,[15] bringing together biochemists and physiologists, anatomists and embryologists to talk with psychiatrists and neurologists. Among the clinicians, the latter were the slowest to perceive a relevance to their specialty, whereas the psychiatrists immediately saw a connection to mental pathology (Scharrer & Scharrer, 1963). The obstacles were seen at a gathering of neurophysiologists, neuroendocrinologists, and neurologists to discuss the evidence for brain-endocrine interconnections. The three-day meeting opened with the observation that "there are endocrinologists here who are not very good neurologists, and there are probably some neurologists here who are not very good endocrinologists; perhaps there are some here who are neither" (Reichlin, 1966, p. 3). Toward the end of the conference, it was admitted that "the endocrinologists view the hypothalamus with the rest of the brain as a sort of cherry on top, and we have no idea … how all the structures and functions are integrated" (Ibid., p. 248). An additional obstacle was found: "It is difficult to argue about results obtained in interdisciplinary studies when there has not as yet been basic agreement on interpretations of data from one discipline" (Purpura, 1966, p. 249). Between those insights, the state of neuroendocrinology at that time included: "the proposal that the median eminence serves as a kind of hypothalamically-innervated neurohypophysis which produces the releasing factors or transmitter substances that guide pituitary secretion, so that the pituitary becomes, in turn, a kind of target endocrine organ. Of related significance has been the identification of these transmitter substances, the nature of which has been under study by many investigators over a period of more than a decade."

> Moving from the endocrine field … into the hypothalamic region overlying the median eminence, it has been fascinating to learn of the ambivalent or uncommitted nature of its embryonic matrix and of the role of developing endocrine substances and other factors in determining its subsequent sexual polarization in maturity. Of additional major significance has been the differentiation in the hypothalamic region of two distinct mechanisms, one of which deals with … ovulation and reproductive tract function, while the second, which is both spatially separate and functionally discrete, serves mating and other reproductive behavior. Unquestionably, there must be close operational relationships between these two mechanisms but they can, nevertheless, be differentially blocked. It has been of the greatest interest to learn of the application of these findings to the development of effective oral contraception, with its broad social implications for population control. Of particular ingenuity is the

[15] Conferences, textbooks, and journals participating in exchange of information in this field are found in Scharrer and Scharrer, 1963, pp. 2–4.

> use being made of the biphasic action of the progestational steroids, to take advantage of their negative phase, so as to block or reduce the excitability of the hypothalamic gonadotrophic mechanism, while leaving unimpaired and, possibly, even enhancing the excitability of the second mechanism serving mating behavior. As a result, all of the drives toward mating are preserved or increased without the consequence of ovulation and potential conception (Magoun, 1966, pp. 1–2).

After commenting on the paradox of pharmacologically simulating pregnancy to prevent conception, the summary continued:

> Stemming also from this work are findings relating to the direct endocrine potentiation of behavior. Dual and reciprocal influences seemingly have been identified; certain of them appear facilitatory, while others reduce excitability. The former initiate the exploratory and consummatory phases of sexual activity, while the latter are responsible for succeeding satiety. These contributions seem to be opening the basic processes serving innate behavior to substantive investigation.
>
> Additionally, throughout the presentations, there have been repeated references to the more general organizational aspects of homeostatic regulation of neuroendocrine processes, with involvement of inverse feedbacks simulating automational models of contemporary technology and engineering systems.

While those exchanges of ideas and information were advancing the concepts of a many-faceted neuroscience, in other quarters a narrower goal was being pursued – identification of the agent(s) of control.

Race to the Prize: From Collaboration to Competition

As Nicholas Wade (1981, p. x) pointed out, "The strange truth is that scientists bear a striking resemblance to ordinary people" and illustrated this by describing the competitive pursuit of the brain's hormones by Roger Guillemin and Andrew Schally. It lasted for 21 years and "involved creating two rival teams of experts and investing time, money, and reputation in a venture which, for the first fourteen years seemed to many onlookers to be headed for humiliating failure" (Ibid., pp. 3–4). The gamble's successful outcome became a foundation of neuroendocrinology and opened doors to medical therapies in spite of its basis on the bizarre concept that the brain could act as a "vulgar" gland. This extended program reached its climax in Stockholm, Sweden, on December 8, 1977, when Guillemin, Schally, and Rosalind Yalow shared the Nobel Prize in Physiology or Medicine.

Roger Guillemin's Program

Following his award, Guillemin delivered a lecture on "Peptides in the brain, the new endocrinology of the neuron" (1978). Describing the peculiar anatomy of the junctional region of the ventral hypothalamus and anterior pituitary, he acknowledged the Scharrers' (1963) concept of neurosecretion and the guess that small peptides were involved. Characterization of those substances was a challenging combination of devising routine bioassays for potency of chemically pure fractions of an enormous number of fresh hypothalamic and pituitary fragments. Late in 1968, 1 milligram was isolated of the first hypophysiotropic peptide, the factor that through the pituitary regulates the thyroid gland (Burgus, Dunn, Desiderio, & Guillemin, 1969). In Guillemin's estimation the isolation and characterization of the thyroid releasing factor (TRF) "separated doubt – and often confusion – from unquestionable knowledge ... and neuroendocrinology became an established science on that event" (Guillemin, 1978, p. 162). The isolation and synthesis of somatostatin (summarized by Guillemin & Gerich, 1976) followed work on the luteinizing hormone that competed directly with Schally's laboratory.

Two years after his Nobel lecture, Guillemin expanded his ideas on the new endocrinology, this time to the brain, in which he brought together research utilizing stimulation and lesion experiments that yielded information about the activities of vasopressin and oxytoxin, which Vincent du Vigneaud and his group (1953) had isolated and characterized as nonapeptides. Guillemin concluded with the hope that the early discoveries would lead to "profound reappraisals not only of mechanisms involved in the functions of the normal brain but also of mental illness" (1979, p. S80).

Andrew Schally's Approach

Subsequent to verification of the direction of flow of blood in the pituitary portal circulation,[16] the proposal was advanced (Hinsey & Markee, 1933; Harris, 1948) that neurohumoral substances from the posterior hypophysis might reach the anterior lobe by way of the portal circulation. That possibility, together with the demonstration by Sawyer, Everett, and Markee (1949) that the central nervous system exerted control of gonadotrophic secretion, crystallized the ambition of Andrew Victor Schally to seek direct evidence of what was then only a speculative proposition. In the McGill University laboratory of Murray Saffran a systematic search was underway for such a hypothalamic hormone. As his doctoral thesis (1957), Schally devised a tissue culture test for measuring the release of the pituitary hormone, corticotropin releasing factor (CRF) that stimulates the adrenal cortex to release ACTH (Saffran, Schally, & Benfey, 1955).

On the award of his degree, Schally accepted an invitation to work in Roger Guillemin's laboratory at Baylor University, Houston, Texas. There rancor and

[16]First depicted elegantly but incorrectly by Popa and Fielding (1930), reversed in amphibia by Houssay et al. (1935) and Green and Harris (1946), and extended to man by Wislocki and King (1936).

resentment gradually developed as productive experimental results were not forthcoming, and in 1962, while Guillemin was working in Paris, Schally moved to New Orleans and Tulane University and the Veterans Administration Research Laboratory. To further differentiate his research from that of Guillemin, Schally shifted to the use of brains from pigs rather than the traditional sheep, and through a VA intermediary persuaded Oscar Mayer, the meat packer, to give him a half-million brains, gratis.

Both laboratories pursued the components of TRF and in 1966, Schally identified the amino acids involved, but when tests showed no biological activity, he switched to the luteinizing hormone releasing hormone (LHRH), the hypothalamic peptide regulating the pituitary hormone (LH) that triggers the production of sex steroids; by 1971 he found the structure of the LHRH molecule.

Through the 70s, Schally and his team continued to search for other hypothalamic hormones, but their activities turned increasingly to the synthesis of LHRH analogs – structurally modified derivatives of the original hormone. He realized that an LHRH antagonist promised to be an ideal female contraceptive, blocking some of the pituitary receptors to decrease the gland's production of hormones critical to reproduction. An LHRH antagonist contraceptive, Schally reasoned, would permit relatively normal circulating concentrations of sex steroids during most of the reproductive cycle.

In addition, Schally turned to exploring agonists, the synthesis of which required fewer changes in the natural molecule. By 1976 he had developed for human use an analog that was 100 times more potent than natural LHRH, but tests revealed negative effects: sex-steroid blood concentrations plummeted. Initially disappointed, Schally soon realized that LHRH agonists provided a revolutionary therapy for cancer. In conclusion, he assured the Nobel audience (Schally, 1978) that in addition to the three hypothalamic regulatory hormones known to be valid (thyrotrophin, gonadotrophin, somatostatin) and at least six more reasonably well established, there would be others, with additional clinical applications. Perhaps the most exalted evaluation of the work of Guillemin and Schally was in the introductory speech at the Nobel award ceremony: "It is justifiable to say that they have uncovered a substantial part of the link between body and soul" (Ibid., p. 24).

ROLE OF THE HYPOTHALAMUS IN HOMEOSTASIS

The usual clustering of feeding, drinking, sex, and sleep as hypothalamic-related activities obscures a more global function of this gland: the conservation of the steady state of the organism's inner environment, or homeostasis. The first American physiologist to concern himself with this essential function was W.B. Cannon. As summarized by Langley (1975, p. 74), Cannon's first outline of the concept and use of the term "homeostasis" was in a paper read at the 1925 Congress of American Physicians and Surgeons in New Haven and published in their *Transactions* (1925), and a full development of homeostasis appeared in 1929 in *Physiological Reviews*. Whereas Claude Bernard propounded the "milieu intérieur" at the cellular level, and Joseph Barcroft wrote of "architectual" features in physiological function, the American's view was specific and based on a firm and ranging body of experimental work.

The Cannon (Peripheralist) Story

Walter Bradford Cannon (1871–1945; Fig. 5.7) enjoyed a career characterized by both breadth of accomplishment in teaching and writing and depth of experimental inquiry surrounding a central theme: the bodily adjustments

Fig. 5.7. Walter B. Cannon, shown with Chim and Panzee at the Franklin Field Station in 1923, made major, well known contributions in several areas of physiology. It is not general knowledge that a mountain peak in the American West was named for him.

needed to preserve a normal state of equilibrium. During his active years that began at Harvard University and ended on retirement from Harvard Medical School (1898 to 1942), Cannon's many students and colleagues[17] learned to appreciate the rare quality and wide extent of his commitment to neuroscience research. Acclaimed in his own country, as were Pavlov and Sherrington to whom he has been compared (Brooks, 1975), Cannon was celebrated 30 years after his death by a conference arranged by one of his most distinguished former students, Chandler Brooks; contributors to the conference provided details of many facets of their hero's life.

Education and Training

Cannon was a New Englander by ancestry and choice, yet was born and reared in the Midwest, in comfortable circumstances. He elected to go East to Harvard College and remained there with few interruptions, becoming one of the first completely American-trained physiologists, thus signaling the rising stature and independence of physiology in the United States. Early in his career, in fact, when weighing the merits of a trip abroad his opinion was that it was not so much for acquiring knowledge as for exposure to the university attitude (Benison, Barger, & Wolfe, 1987).

Research Career

When the ailing Henry Bowditch retired in 1906 as head of the department of physiology that he had so successfully launched, Cannon became his successor after a painful interregnum and guided the department to even greater renown. His wide knowledge and free rein allowed his colleagues the pursuit of collaborative as well as individual interests with the result that productive foci of neuroscience research developed within the department around luminaries such as Alexander Forbes, Hallowell Davis, and others.

As Horace Davenport noted (1975), the initiation of Cannon's career was the perfect example of the right idea given to the right students at the right time: Bowditch suggested to Cannon and Albert Moser that they study deglutition with the new Roentgen-ray method. Going far beyond Beaumont's observations (1833) on a human patient/subject (see p. xvi), the result was a thorough examination in laboratory animals of activity of the gastrointestinal tract and the effects of sensation and emotion on it. Those experiments served as the route to the study of the autonomic nervous system and its role in maintenance of the inner environment, which became the subject of one of Cannon's classic publications, *Bodily Changes in Pain, Hunger, Fear and Rage* (1915). A third phase of the work of Cannon and his associates was to

[17] A list of 359 students and staff who were at the Department of Physiology while Cannon was on the faculty or chairman appears in Brooks, Koizumi, and Pinkston, 1975.

demonstrate the action of the sympathetic nervous system in coping with stress, achieved by the successful maintenance under controlled laboratory conditions of animals from which the sympathetic chain of ganglia had been removed bilaterally. That work, summarized in *The Wisdom of the Body,* (1932) led naturally to the adrenal medulla and humoral transmission of the nerve impulse.

First in collaboration with Z.M. Bacq from Belgium (1931) and continued with Arturo Rosenblueth from Mexico, Cannon demonstrated in unanesthetized cats with previously denervated heart and under normal stimulation the cardiac acceleration that O. Loewi (1921) had found in the isolated toad heart. The productive collaboration with Rosenblueth, a theorist and mathematically oriented physiologist, led to elaboration (Cannon & Rosenblueth, 1933), of an erroneous hypothesis[18] that two neurotransmitters, "sympathin-E" (excitatory) and "sympathin-I" (inhibitory), combined with a mediator (epinephrine) at the receptor level and circulated in the blood stream. Demonstration that the actions of sympathins E and I were identical with those of epinephrine and norepinephrine respectively came later (Bacq, 1934).

Evaluation by His Students

Cannon's ranging contributions to neuroscience formed primarily a legacy of rigorous demonstrations of the many-faceted interconnections involved in homeostasis. He chose thirst as the most effective and ubiquitous of the peripheral stimuli that indicate body needs and made it the subject of his Croonian Lecture (1918), in which he displayed his rare ability to pull together stark data into meaningful interpretation, for he had been working on problems related to traumatic shock in war casualties. Cannon was dedicated to imparting to students a useful, integrated knowledge of physiology; from early in his teaching he prepared – and continuously revised through eight editions (1910–1931) – *A Laboratory Course in Physiology,* that reflected new discoveries and techniques. He wrote, published, and lectured in professional and popular vehicles on pressing issues, for example, on characteristics of anti-vivesection literature in *Scientific American* (1914). A religious man, he did not hesitate to make known his thoughts on evolution and moral values, citing (1945) rigorous scientific conduct, sharing observations and collegial relationships, and freedom of investigation. Recognition of Cannon, both the man and his work, continued after his death in 1945. In addition to the conference described above organized by his one-time student, Chandler Brooks, a second honor was the choice of Cannon's autobiography, *The Way of an*

[18] According to Bacq (1975, p. 81, fn), that error and its tenacious support may have been the basis of Cannon not being included with Loewi and Dale in the Nobel Prize in physiology or medicine of 1936.

Investigator (1945) as a souvenir commemorating the XXIV International Congress of Physiological Sciences held in 1968. The final tribute is the preparation of a definitive biography by coauthors who knew him well and had access to an extensive archive at his Alma Mater (Benison, Barger, & Wolfe, 1987). Celebrated in his own time, Cannon is prominent not only in the neural sciences but also in the behavioral sciences as shown in Part 2.

Chandler Brooks and the Autonomic Nervous System

Born in the undeveloped state of West Virginia early in the century, Chandler McCuskey Brooks (1905–1989; Fig. 5.8) was reared in a Presbyterian minister's God-fearing family. His father was also an amateur naturalist who inculcated a love of nature and order into his young son. The college preparatory years were spent in Massachusetts; after Oberlin, Brooks gained his Ph.D. in physiology at Princeton under Philip Bard in 1931. He accompanied Bard to Harvard as a Fellow in Cannon's department and commenced studies related to the autonomic nervous system. Two years later, and again with Bard, he moved to the physiology department at Johns Hopkins University School of Medicine. Following a Guggenheim fellowship with J.C. Eccles in

Fig. 5.8. As a self-proclaimed generalist, Chandler McC. Brooks considered his work an extension of the Cannon school of the autonomic nervous system. Photograph from Seller, 1991, facing p. vii. With kind permission.

New Zealand in 1948, an association that diverted him from the autonomic nervous system, Brooks was appointed chairman of the Departments of Physiology and Pharmacology of the Long Island College of Medicine, which became the Downstate Medical Center of the State University of New York at Brooklyn.

By mid-20th century the earlier American "schools" of excellence in neuroscience grouped around outstanding figures as department heads – Harvard (Cannon), Yale (Fulton), Johns Hopkins (Bard), Northwestern (Ranson), Washington University (Erlanger) – had diminished or disappeared and not yet been replaced. Going outside that mold, to fulfil his responsibilities to train broadly based physicians at Downstate, Brooks formed a faculty of wide-ranging research interests. This large view fitted with his own research career as a generalist. He reported (1986, p. 19) that his bent was to look "at the totality of things ... the objective of the biologist is to understand the nature of life." The generalist role allowed him the freedom to pursue tangential directions, as he did at the height of interest in the reticular formation arousal reaction by undertaking research on sleep: a mattress company supported a series of comparative studies that led to publications and interviews and ultimately to the Sleep Research Foundation. From 1966 to 1972 he was dean of the School of Graduate Studies at Downstate and briefly, acting president and was made Distinguished Professor in 1971.

Founding of the *Journal of the Autonomic Nervous System*, in 1971, represented to Brooks a culmination that "fitted into my heritage from my thesis on through the Cannon days and I look on myself as ... carrying on the Cannon tradition – the tradition of the autonomic nervous system" (Brooks, 1986, p. 20). On retirement in 1981, he turned to a serious quest for answers to fundamental questions and became a Fellow of the Center of Theological Inquiry at Princeton, New Jersey, a status which allowed him to continue extensive lecturing and writing. In a review of the hypothalamus that traced the surges of interest in its myriad functions, Brooks (1988, p. 666) asked: "Is there something more?" and answered that the gland holds a key to the " 'soul' ... an admixture of mind and emotion ... ." This statement may be interpreted as a move toward the center by a neuroscientist dedicated to a peripheral system.

S. Walter Ranson's (Centralist) Contribution

One of the troika of outstanding centers of neuroscience activity in Chicago during the 4th and 5th decades of this century was at Northwestern University Medical School in the Institute of Neurology, which flourished from 1928 to 1942 under the firm hand of its founding director, Stephen Walter Ranson (1880–1942; Fig. 5.9). During those 14 years that ended with Ranson's unexpected death, the institute was "humming like a beehive" in exploitation of

Fig. 5.9. Stephen Walter Ranson, shown informally at about age 35, shared with his close friend Joseph Erlanger an old-world philosophy of tight control of their research programs.

the novel stereotaxic instrument and accumulating evidence that related to the role of the central nervous system in maintenance of homeostasis. The significance of the work of Ranson's group may be attributed to its leader's rigid sense of direction.

Early Life and Education

Born into a conservative, intellectual family in Minnesota, Ranson grew up immersed in medicine and science, where his serious view of life appeared at an early age. At the University of Minnesota he favored psychology, but association with the comparative neuroanatomist, J.B. Johnston, bent him toward the anatomy of the nervous system. His last college year was spent as a student of H.H. Donaldson at the University of Chicago, who became his mentor for the Ph.D. (1905) and work on spinal nerve degeneration. Two years later, he gained the M.D. from Rush Medical College and immediately joined the anatomy department at Northwestern. As already mentioned, in 1924 Ranson moved with a graduate student, Joseph Hinsey, to St. Louis to become Professor of Neuroanatomy and Histology and a bright light in the neuroscience firmament there. The interlude lasted only 4 years, however, for Ranson and Hinsey returned to Northwestern and inaugurated a research program in the central nervous system at the Institute of Neurology created for that purpose and housed in the tower of a new building.

A New Type of Textbook

Ranson's research career was marked by an ever-continuing search for the physiological importance of structural observations and by a desire to elucidate their clinical implications, qualities wholly original in outlook and unusual in that era in a neuroscientist untrained in physiology. That attitude was apparent in the first edition of the textbook, *The Anatomy of the Nervous System from the Standpoint of Development and Function* (1920), which was in its seventh rewriting when Ranson died 22 years later. His "functional interpretations of structural background" (Arey, 1943, p. 616) "placed him in the forefront of leadership of that large group of American anatomists to whom a distinction between anatomy and physiology is only a pedagogic discipline to be forgotten in investigation" (Magoun, 1942, p. 3).

A Coordinated Research Program

Ranson's initial investigations at the Institute of Neurology reflected his interest in the prevalence of unmyelinated sensory fibers in dorsal root ganglia and spinal cord and the conduction of pain. Studies of limb reflexes and muscle tonus were aided by revival of the British stereotaxic instrument (see p. 115) and led to the fundamental concept of maintenance of excitation within the neuraxis by reverberating neuronal circuits (see Fig. 5.10). A chance observation that a "hypothalamic" cat can walk focused Ranson's attention on higher brain centers (Hinsey, 1943) and the hypothalamic region and its participation in a broad range of homeostatic adjustments, e.g., the innervation of the hypophysis and water exchange. From the rapid succession of papers it seemed that "the hypothalamus stands at the entrance to the lower brain, guiding the tempo of the internal realm of our existence, with one rein to the pituitary gland and others to the sympathetic system" (Magoun, 1942, p. 2).

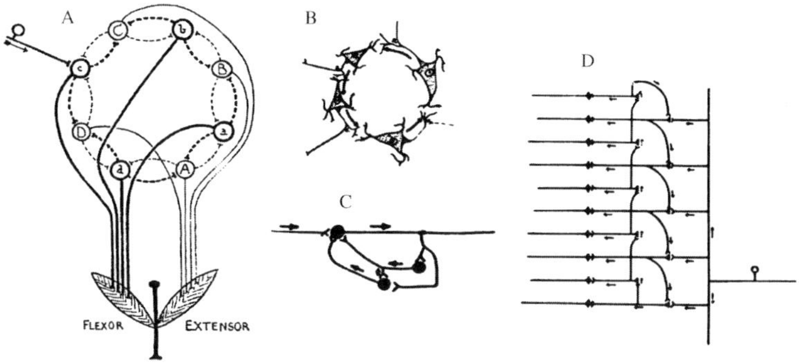

Fig. 5.10. Diagrams of self-reexciting or reverberating neuronal circuits maintaining prolonged excitation in the central nervous system. A from Graham Brown (1916); B from Kubie (1930); C from Lorente de Nó; D from Ranson and Hinsey (1930). Composite from Magoun, 1981, p. 192, Figure 11.

During the latter half of the 1930s, as the institute's investigative program became more and more concentrated on the hypothalamus, Ranson developed a proprietary interest in research in that region of the brain, with scorn for publication of whatever he considered shoddy or erroneous. He was particularly skeptical – Fulton accused the group of bias (letter Fulton to Ingram, 24 March, 1939) – of the claim of the Swiss physiologist, Walter R. Hess, of a diencephalic center stimulation of which produced sleep in cats.[19] Ranson's doubts were based on not being able to confirm Hess's published results, his conviction that they were due to electrolytic destruction, not stimulation, and his own studies (Ranson, 1939) which demonstrated that somnolence in cats and monkeys followed a reduction or impairment of hypothalamic activity. From studies for which he was awarded the Nobel Prize in 1949, Hess had differentiated the gland into a rostral region that operated via the parasympathetic nervous system and a caudal region exerting its influence through the sympathetic system. Although some support for this concept could be derived from the homeostatic role of the gland in temperature regulation (Clark, Magoun, & Ranson, 1939), the behaviors mediated through the hypothalamus were distributed 180 degrees out of phase with those proposed by Hess.

The Northwestern Team

Although reticent and suspicious, the institute's director related well with the many fellows and students attracted by the opportunity to learn the secrets of a cutting-edge methodology and by Ranson's reputation in an emerging field; he assimilated them as collaborators in some aspect of the larger, closely supervised program. In the use of the Horsley-Clarke instrument, Ranson seemed content to watch and comment on the procedures[20] but at the same time he was known as an "unremitting worker of keen mind and quiet, unassuming dignity" (Magoun, 1942, p. 3). His leadership and the research program received justified recognition by the dedication to him of the volume of papers on *The Hypothalamus and Central Levels of Autonomic Integration*, presented at the annual meeting of the Association for Research in Nervous and Mental Diseases in 1939. The resilience of both the work and the topic was manifested 30 years later by another publication, *The Hypothalamus,* the title page of which reads: "This volume is dedicated to the memory of James W. Papez, Stephen W. Ranson, and John D. Green, whose pioneering studies and concepts have shaped its frame" (Haymaker, Anderson, & Nauta, 1969). A lectureship and an endowed chair were named for him by

[19] The details are constructed from the Fulton correspondence at Yale University Library, reproduced with permission in Windle, 1981, pp. 222–228.

[20] Both Magoun and Hinsey in their interviews remarked that they had never seen Ranson actually manipulate the instrument.

Northwestern University; no formal archive exists.[21] There is, however, as with Cannon, a commemorative publication compiled and edited by a former student, William F. Windle (1981). The students and colleagues who reminisced about their experiences with Ranson included practitioners (neurosurgeon Loyal Davis and psychiatrist Charles Fisher) and heads of basic science departments: among many others, there were Joseph Hinsey and "Rex" Ingram, John Brobeck at the University of Pennsylvania who helped to elucidate neural factors in obesity; John Brookhart continued studies on muscle-nerve at the University of Washington; William Windle at the National Institutes of Health, Bethesda, Maryland, on developmental aspects of the mammalian nervous system; and Horace Magoun at the University of California, Los Angeles, delineating the role of the reticular core in arousal mechanisms.

After Ranson's sudden death in 1942, William Windle was named director and Magoun "was invited to move from [Ranson's] well-appointed institute… down among the bare bones of the anatomy department" (Magoun, 1975, p. 522). When Windle left 4 years later for greener fields at the University of Washington, Wendell Jordan Krieg was made director and unsuccessfully endeavored to sustain the institute in the face of dwindling administrative support; perhaps one of his last actions was to invite Giuseppe Moruzzi at Pisa as a National Research Council fellow to the institute. When Moruzzi arrived late in the year, there was no space available until Magoun invited him to share his resources in the basement laboratory; the outcome of their collaboration is described in Chapter 6.

Joseph Hinsey's Imprint on the Institute

The high productivity of Ranson's institute and his strict standards in application of new methods to difficult problems rested heavily on the shoulders of the members of the team who were invited to stay on after earning their degrees. One of those was Joseph Clarence Hinsey (1901–1981; Fig. 5.11), a major influence during the formative years of the institute and with whom Ranson co-authored 56 papers in 6 years.[22] Hinsey had transferred to Northwestern University as an undergraduate intent on entering its medical school when he graduated Phi Beta Kappa, but was diverted by an assistantship in zoology. Accepted into Ranson's program in anatomy as a Ph.D. candidate, he joined him at Washington University with the manual of the

[21] Photocopies of correspondence and a collection of about 2000 offprints sent to him are deposited at the Neuroscience History Archives, University of California, Los Angeles.
[22] For a Ranson bibliography, see Magoun, 1942, and Sabin, 1945. Ranson was extremely proud of the institute's productivity, and annually sent bound copies of offprints to eminent figures in the field, a public relations gesture that accounts for the unusual numbers of offprints from authors who reciprocated.

Fig. 5.11. Although Joseph C. Hinsey accomplished important collaborative work as a doctoral student with S. Walter Ranson, his major contribution came as an administrator and educator in medical sciences, notably at Cornell University Medical College.

dissection of the fetal pig he had been instructed to write. "I was lucky to be able to show up with the manual .... After three or four days [of assisting in the class] he told me I was getting along so well that he was putting me in charge" (letter Hinsey to H.W. Magoun, 30 June, 1979).

Hinsey gained his Ph.D. in 1927 from Washington University and the next year, he was invited to go with Ranson to Northwestern University Medical School to help set up the Institute of Neurology created there to lure Ranson "home." On return to Chicago, they shared an office, laboratory, and the research program on postural contraction, as already described. Two years later, however, wanting to stand on his own, Hinsey accepted the repeated offer of Stanford University, in spite of Ranson's efforts to retain him with visions of discoveries with the new Horsley-Clarke apparatus. He departed in the summer of 1930, just before the new unit was delivered and "Rex" Ingram arrived and was assigned the task to get it up and running (see p. 115).

After 6 years at Stanford, Hinsey became department chair (and Herbert Gasser's successor) in physiology at Cornell University Medical College in New York, then transferred to anatomy and again replaced Gasser, 1942 to 1953, as dean of the medical school, when Gasser went to Rockefeller

Institute. The warm friendship between Gasser and his protégé had commenced during the latter's sojourn in St. Louis, when in spite of Ranson's advice, Hinsey suggested to Gasser their collaboration using the new cathode ray oscilloscope to investigate the nerve fibers involved in dilatation of blood vessels in previously deafferented limbs upon stimulation of spinal dorsal roots. As Hinsey recalled, "... his response was to ask when I could bring the animals over and begin" (Hinsey, 1981, p. 73). The experiments continued after Hinsey returned to Chicago; he prepared the cats, shipped them to St. Louis, and he and Gasser blocked out several days of uninterrupted work. Years later, Gasser presented the Peter Coin[23] to Hinsey as representing his favorite, most productive student who was also a teacher.

In retrospect, Ranson's most widely recognized contribution during the last productive decade was his revival and demonstration of the great usefulness of the Horsley-Clarke instrument. "[It] became the physical vehicle of the fame of that institute" (Krieg, 1975, p. 2). From its adaptation to the rat, cat, and monkey by the group at Northwestern, its utilization then spread to a wide range of experimental animals and its use in human brain surgery became commonplace. It was the unanimous belief of his peers that although Hinsey departed before the instrument was in full use, his role in planning the program for which it was imported was indeed significant. That the stereotaxic instrument was in the hands of one of the century's most productive and inspired groups of experimental biomedical scientists adds to the renown of Ranson's short-lived institute that flared and died with him, but laid the groundwork for many who trained there.

[23]For information about the Doctor Robert Peter Coin and its tradition, see Hinsey's Preface to Gasser's autobiography (Gasser, 1964).

Chapter 6
The Peripheral Nervous System

During the 100 years following Galvani's 18th-century studies, most successful experimental investigations of the vertebrate nervous system were confined to the muscular response to activation of peripheral nerves. In fact, the contraction of the leg of a pithed frog became a standard test of nerve activity and history has many examples of physiologists backing into the nervous system by way of the musculature; 20th-century "neuroscientists" were no exception.

European Roots

In 1847, a quadrumvirate of young, "materialistic" physiologists – Hermann von Helmholtz, Émil Du Bois-Reymond, E.W. von Brücke, and Carl Ludwig – meeting in Berlin, resolved "to expunge from physiology all evidence of vitalism … and to reduce all biological phenomena to the principles of physics and chemistry" (Clarke & Jacyna, 1987, pp. 209–210). Although they had obtained medical degrees for practical reasons, they had little clinical experience to distract them and devoted their full time to research and training their students and assistants. Among their many independent investigations, they made great progress in exploring the electrical features of the nerve impulse and contraction of muscle. Their experimental findings were influential in stimulating younger investigators, such as biochemist Ludimar Hermann and physiologists Ernst Overton and Julius Bernstein, to study the propagation of action currents and the role of electrolytes. And in England, the high-water mark of the Sherrington school of neurophysiology, first at Liverpool and then Oxford, was the study of spinal reflexes, utilizing the string galvanometer to record contraction of the muscles involved. Somewhat later at Cambridge, electrophysiologist Edgar Adrian used the capillary electrometer – until biophysicist Bryan Matthews developed an oscilloscope – to record nerve impulses. The resolve of the German scientists to find a physiology based on chemistry and physics reflected the afterglow of the "romantic" biology of a few decades earlier[1] and cleared the way for at least two avenues of progress that were much later subsumed by neuroscience, one leading to neurophysiology and the other to neurochemistry.

PHYSIOLOGY OF THE AXONS

In contrast to the vigorous growth of comparative neuroanatomy in American studies between 1870 and 1920, neurophysiology made little progress, nothing comparable to the achievements at Oxford and Cambridge some of which have been mentioned. That imbalance was righted after the First World War by American physiologists entering the field, most of whom had benefited from a sojourn in English or Continental laboratories. Laboratories in Philadelphia, Boston, Chicago, and St. Louis were working on peripheral nerve and addressing the intricate engineering task of devising apparatus to faithfully detect and record the relatively faint electrical signals of nerve activity. By the 4th decade, the stage was set for a bright moment in American neuroscience that thrust the neurophysiologists to the forefront and created a unique American school known as "the axonologists." The original participants in the movement represented diverse backgrounds, locales, and research interests, but they all worked on peripheral nerves. They were peers in U.S. biomedical ranks, each an established investigator, and seem to have had a keen appreciation of the benefits accruing to an exchange of information about what interested them. The collegiality of the annual dinners and discussions became legendary and necessitated no formal organization as they took turns assuming responsibility for the arrangement and content of the meetings.

Dramatis Personae

Alexander Forbes

The acknowledged leader of the American school of neurophysiology, Alexander Forbes (1882–1965) combined his New England extraction and education with natural courtesy, curiosity, and ability. Those characteristics gained him respect in several fields – sailing and navigation, aerial exploration and mapping, radiometry, and electrophysiology. On completing his M.D. at Harvard, Forbes spent 2 years (1910–1912) with Sherrington at Liverpool in study of spinal reflexes with myographic recording. During the First World War, Alex gained familiarity with electronic devices of the time and, on returning to Harvard, applied thermionic valve techniques to increase the amplifying capacity of the string galvanometer by some 50 times. By 1920, he was able

[1] For a brief discussion of the "organic physicists" and this interesting movement, see Clarke and Jacyna, 1987, p. 208.

to obtain better records of nerve action potentials than had been recorded before (Forbes & Thacher, 1920), probably the first application of a vacuum tube amplifier to biology (A.M. Grass, 1980). In 1921, he returned to England to spend the summer with Adrian at Cambridge in study of the electrical activity of the nervous system. A year later he published a notable review, "The interpretation of spinal reflexes in terms of recent knowledge of nerve conduction" (Forbes, 1922), which integrated the earlier myographic contributions of the Sherrington school with the neurophysiological observations of the Adrian group in a remarkable synthesis. Forbes had "physiological insight" (Eccles, 1970, p. 394), and his study (Forbes, Campbell, & Williams, 1924) showing that the muscular response to electrical stimulation was the same as that to natural stimulation was fundamental to the development of the evoked potential technique.

Analogously, the activities of others in this field were initially concentrated on electromyography. As already noted (see Chapter 4), after spending a considerable part of the 1920s in Sherrington's laboratory at Oxford, John F. Fulton (1926) published a monograph on muscular contraction in which he emphasized an advanced feature of the ubiquitous myographic records, called "the Oxford angle," which was discovered to be an artifact, to Sherrington's embarrassed chagrin. Also among those working with muscle, Joseph Erlanger's early research at the University of Wisconsin was concentrated on electrocardiology. Wallace Osgood Fenn (1893–1971), at the University of Rochester, New York, before his later outstanding contributions to respiratory and electrolyte physiology, had investigated the energy-work relations of muscular contraction. And when physiological psychologist Donald Lindsley obtained a National Research Council postdoctoral fellowship to work in neurophysiology at Harvard in 1933, he was assigned a project in human electromyography, recording motor units at Massachusetts General Hospital, under the mentorship of Hallowell Davis.

Detlev Wulf Bronk (1897–1975)

Previous to becoming an international statesman of science, Det Bronk (Fig. 6.1) spent a National Research Council fellowship in 1928 at Cambridge with Adrian, who was then extending his work on the EMG with concentric electrodes and attempting to record action currents from single nerve fibers. As Bronk developed this painstaking method, it consisted of gingerly cutting across a mixed nerve bundle until only a single fiber remained functional. The technique was demonstrated at the meeting of the Physiological Society on October 13, 1928, and the report (Adrian & Bronk, 1928) was one of four (Adrian & Bronk, 1929; Bronk, 1929a,b) with Bronk's name attached that recorded significant experimental results during the one-year tenure of the fellowship, a manifestation of high motivation and energy. Among the many new

Fig. 6.1. Detlev W. Bronk's administrative accomplishments peaked in his transformation of the Rockefeller from an institute to a university.

findings was the increase in frequency but not amplitude in forced breathing in rabbit. Shifting to hind-limb reflexes in cat, they found groups of muscle fibers that were innervated by a single neuron. The resemblance of the impulses from sense organs and motoneurons suggested that both shared "some property common to axon or dendrite terminations in general" (Adrian & Bronk, 1929, p. 151).

On returning to the University of Pennsylvania in 1929, as director of the Eldridge R. Johnson Foundation for Medical Physics, Bronk's later work with his associates confirmed and extended the discoveries for which Adrian shared the Nobel Prize with Sherrington in 1932 and helped elucidate the functions of the carotid sinus and some aspects of sensory physiology. As a leader in establishing biophysics as a discipline, Bronk integrated physics, chemistry, and mathematics into the physiology of life processes.

Bronk's executive ability was soon manifest, and he held successive influential positions: the presidencies of Johns Hopkins University, the Rockefeller Institute for Medical Research (which he restructured as Rockefeller University), and the U.S. National Academy of Sciences (1950 to 1962); he was also chairman of the policy-making board of the National Science Foundation

Fig. 6.2. The career of Herbert S. Gasser, right above, photographed in 1929 by F.O. Schmitt, as researcher and administrator (and a talented pianist) included sponsorship of R. Lorente de Nó. His protégée, Helen Tredway Graham (1890–1971), received support for work on histamine from the National Institutes of Health at age 79.

(1956 to 1964) and organizer of the Space Science Board in 1958. "[A]s a mover and shaker in the American scientific community Bronk was without peer" (Anon., 1984, p. 24). Bronk and his family lived part of the year in a gated enclave at Woods Hole, Massachusetts, where he had anchorage for his yacht, yet he was not above searching on hands and knees for an electrical outlet beneath the lectern during a scientific session he was moderating.[2]

Herbert Gasser

In 1916, at Washington University, St. Louis, Herbert Spencer Gasser (1888–1963; Fig. 6.2) attempted to make the Einthoven string galvanometer sensitive enough to record nerve impulses. While completing the M.D. at Johns Hopkins, he and H. Sidney Newcomer, a fellow student with a background in biophysics, had developed a 3-stage amplifier using thermionic valves. Trying it out in St. Louis, they obtained records from the dog phrenic nerve (Gasser & Newcomer, 1921). Although the amplification was sufficient

[2] Personal communication.

to detect nerve action currents, the lag in the string galvanometer greatly distorted the fast potential changes. After persuading Joseph Erlanger (1874–1965), then chairman of the Department of Physiology at Washington University School of Medicine, that the nervous system was worth investigating and learning of the availability from Western Electric of the inertia-free, cathode-ray or "Braun" tube, Gasser and Erlanger published "A study of the action currents of nerves as disclosed by the cathode ray oscillograph" (1922). Twenty years later, in 1944, the two shared a Nobel Prize for this achievement, the first awarded for neuroscience in America. Not only had an era of new discoveries of the nature of axonal activity opened but also the use of the cathode-ray oscilloscope in biology had been inaugurated. Erlanger and Gasser (1924) with George H. Bishop subsequently, found that the compound nature of the action potential recorded from bullfrog sciatic nerve to be a summation of potentials from several fiber bundles and that bundled potentials travel at specific speeds (Erlanger, Bishop, & Gasser, 1926).[3]

In 1966, at a memorial program for Erlanger, George Bishop remarked:

> These two men were not particularly sophisticated in either electronics or physics, and their troubles were varied and cumulative. Noise in the amplifier, noise in the environment, vibration from trucks and sparks from the trolley line called for shielding and insulation, Switches had to be devised which would operate in fractions of a thousandth of a second, without chatter etc. Finally, no camera would photograph the dim trace obtainable at the speed required to record the nerve impulses. The first records were obtained by holding a photographic film against the end of the [Braun] tube, and repeating the stimuli until an accumulation of superposed traces induced a bleary line on a somewhat fogged background. I can only compare their progress to the treck [sic] of the pioneers in oxcarts across the plains and mountains of the West. They were pioneers of electrophysiology and they encountered every conceivable obstacle but the Indians. Far more time was spent on reconstruction of apparatus than on the recording of nerve, when one good record occasionally made a successful day (unpublished manuscript, Washington University Medical Library, with permission).

The results were summarized (Erlanger & Gasser, 1937) with better records than the originals obtained on antiquated apparatus.

Establishing the "Axonologists"

The fabulous new "plant" of Gasser and Erlanger, the cathode-ray oscilloscope (CRO) and its attending amplifier stages, were exploited in a frenzy of activity and published findings by the principals and at other laboratories as

[3] Chase and Hunt (1995) have prepared a lucid and detailed description of Gasser's work and it's significance.

well. With novel apparatus and observations, there was a great deal to talk about, and Ralph Gerard, then at the University of Chicago, recognized a critical mass and made sweeping plans. When the American Physiological Society (APS) met in Chicago in 1930, he "invited everyone working on nerve who was present at the meeting" (Gerard, 1975, p. 467) for dinner at the Faculty Club and on to his home for discussion of their common interests. Fifteen were tapped,[4] but only ten attended: Gasser, Erlanger, Bishop, Bronk, Hallowell Davis, Wallace Fenn, Forbes, Grayson McCouch, Francis Schmitt, and Ralph Lillie, plus their host, Gerard.[5] No records were kept, and that inveterate diarist, John Fulton, was abroad, so all we know is that the evening's collegiality was so timely that the dinners were repeated for the subsequent 12 years.

The annual gatherings of "axonologists" (a name suggested facetiously by Forbes) provided occasions for debates on the nature of nerve impulse transmission and served to elevate the field to a position of dominance in physiology. By the mid-30s,

> the 'Axonologists' were the important people, and almost strutted through the corridors, being very conscious that they alone were at the frontiers of physiological discovery .... [T]he axonologists, with the neurophysiologists in general, enjoyed the greatest battles between the 'Electronists' who were sure that nerve messages were transmitted electrically, and the 'Neurohumorists' who were equally certain that transmission was basically chemical. There would be fantastically excited discussions, all the more fun since some of the protagonists had languages other than English as their native tongue [e.g., Rosenblueth, Nachmansohn, Lorente de Nó] and with the increasing heat of argument the native accent emerged more and more (Alan C. Burton quoted in Fenn, 1963, p. 93).

Within very few years, action potentials became the "hot" field for ambitious physiologists. With the growing number of invitees, the group became too large and unwieldy and lost its "private informality" and so the meetings were discontinued by agreement after the last annual dinner of core members, in Boston in 1942. The tradition of assembling just before the annual APS meeting was preserved by a more formally constructed program on some aspect of neurophysiology, open to all. Succeeding accomplishments of most of these early figures will be presented under later headings.

Basis of Neural Inhibition

The question of inhibition – was it merely cessation of excitation or did it have a self-constituted mechanism – had occupied the thinking of neuroscientists since Sherrington proposed a competition of central states of excitation

[4] Their names are listed on the invitation in the Detlev Wulf Bronk Collection of the Rockefeller Archives Center.

[5] For more about the this famous group, see Fenn, 1963; L.H. Marshall, 1983.

and inhibition that acted on spinal reflexes (Hoff, 1962). Head and Holmes targetted the corticothalamic system as the pathway of cortical inhibition of lower centers removal of which allowed hyperpathia (Walker, 1938, p. 250). American evidence came from studies on cortical control as well as spinal reflex activity.

At the Cerebral Level

John Fulton (1926), reasoning from the paralysis seen after his many ablation experiments, believed that the neocortex could both inhibit and excite lower centers. Support for that view was from work of Sarah Tower and Marion Hines (1935). They found no release phenomena after pyramidal transection and in further experiments (Hines, 1936, 1937) two effects of inhibition, paralysis and spasticity, were distinguished and related to a narrow strip (4s) of frontal motor cortex (see Chapter 4).

By the mid-40s, researchers were finding inhibitory sites further afield. Magoun and Rhines (1947), stimulating centromedial points in the bulbar brain stem, identified inhibitory sites that corresponded closely with Wang's vasodepressor regions and Robert Pitts's end-of-inspiration center.

At the Spinal Cord

When Herbert Gasser was appointed director of the Rockefeller Institute for Medical Research in the 1930s, his main research interest lay with that of the axonologists – in the peripheral nervous system. Nonetheless, he assembled a group of electrophysiologists whose research in the spinal cord could take full advantage of the superb electronic equipment designed and built at Rockefeller by the Dutch electrical engineer, Jan Toennes (Patton, 1994). The institute's preeminence in the field was assured by the presence of Rafael Lorente de Nó, Harry Grundfest, David Lloyd, and Birdsey Renshaw, in addition to Gasser himself.

As part of the group around Alexander Forbes at Harvard and aiming toward his dissertation, Birdsey Renshaw (1911–1948) adapted an experiment by Lorente de Nó, measured homolateral synaptic delay times at spinal motoneurons, and proposed that their briefness (0.5 millisecond to 0.9 millisecond) indicated the dorsal root-ventral root reflex arc had to be monosynaptic (Renshaw, 1940). He developed in meticulous detail the fashioning of satisfactory microelectrodes from ultraclean Pyrex pipettes filled with Ringer's-agar into which dipped a silver-silver chloride wire. The action potentials from rabbit or cat hippocampus and cortex were interpreted (Renshaw, Forbes, & Morison, 1941) in terms of the membrane hypothesis (see below).

On receiving his degree, Renshaw joined Gasser's team and continued his investigation of spinal reflexes. He found that certain motoneurons "condition the reflex discharges of other motoneurons. The conditioning effect is often

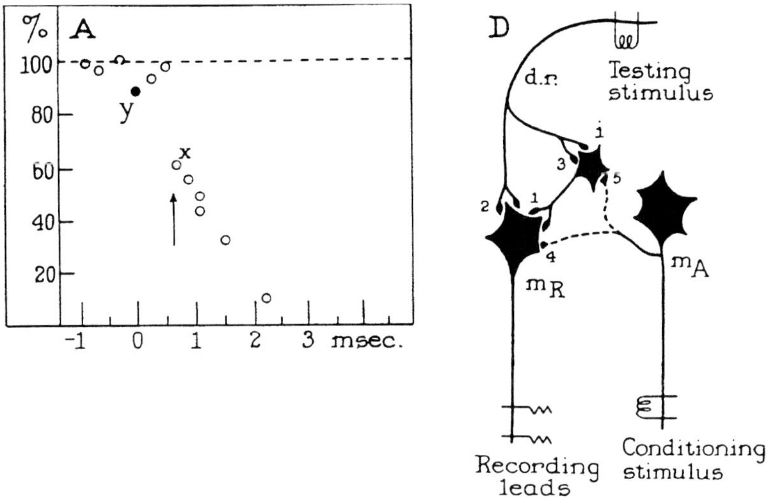

Fig. 6.3. Details of an illustration from Renshaw, 1941, p. 176, Figure 6, demonstrating recurrent inhibition (A, at arrow) and (D) all possible synaptic connections and neurons which might have been involved. Renshaw cells are small, internuncial, ventral horn cells receiving "their afferent stimulation from axonal collaterals of neighboring motor neurons and in turn inhibit the same or other ventral horn motor neurons, thus constituting a negative feedback mechanism" (Lockhard, 1992, p. 233).

inhibitory" (Renshaw, 1941, p. 167). That work marked the discovery of recurrent inhibition mediated by spinal neurons possessing axonal collaterals (Fig. 6.3). After interruption by the Second World War, Renshaw (1946) studied the central effects of centripetal impulses. He predicted the existence of internuncial cells – later named for him – that respond to a single ventral root volley with a train of high-frequency spikes instead of the single spike characteristic of most motoneurons (Eccles, Fatt, & Lundgren, 1957).

The advantage of the monosynaptic spinal reflex as a tool for studies free of the influence of interneurons was exploited by David P.C. Lloyd (1911–1985; Fig. 6.4) in a series of experiments that helped to link the early myographic analyses of the reflex with the findings made possible by the micropipette electrode studies of membrane potentials (Patton, 1994). He demonstrated the direct central inhibitory action of afferent fibers (Lloyd, 1941), identified the origins and projections of afferent nerves of different sizes (Lloyd, 1946a), and determined, using 2-neuron reflex arcs, the reciprocal reflex patterns in the lower spinal cord (Lloyd, 1946b). In his words, "Without the necessity for other than direct reflex connections, the myotatic unit exhibits, complete within itself, the elementary mechanism of reciprocal innervation" (p. 444).

In 1951, a culminating achievement in this field – discovery of the biophysical basis of neuronal inhibition – was announced by John C. Eccles

Fig. 6.4. Canadian by birth, David Lloyd joined Gasser's group at Rockefeller in 1939 and advanced neurophysiology using the monosynaptic ventral reflex arc as a tool. In 1970, he left the institute to live in England. Photograph from Patton, 1994, p. 196, with kind permission.

(1903–1979), then at the medical faculty in Dunedin, New Zealand. Applying the technique of intracellular recording from spinal motor neurons that in the hands of Ling and Gerard (1949), had proven successful with muscle,[6] Eccles inserted a fine glass microelectrode and found the inhibitory postsynaptic potential to be associated with an increased charge on the neuronal membrane, i.e., with its hyper-polarization, the inverse of neural excitation, in which the membrane is depolarized (Brock, Coombs, & Eccles, 1951). With mandated retirement approaching, in 1966 Eccles accepted an invitation to become director of the neurobiology unit of the Institute for Medical Research, newly established by the American Medical Association in Chicago. Two years later, he became distinguished professor of physiology and biophysics at the State University of New York, Buffalo until his retirement in 1975. During that period he wrote four monographs and contributed at length to at least three published conferences, expressing his views on the mind/brain question, among other issues.

Elucidation of the phenomenon of inhibition at the cellular level, however, did not solve the problem of inhibitory effects in tissues. An intramural

[6]Ling's reminiscences of this work, recorded in 1986, described the solution to the problem of drifting baselines encountered by Judith Graham, his predecessor in Gerard's laboratory.

conference in 1967 brought together investigators at the University of California at Los Angeles "to discuss the inhibitory effects (and their electrophysiological correlates) which can be produced by experimental manipulation of prefrontal cerebral cortex, caudate nucleus, and basal forebrain" (Schlag and Scheibel, 1967, p. v). The results were expressed succinctly, again in the Foreword: "The relationship between a nerve cell which becomes hyperpolarized, a motoneuron which fails to respond, endocrine secretion which is interrupted, or an animal which stops bar-pressing, is not clear at the present time."

MOSTLY MEMBRANES

The history of the studies of membrane morphology and function illustrates the powerful contribution that physics and chemistry have made to the knowledge of many biological puzzles, in concordance with the four 19th-century Germans' construction of biological science, as mentioned. That history also furnishes examples of science on hold, waiting for new technology, in this case diffraction X-ray, electron microscopy, voltage and patch clamping, and many other methodologies, before further progress became possible. At the opening of the century, the concept of an inscrutable nerve impulse accompanied by an electric current was generally accepted by physiologists. Decades elapsed before biophysicists elucidated the role of membranes in the generation and propagation of the nerve impulse; by midcentury the solution was in place with the focused work of British investigators.

The Myelin Story

The nature and formation of the myelin sheath was first unlocked in the peripheral nervous system of vertebrates before its counterpart in the brain was deciphered and only after a long interaction of biochemical studies on nerve with investigations of structure. Peripheral myelin is also older phylogenetically than central myelin, as Flechsig proposed in 1901.

Biophysics Around the Axon

The early work of the American biologist, Francis Otto Schmitt (1903–1995) and his associates, first at Washington University, St. Louis and after 1942 at the Massachusetts Institute of Technology, on the ultrastructure of neuronal membranes revealed early indications of myelin's lamellar structure and has been considered the beginning of the modern study of myelin (Terry, 1965). It was also evidence of the advances in physical and biochemical methods made between the two World Wars (Schmitt & Bear, 1939).[7] Only after the advent

[7] Schmitt's later involvement in the Neurosciences Research Program is described in Volume 2.

of the electron microscope and thin sectioning could Schmitt's student, B.B. Geren, in her classic paper (1954) show the lamination of myelin around the peripheral axon (Fig. 6.5) and the connection of the axon with the Schwann cell by the mesaxon, first seen by Gasser (1952), leading to a hypothesis on the formation of myelin from the Schwann cell. The previous year, J.B. Finean in Germany had described myelin as two bimolecular layers of lipid alternating with protein layers. Geren went on to study the myelin loops at the nodes of Ranvier in mouse (Uzman-Geren & Nogueira-Graf, 1957) as Robertson (1957) did in frog nerves. The role of nodes in enhancing conduction velocity and efficiency attracted the attention of theorists and experimentalists alike to studies of saltatory conduction and membrane mechanisms in not only peripheral nerves but also in the central context. The myelinated axon has been called the superhighway of the nervous system (Bunge, 1968, p. 197).

During the 1950s, the dissimilarities of the peripheral and central myelins and the morphogenesis of their sheaths were in the limelight. Building on the original work of Del Rio-Hortega in Ramón y Cajal's laboratory and the acceptance of glia as a type of nerve cell, an American pathologist, Sarah Amanda Luse (1918–1970), suggested from her e.m. studies (1956) that axons of the central nervous system are ensheathed with myelin derived from oligodendroglia. Although the "jelly-roll" theory of peripheral myelination had been accepted (Terry, 1965), two additional theories regarding the mechanism of

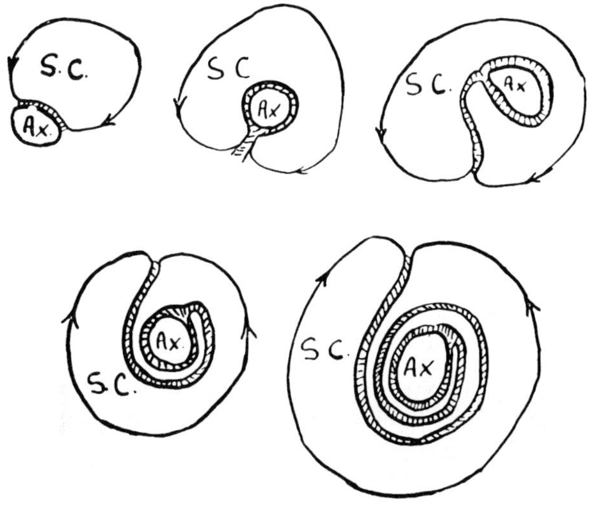

Fig. 6.5. Diagram of myelin formation in "jelly-roll" fashion from B.B. Geren (1954, p. 560). Ax-axon, SC-Schwann cytoplasm. Earlier, Herbert Gasser (1952) had found the sheath well defined and attached to the axon by gossamer ligaments which he called "mesaxons." Electron microphotographs by M.B. Bunge (1962) and her colleagues demonstrated the connection between the sheath and oligodendroglia.

central myelin formation were put forth from abroad. These were sorted out in a masterly review (R.P. Bunge, 1968), which emphasized the characteristics of adult vertebrate CNS myelin and its internodal conduction efficiencies in relation to the known functions of PNS myelin and argued for better coordination of knowledge of conduction in the white matter with the cytoarchitecture of the membranes involved. Two decades later, sequencing of myelin constituents, principally myelin basic protein (MBP), had revealed numerous heterogeneities, including but not limited to, molecular size and configuration, animal species, brain sites, and rate of metabolic turnover (see P. Morell, 1984; Kies, 1985). Current research on the MBP gene shows a relationship with the rodent immune system and neurons in the central and peripheral nervous systems; very recent findings of potential importance in understanding multiple sclerosis in humans have come from the team of A.T. Campagnoni at the University of California at Los Angeles showing that major myelin proteins are expressed in the human thymus (Pribyl et al., 1996).

"Kacy" Cole's Legacy

Coopting the title of his autobiography to introduce the section on membranes recognizes the scientist, Kenneth Stewart (Kacy) Cole (1900–1984; Fig. 6.6), on whose slender shoulders rests the American contribution to one of

Fig. 6.6. Well into retirement at La Jolla, California, Kenneth Cole, photographed in front of the enlarged action current in 1979, relished the widespread use of the "voltage clamp" which he invented, enabling a stable membrane current against which the impulse action current was measured.

the real breakthroughs in 20th-century neuroscience. Tall and slightly stooped, he claimed to be an electrician throughout his career (Cole, 1979, p. 1) and a "Woods Hole addict" where he could work relatively quietly in modest space. His staccato prose abounds in seeming nonsequiturs and evidences of a wry humor.

Early Years

Born in the Midwest and raised in an academic atmosphere, Cole rebelled by taking jobs as deck hand or machinist on ships sailing the Great Lakes. He matriculated at Oberlin College and went on to gain his Ph.D. from Cornell in 1926, then to Harvard (and switched from physics to biology) and a year in Leipzig with the physical chemist, Peter Debye. On his first job as a physiologist – at Columbia University's College of Physicians and Surgeons working for H.B. Williams[8] – Cole did "all sorts of odd jobs" as consulting physicist to the hospital, and indulged his love of sailing.

In late 1932, Cole was asked to help organize a symposium on nerve and muscle and put his stamp on it with the title *Cold Spring Harbor Symposium on Quantitative Biology*, the first in a series which became a preeminent forum of information exchange among physicochemists interested in vital phenomena. Howard James (Bim) Curtis (1906–1972) joined the group late in 1935 and a "feverish rush" ensued to complete experiments to present at the following summer's symposium, which was to be marked by Zed Young's lecture on the use of the squid unmyelinated giant axon, a rapid conductor because of its size. When Cole asked why it had escaped prior notice, Young is reported (Cole, 1979, p. 10) to have replied that near the end of his work at the Naples Marine Station he had found the monograph by L.W. Williams, *The Anatomy of the Common Squid, Loliga pealii, Lesueur* (1912), in which the axonal system was described. By coincidence, the widow of the author, a Harvard anatomist, had been Cole's landlady for 2 years and they had discussed the work, but "I didn't see the monograph for 10 years."[9]

Later Career

During the war years, Cole enlisted his executive talents directing the Oak Ridge staff on the south side of Chicago's Midway. That organization disintegrated with the release of the atomic bomb and the fallout was taken up by the University of Chicago as institutes, with the Institute of Biophysics under Cole's direction. George Heinemann Marmont (1914–1983), a mathematician

[8] The same Williams who had helped Alexander Forbes with his amplifier in 1920.

[9] In his autobiographical contribution to *The Neurosciences*, Young (1975) did not mention Williams nor, inexplicably, did Cole (1979, p. 10) in his volume, *Membranes, Ions and Impulses* (1968).

and biophysicist, joined the institute and the studies on nerve were resumed. In 1949, Cole moved to the Washington area to become technical director of the Naval Medical Research Institute (NMRI) where he built up a complex of strong research groups in medical fields related to possible naval emergencies, such as diving physiology.

Cole made a final transfer in 1954, but only across the road. As the postwar fortunes of the NMRI were declining, not only in funding but also from the ravages of McCartheyism, those of the National Institutes of Health, on the west side of Rockville Pike, Bethesda were entering an ascendancy. Cole was asked to create a new laboratory of biophysics in the National Institute of Neurological Disorders and Blindness, giving him the opportunity to resume the "squid work" he no longer had the funds to support. With John Wilson Moore from NMRI and others, Cole's research on biologic membranes continued for 23 years until his retirement and departure for California. At the NIH, Cole's group was the front-runner among several independent investigators on the same campus who were interested in the same general field: Abraham M. Shanes arrived in 1950 and worked on the nerve-muscle junction and Ichigi Tasaki (1974) on sense organs; Daniel L. Gilbert (1925–2000) joined Cole's group in 1962 to pursue the physiology of biological membranes and toxicology of oxygen.[10]

His Research Contribution

Cole's research interests reflected his point of view as a hands-on physicist, in accord with his early training, and when he decided to mix biology with physics he became one of the first "electrophysiologists." By focusing on biologic membranes, however, he went beyond the axonologists' usual field – the nerve impulse – and into the wider area of excitable systems. The summer of 1938 Cole invited Alan Hodgkin, in Gasser's New York laboratory on a Rockefeller fellowship, to visit Woods Hole. During the winter months Cole and Curtis had been using Nitella and had published (1938) the paper that was "a pride and joy to me. If (as we have been accused of doing) we started a new era of axonology, it was in this paper that we had to be original" (Cole, 1979, p. 11). With the seasonal return of the squid, and having improved their equipment, they were repeating the experiments in which they measured the impedance change (ΔZ) during passage of an action potential. In Cole's words, "Hodgkin visited when we had ΔZ on the scope. He was as excited as I've ever seen him, jumping up and down as we explained it. He also appreciated

[10] *Editor's note:* Even Lorente de Nó, champion of electrical transmission of the nerve impulse, in the late 1960s considered a move to NIH, but reneged at an advanced stage in the negotiations after discovering that he would not have control of his laboratory budget.

the importance of the resting membrane resistance … ." In a collaborative paper (Cole & Hodgkin, 1939) they reported the accomplishments of that short visit.

Those events were also recalled by Hodgkin in Stockholm in 1963 (Sourkes, 1966, p. 410):

> I arrived [at Woods Hole] in June 1938 and was greeted by a sensational experiment, the results of which were plainly visible on the screen of the cathode-ray tube. Cole and Curtis had developed a technique which allowed them to measure changes in the electrical conductivity of the membrane during the impulse; when analyzed, their experiments proved that the membrane undergoes a large increase in conductance which has roughly the same time course as the electrical change.

And the data seemed to fit Bernstein's and Lillie's ideas of a transient membrane breakdown during the passage of the action potential,[11] a theory later refuted by further evidence from Cole and from Hodgkin that the action potential exceeded the resting potential by 40 mV to 50 mV and that "some process giving a reversal of E.M.F. [electromotive force] was required" (Sourkes, 1966, p. 411).

By this time (1938 to 1940) Cole had a creative relationship with his British friends, flowing from their common experimental experiences, and was receiving more support from his overseas colleagues than from his American peers. He recalled (unpublished communication, 1982) being subjected to a "Star Chamber" questioning of his data at Rockefeller Institute in the late 1930s and the accusation by Lorente de Nó that he was deceiving his readers; "Even Gasser was undecided," an especially unhappy blow.

When laboratory work resumed after the war, Cole (1949) and his associate, Marmont (1949) impaled squid axons with long metallic electrodes and commenced using electronic feedback to apply current uniformly to the membrane. With that technique, they were able to measure the experimental variable, the current passing through the membrane. After Hodgkin discussed improvements with the inventors, the method was used at Plymouth the summer of 1948, yielding data for analyses that led to the publications of 1952 on the ionic basis of the nerve impulse and the Nobel Prize in 1963. When Cole visited his British colleagues in late 1948, he realized that "[i]n a little over a year they had corrected most of my difficulties, caught up and run past me. I was pleased that we had been their starting point and that they had confirmed my results, although I was not entirely happy to have the concept and the technique dubbed the 'voltage clamp' " (Cole, 1979, p. 17). Not until

[11] Ralph Lillie's iron-wire model of the nerve impulse and its theoretical usefulness were admirably described by Andrew Huxley, 1995, p. 7.

1973 were gating currents detected (Armstrong & Bezanilla) and 3 more years elapsed before German workers (Neher & Sakmann, 1976) measured single-channel currents.

Invertebrates As Neuroscience Paradigms

The comparative neurobiologists, by confining their experiments to the invertebrates, have the advantage of studying larger neurons and fewer of them in a given volume of tissue than is the case with those who utilize vertebrate models (R. Chase, 1979). This relative but illusory simplicity created an opportunity for multiple comparisons, within and between species, an approach that was made a special crusade by one of America's preeminent neuroscientists, T.H. Bullock (1984). In contrast to Walter B. Cannon's exhortation, "Observe, observe" (see Fig. 6.7), Bullock's "Compare, compare" is more fundamental: Compare first, then observe the differences and similarities to arrive at discoveries. To realize the analytical power of the comparative approach, Bullock recently argued (1995) that it is incumbent on neuroscience to continue to characterize in detail the differences among animals – anatomy, physiology, behavior – with the purpose of answering the question: Is a brain "better" because it has more connections or is there some qualitative difference? Whatever the answer, if any, two invertebrate animals have and will continue to contribute immeasurably to knowledge of brain and behavior.

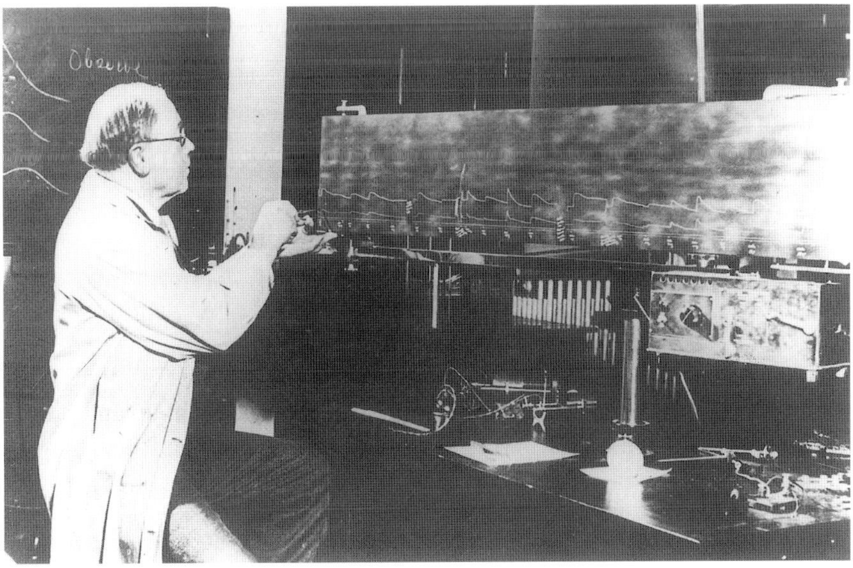

Fig. 6.7. A famous image of Walter B. Cannon with his long-paper kymograph and blackboard admonition. Photograph from the American Physiological Society, with kind permission.

Squid Giant Axons

After the English anatomist-physiologist, John Z. Young, had decided the giant fiber in squid he had discovered in 1929 was a nerve and not a vein (1975, p. 19), he visited the United States in 1935. During the summer at Woods Hole, Det Bronk and Ralph Gerard helped him and Keffer Hartline study action potentials in these giant axons,

> but every time we passed a[n electrical] stimulus to the nerve the oscilloscope trace went sailing away .... So one day, when Det and Ralph were out I said to Keffer, 'Why don't we do away with all this [electronic] equipment and stimulate it with sodium citrate?' So we hooked a fibre up to an amplifier and loudspeaker and put some citrate on the end. Out [of the speaker] came 'buzzzzzz' [of evoked potentials] – one of the best sounds I have ever heard (Ibid., p. 18).

Those were the days – the 1930s – when the excitement generated at actually hearing nerve action currents was a general experience. Adrian and Bronk (1929) had explicitely described the audio output as more revealing than the capillary electrometer tracings. Moruzzi mentioned this in association with his work with Adrian (L.H. Marshall, 1987), and the first printed description of sensory evoked potentials recorded within the cerebrum (see Chapter 7) documented the loudspeaker outputs in detail. Audiences were entranced by the staccato ping of the action current "spikes" as the directional neurons responded to their appropriate stimulus, demonstrated by Hubel and Wiesel and soon repeated by others.

Aplysia Ganglia

This fascinating, gentle creature was first used by Madame A.G. Arvanitaki at her laboratory in the Oceanographic Institute of Monaco, just before the Second World War and in 1949, she described the "spontaneous activity, photosensitivity, and synaptic connections of *Aplysia* neurons. It soon became clear that the preparation had unique advantages" (Kennedy, 1975, p. 52) for analyzing the organization of neuronal networks. The contagion began to spread when *Aplysia* were introduced to Albert Fessard's laboratory in Paris, had an American introduction at Arvanitaki's demonstration in Woods Hole (1960), and Eric Kandel (who shared a Nobel Prize in 2000) spent a year with Ladislav Tauc at Paris. "Then the [American] rush was on, and by the late "sixties an uncounted number of laboratories were filled with Aplysiasts [and] a brisk cottage industry of *Aplysia* collection sprang up on the West Coast" (Kennedy, 1975, p. 52). At this point, the main interest was the connectivity patterns of neurons in the abdominal ganglion (see Kandel, 1969). Kandel's group was the first to interrelate neurophysiological studies with feeding behavior and other investigators utilized a series of animals selected for their widely different patterns of behavior.

Championing Diversity

As an index of the growth of research in the field of invertebrate neurobiology, at the 1977 meeting of the Society for Neuroscience, the number of papers in that field "constituted the third largest of 41 topical categories representative of the Society's interests" (R. Chase, 1979, p. 103). Among those dedicated enthusiasts, Theodore Bullock and Susumu Hagiwara (1922–1989) at the University of California, (both at Los Angeles and San Diego) made important contributions in their separate fields. Hagiwara is noted for his studies of ion channels and excitability, especially calcium ions, across many invertebrate forms. So too, Bullock's career has encompassed a wide range of invertebrates and he has pondered deeply about the results. In the inaugural Jacques Loeb Lecture at Woods Hole, Bullock (1970, p. 566) argued his thesis that a probabilistic theory of neuron activity is plausible "for some parts of the nervous system in some animals." Close attention to noise and variable patterns of firing can promote the movement away from the belief in fixed responses of hard wiring, descriptive of the thinking during the first half of the 20th century, to the second-half concept of a dynamic neuroscience characterized by simultaneous existences of both fixed and variable processes.[12] Later, he characterized this as one of the "quiet revolutions" in neuroscience provided through examination of differences between and among taxa, individuals, and developmental stages and stated that although "the neural basis of humanity" (Bullock, 1984, p. 473) is nowhere near being understood, "we can work on many fronts" in all sorts of creatures.

Bullock's cheerful optimism about the investigative value of invertebrates was justified a quarter-century ago and was sustained. The early authoritative, English-language monographs in the field (Bullock & Horridge, 1965; Wiersma, 1967; Lentz, 1968) were joined by specialized literature comparable in scope to that in the vertebrate domain: behavior (Wiersma, 1975), arthro pods (Hoyle, 1977), neuropharmacology (Leake & Walker, 1980), electrical conduction (Shelton, 1982), cell culture (D.J. Beadle, Hicks, & Middleton, 1982), and biochemistry (Lunt & Olsen, 1988). As the 20th slid into the next century with increasing pressure – and hope – toward a cure for human paralysis, most appropriate to the bright outlook is a text on nervous system regeneration in invertebrates (Moffett, 1996).

In summary, the first half of the 20th century in both America and abroad saw a common full flowering of the nerve impulse, from its chemical and physical elements to its physiology as a propagated wave. The neurotransmitters became agents in their own domain and the biologic membrane's role was established. Anatomy did not disappear but became more particulate as

[12]Relatedly, the trophic theory of Dale Purves mirrors this thesis and is discussed in the next chapter.

simpler invertebrate species swam into view. With adequate instrumentation the time had come to carry investigation of how the brain works into the central nervous system.

NEUROCHEMISTRY AND NEUROPHARMACOLOGY[13]

The second track that emerged from 19th-century European bioelectric science became a recognized neurochemistry as it focused on the nerve action potential before spreading to the central nervous system. In broad brush, the biochemistry of the nervous system may be traced in three phases (D. Tower, 1958; McIlwain, 1966): chemical composition of the brain (Hensing, early 18th century and Thudichum, late 19th century); metabolic studies of substances found within brain tissue, which did not of themselves add to understanding brain function; and studies of specific substances acting on the brain, the latter with spectacular results. In England, as the 20th century commenced, the systemic effects of extracts containing what became known later as neurotransmitters, epinephrine (adrenaline) and acetylcholine, held the attention of Thomas Elliott at Cambridge and Henry Dale at University College London. The latter quoted Du Bois-Reymond's 1877 remark on the possibility, based on chemical stimulation of muscle, that the phenomenon "is chemical or electrical" (Dale, 1937–1938, p. 4P). The great debate on the nature of nerve impulse conduction was beginning to take shape, a debate informed by the parallel development of biochemistry and physiology. The two disciplines shared a common goal – to understand the functions of the nervous system – and as D. Tower (1958, p. 25) argued, it is artificial to separate their history.

American Focus on the Biochemical Basis of Mental Health

The combination of biochemistry with human behavior commenced its move to become an independent subdiscipline in America just before the close of the 19th century, when a few enlightened superintendents of hospitals for the insane were convinced that more than custodial care of the inmates would bring rewards in preventing or alleviating mental disease. That humanitarian attitude was prevalent in Europe and encountered little obstacle to its spread in the United States (Tourney, 1969). A major reinforcement was the severe scolding for their lack of interest in experimental studies given the psychiatrists in 1894 (p. 26) by the popular Philadelphia neurologist, S. Weir Mitchell.

Beginnings in Mental Hospitals

Coincidentally, that same year, one of the first nonclinical laboratories for biochemical research was opened, at Harvard's McLean Hospital in Waverly,

[13] *Editor's note:* Only an outline of this section was completed by the author.

Massachusetts, in a building separate from the hospital and shared with pathologists and physiological psychologists; the superintendent had studied with Wundt and Kraepelin in Europe. The biochemical laboratory's subsequent scientific eminence dated from Otto Folin's directorship beginning in 1901, and continued in neurochemistry with Jordi Folch-Pi as director in 1944 to his retirement in 1977. In that laboratory Alfred Pope and his group developed microtechniques for quantitative analyses of enzyme activities and cellular components of identified neocortical layers (Pope, Caveness, & Livingston, 1952), a neurocytological achievement that merged anatomy and chemistry at the micro level.

Jordi Folch-Pi (1911–1979)

Born in Barcelona and retaining his Catalan intensity, Folch-Pi (Fig. 6.8) was determined to address the research needs in his field and establish a community of neurochemists. The "Folin-Wu procedure" of total protein extraction had become standard, and so he took the next step toward elucidation of nerve membrane mechanisms and isolated the proteolipid complex in central nervous

Fig. 6.8. Jordi Folch-Pi is credited with outstanding research (e.g., the "Folch procedure" of total lipid extraction), promotion of education in the new field of neurochemistry, and its furtherance nationally and internationally. Photograph from *J. Neurochem.*, 35:1, 1980, with kind permission.

system myelin. Folch-Pi's extramural contributions included prominent participation in the formation of both international and national societies for neurochemistry and as an editor of *Journal of Neurochemistry*. He had "universal acclaim as one of the historical founders of the biochemistry of complex lipids, and of neuro-chemistry as a distinct and significant field of learning" (Pope, Lees, & Hauser, 1980, p. 3).

A slightly older pioneer in American neurochemistry, Heinrich B. Waelsch (1905–1966; Fig. 6.9) was born in Brno and educated (M.D. and Ph.D.) at the University of Prague. Emigrating to the United States in 1938, his first employment was at Columbia University, where, in 1954, he was made professor of biochemistry, and when he died, he was director of the New York State Research Institute for Neurochemistry and Drug Addiction as well as chief of research (pharmacology) of the New York State Psychiatric Institute. Waelsch's research interest was metabolism of amino acids, especially glutamic acid, polypeptides, and proteins in brain. He was very active in the International Brain Research Organization (IBRO), attempting to prevent "research without an echo" on the part of foreign students returning to their home countries without equipment or assistants with which to continue their research. Waelsch led the organization of five international, multidisciplinary symposia

Fig. 6.9. In addition to his investigation of amino acids in brain, Heinrich B. Waelsch was committed to international efforts to promote brain research, serving as secretary general to IBRO, 1963 to 1965. Photograph ca. 1960, kindness of Salome Waelsch.

between 1954 and 1962 which "had a major influence in development of neurochemistry in proper perspective throughout the world" (W.M. Sperry, 1966, p. 1262).

Percival Bailey's Showplace: the Himwich Team

Responding to public pressure in the aftermath of the Second World War, the Veterans Administration set up laboratories of neurochemistry and neurological sciences to address some of the basic issues of mental and neurological disability. The circumstances surrounding the establishment of one of those new dedicated laboratories vividly illustrate the rewards of well placed political connections. When Percival Bailey in Chicago was asked by then Governor Adlai Stevenson to upgrade the psychiatric care system in Illinois, in 1951 Harold and Williamina Himwich were persuaded to establish the Thudichum Psychiatric Research Laboratory at the Galesburg State Research Hospital. "At that time, Percival Bailey was very strong in the state, and he saw to it that the hospital was to be made into a showplace .... The main idea of the research was, of course, to find treatments – or at least ways of alleviating – mental illness" (W.A. Himwich, 1980, p. 20). Funding, installation of air conditioning for the animals so they would not die during the hot summer, and even a new building were not a problem "as long as Percival Bailey was alive because he handled it all" (Ibid., p. 27).

Williamina Elizabeth Armstrong Himwich (1912–1993; Fig. 6.10; right), with Scotch and American Colonial antecedents, was educated in the Midwest and held a doctorate in nutrition from Iowa State College. At the end of the Second World War, she and her husband, Harold, were at Edgewood, Maryland in the Army Chemical Center where they worked on organophosphates and related enzymes. In 1949, Williamina became a research medical bibliographer in toxicology at Johns Hopkins School of Medicine under contract with the National Library of Medicine. and, in association with Eugene Garfield (Wortis, 1994), an expert on the interface between medical indexing and computers. Thirty years after moving to Galesburg and raising three sons while contributing a steady stream of research reports on topics of brain metabolism, she joined the National Library of Medicine to work in the specialized information service dealing with the current literature in neurotoxicology.

Harold Edwin Himwich (1894–1975; Fig. 6.10; left) was a second-generation American with a strong Eastern European medical-family background. After his medical degree from Cornell Medical College, he progressed from intern to house physician at Bellevue, then held successive research positions at Cornell, Harvard, and Yale, where he and Nahum (1929) discovered that the respiratory quotient of the brain is unity, indicating that glucose is the main energy source of that organ. In 1935 the couple moved to Albany, where he was professor of physiology and pharmacology, and then to Edgewood,

Fig. 6.10. Although Harold and Williamina Himwich pursued their own lines of productive research, they also succeeded in raising a family and serving their profession of biological psychiatry. Photographs from Wortis, 1975, p. 681; 1994, p. 291.

Maryland. Harold was glad to answer Percival Bailey's bid to Galesburg State Hospital to escape from participation in development of chemical weaponry (Wortis, 1994). His interest was focused then on the pathophysiology of psychoses, an unpopular position at the time of analytical psychiatry; he and his wife each had distinguished careers and were elected separately to the presidency of the Society of Biological Psychiatry. With an enormous published output,[14] Harold Himwich investigated many chemical constituents of body fluids from mental patients and the effects of drugs on them.

In Graduate Schools

The evolution of neurochemistry may be seen in a succession of American texts used in teaching. In 1937 Irvine H. Page, a physician writing from the Rockefeller Hospital, published *Chemistry of the Brain* in which he proclaimed that "Probably no other organ … may be studied with greater facility than the brain" (Page, 1937, p. ix): blood samples to and from the brain are collected easily; it is bathed in cerebrospinal fluid which isolates it and can be itself studied; and it has unique qualities of metabolism. A decade later, David Nachmansohn centered on the nerve with *Chemical Mechanism of*

[14] Perhaps one of the most accurate and complete bibliographies of a scientist, as it was prepared by his wife who had an outstanding reputation as a bibliographer of medical sciences.

Nerve Activity (1946), and in little more than another decade, the intrusion of molecular biology was made apparent by the amplification of the title to *Chemical and Molecular Basis of Nerve Activity* (1959).

Concurrently with the rise of interest in brain chemistry, there developed in American medical schools departments of physiological chemistry and pharmacology independent of biological chemistry, and by midcentury biochemists working in the nervous system were calling themselves neurochemists. Banding together for more specialized meetings than those of the biological chemists, in 1969 the American Neurochemical Society was organized (the international society had been formalized 4 years earlier), with Folch-Pi as president. The rise of neurochemistry in academia was fueled in part by the postwar surge in new methodologies and instruments as well as biomedical funding. Additionally, it was concomitant with the excited debate over the nature of impulse transmission at the synapse – "spark" or "soup" – electrical or chemical, which in turn stemmed from the discoveries of neurotransmitters and the concept of adrenergic and cholinergic transmitter release systems, a British proposal (Dale, 1936–1937) that had been adopted by the field.

Neuropharmacology and Neuropsychopharmacology

There are no techniques specific to neuropharmacology by which this subdiscipline may be identified (Tourney, 1969; Cooper, 1987). The patch-clamp apparatus seen in every pharmacology laboratory with which fluxes through membrane channels are measured was borrowed from neurochemistry; any investigator interested in a neuroactive drug may be called a neuropharmacologist. Neuropharmacology and the search for new drugs was revolutionized by the gradual realization that almost all drugs act with the mediation of receptors, a vast elaboration of the concept initially suggested by the British pharmacologist, Langley (1905). The early century American neuropharmacologists, however, did not pursue that idea for their interests veered toward exploration of the physiologic consequences of drug action in the organism.

By the 1960s, the empirical demonstration of the effectiveness of psychotropic drugs had made them competitive with the psychoanalytic approach to mental illness, and the neuropsychopharmacologists were emboldened to form an American College to "catalyze developments in diverse areas of clinical and basic neuroscience ... ." (Meltzer, 1987, p. ix). Through the publication of occasional review volumes, in 1968, 1976, and 1987, the college has charted worldwide progress in its subdiscipline: chemical neuroanatomy and toxicology of the CNS, electrophysiology of drug action, and biology of learning and memory; research with receptors, the neuropeptides, neuroendocrinology, and animal models; contributions of molecular biology to the understanding of the earlier areas. The last edition, *Psychopharmacology: the Third Generation of Progress*, highlights the continued sophistication of methods and progress in translating basic research findings into clinical utility. The members of the American College of Neuropsychopharmacology are cognizant of their indebtedness to the pioneers in their specialty.

Abel and Cannon

As the first professor of pharmacology in the Western Hemisphere (Davenport, 1982), John Jacob Abel (1857–1938) brought great renown to his Alma Mater, the University of Michigan, where he trained with Vaughn and Sewall, and at the Johns Hopkins School of Medicine, the scene of much of his research on epinephrine, the vasopressor effect of which had been first noted by Oliver and Schäfer (1895). Abel commenced work on the drug in 1895 and published his "definitive" paper 4 years later (Abel, 1899) on the first hormone to be isolated (Hartung, 1931). A clever Japanese chemical engineer, Jokichi Takamine, visited Abel's laboratory (by then at Hopkins) probably in 1900, and applied for a U.S. patent and trademark on Adrenalin late that year; the correct elementary formula was determined by T.B. Aldrich (1901) who had been an assistant to Abel for 5 years at Hopkins and then was employed by Parke, Davis and Co., the owner of the patent. A half-century elapsed from Abel's first experiments before the contamination of natural epinephrine with norepinephrine was recognized (Goldenberg, Faber, Alston, & Chargaff, 1946). A recent history of the development of American pharmacology singled out Abel as the shaper of that discipline, not as much for his discoveries as for initiating publication of two journals and organization of two professional societies. In addition, "His laboratory was the key institution in the education of the first generation of home-grown American pharmacologists" (Parascandola, 1992, p. 61).

As noted, Cannon at Harvard in the mid-30s had embarked on a search for the chemical transmitter in the sympathetic nervous system and postulated two forms of what he was pretty sure was "adrenin" (Davenport, 1982). Comparing the excitatory or inhibitory effects of stimulating sympathetic nerves to specific organs with the effects of injecting the natural product, he concluded erroneously that he was dealing with two products, sympathins E and I. That misinterpreted work, however, was only a small part of Cannon's career-long contributions to the physiology and pharmacology of the autonomic nervous system. By midcentury, the earlier, relatively crude studies of epinephrine had been superseded by the recognition of a vast number of substances that controlled or moderated neural action.

Metabolism of the Neurotransmitters

As part of the large group that moved in 1949 to Bethesda with James Shannon (later director of the National Institutes of Health), Julius Axelrod became intrigued with the sympathomimetic amines (Axelrod, 1975), the first of which had been Abel's epinephrine. Finding a neglected line of inquiry, with his associates, notably B.B. Brody, he explored the metabolic pathways and their related enzyme catalysts and found that neurotransmitters in the body may be metabolized and removed by natural processes, an idea that is the foundation of biochemical pharmacology (Ibid., p. 192). Using radioactive noradrenaline, it was shown that in nerve terminals the released neurotransmitter is in a continuous state of flux, through removal by re-uptake or breakdown by monoamine oxidase or catechol-O-methytransferase. For these and other insights, Axelrod was made a Nobel laureate in 1970.

Chlorpromazine (CPZ)

For a second time during the 20th century, mental hospitals played an important part in the history of neuroscience in America and elsewhere. In the early century, as already noted, superintendents had led the change from custodial barracks to treatment centers. At midcentury, the use of a new tranquilizer reduced the role of hospitalization in the care of large numbers of human beings with brain dysfunctions. That history also illustrates the pivotal place of drug manufacturers in the infrastructure of pharmacologic science. This dual success story with a dark side (Swazey, 1974) commenced in the early 1950s in France when a pharmaceutical company found that a relaxing "cocktail" containing a phenothiazine amine also had sedative properties that calmed and made more accessible patients with certain types of mental disorder, especially schizophrenia. The marketing success of Thorazine in the United States was more than matched by CPZ's spectacular effect on mental hospital populations in the United States [a decrease of 39 percent in 20 years (Fig. 6.11)]. The reduction of numbers of hospitalized patients, however, also exacerbated the problems of homelessness and other social ills.

The introduction into pharmacologic sciences of the phenothiazine amines instigated an explosion of therapeutic testing and a hunt for synthesized configurations with tailored characteristics, fewer side effects, for example. The psychotropic drugs represented in the '50s and '60s an enormous contribution to human well-being as well as a great boost in a subdiscipline of neuroscience. It was only a small step, however, when viewed in relation to the enormous stride taken by the relatively new field of immunopsychopharmacology which came later (see Part 2, The Behavioral Sciences).

Drugs as Tools

While a student at Yale Medical School in the late 1940s, Daniel X. Freedman (1921–1993; Fig. 6.12) conceived of using neuroleptics as a reversible method of probing the mechanisms of the nervous system. Later, at the National

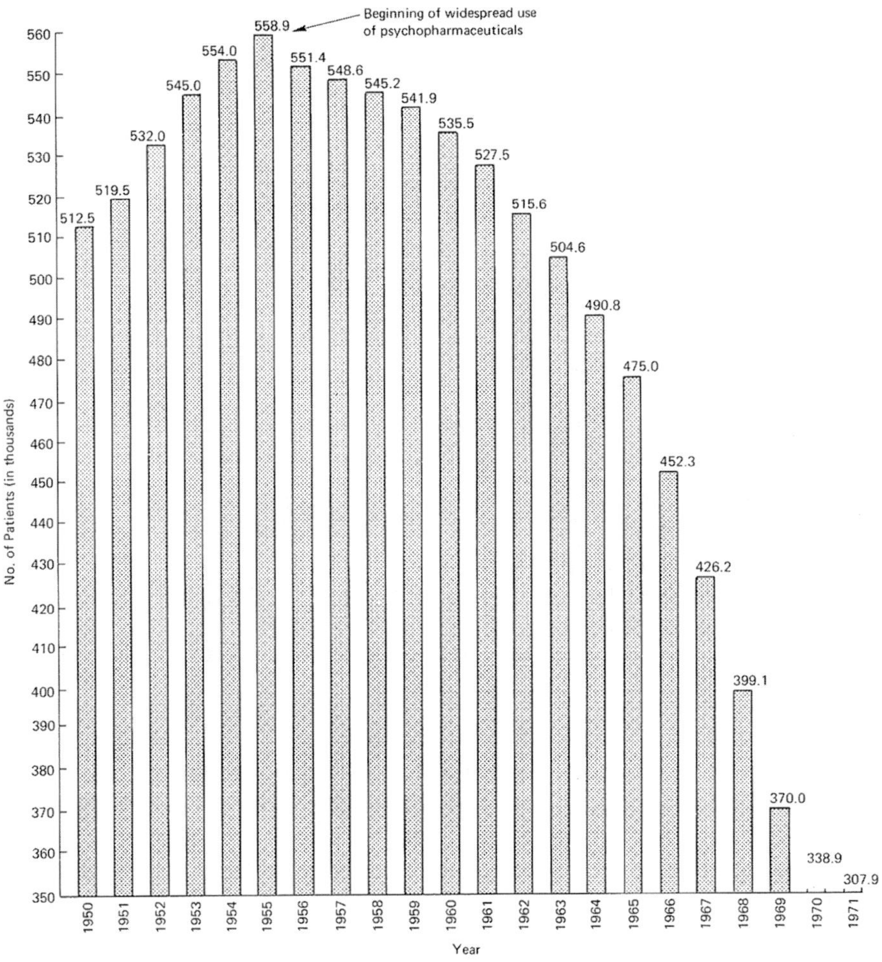

Fig. 6.11. Number of resident patients in U.S. state and locally governed mental hospitals from 1950 to 1971. Chart prepared for publication in the first of four historical manuscripts sponsored by the NRC Committee on Brain Sciences and funded by the National Institute of Mental Health in an effort to narrow the gap between science and society. From Swazey, 1974, p. 241, Figure 6.

Institute of Mental Health, he and collaborators focused on LSD, showing the dependence of the intensity of its effects on the amount of the neurotransmitter, serotonin, present at synapses. Freedman retained his research interest in LSD throughout his career and at the same time juggled administrative and clinical responsibilities, service on national and congressional advisory groups, and almost three decades of editorship of *Archives of Biological Psychiatry*. Through his serious commitment to the journal, Dan Freedman

Fig. 6.12. The intensity brought by Daniel X. Freedman to all his activities included research and promoting biological psychiatry.

greatly influenced the course of psychopharmacology and biological psychiatry in the United States by his pioneering role. The new-found knowledge of biological psychiatry was heralded in that era by Ralph Gerard, with his customary eloquence: "No twisted thought without a twisted molecule!"[15]

[15] From S.J. Novak (1991) based on interviews with D.X. Freedman.

Chapter 7
Into the Central Nervous System

Until the 1930s, the electrophysiology of the nervous system had been studied in the United States for the most part away from the brain, in the spinal cord and autonomic and peripheral nervous systems. That focus was moved forward into the central nervous system by an enterprising graduate student at the University of Chicago working in a laboratory headed by a mature neurophysiologist with a propensity for trying out imaginative ideas. Together they took the axonologists' stimulation-recording techniques into the brain where the full potential of the methodology's versatility would be realized and a new world of discovery opened.

EVOKED POTENTIALS

The neuroscientists who introduced the rubric "evoked potentials," which eventually became a domain *per se* with its specialized journal, monographs, and conferences, were men whose collaboration commenced as a brief excursion into axonology (Gerard & Marshall, 1933). Wade Hampton Marshall (1907–1972; Fig. 7.1) had come to Gerard's laboratory at the University of Chicago with a solid background in physics and was competent to build amplifying and stimulating circuits tailored to his projected experimental goals. Those primary skills were fine-tuned several years later during a summer of contact with Albert Grass at Harvard's Physiology Department, before moving to Johns Hopkins Medical School as a fellow of the National Research Council (NRC) with Bard in physiology and then to establish a neurophysiology laboratory at the Wilmer Institute of Ophthalmology. In 1947 Marshall established a neurophysiology laboratory at the National Institutes of Health, Bethesda, Maryland, the first to be organized there, of which he was chief until his retirement in 1970. The sustained high productiveness of that laboratory during the 1950s and '60s was a product of liberal leadership and a steady stream of research associates – newly graduated M.D.s satisfying

Fig. 7.1. Wade H. Marshall constructed as well as utilized the new electronic gear in study of the nervous system. Between 1935 and 1950, he established three productive laboratories in succession: in the physiology department and the Marburg Institute at Johns Hopkins School of Medicine and at the National Institutes of Health.

their military obligations while training in research, a federal program that produced many competent researchers.

A Premature "Electrical Atlas of the Brain"

Recalling his initial reluctance to explore the "morass" of the central nervous system, Ralph Gerard described how finally he and Wade Marshall, with help from a visiting psychiatrist, Leon Saul, explored the electrical activity of the cat's brain with oscilloscope (CRO) and loudspeaker. "[W]e followed 'evoked potentials' – which we named – from sight, sound, touch, and proprioception into all sorts of regions where sensory impulses were not supposed to go … ." (Gerard, 1975, p. 468). They found a great variety of "spontaneous" rhythms as the electrode was moved slightly and published (Gerard, Marshall, & Saul, 1936) nine page-length tables of coordinates, structures, and spontaneous and evoked responses of electrode placements, both CRO-recorded and auditory, such as howling, a terrific roar, a very high, siren note, hoarse blasts, crashes, a chugging rhythm, and so on. Opposite each table there was an enlarged microphotograph of cat brain sections on which had been superimposed a numbered millimeter grid. The text explained that "Dr. Ranson had given assistance in the use of the instrument, and, especially, made available to us serial sections of cat's brain … ." (Ibid., p. 677). The interpretation of such data was, to say the least, baffling and, while Gerard was in Europe for the summer, Marshall hastily completed his doctorate and left for George Washington University.

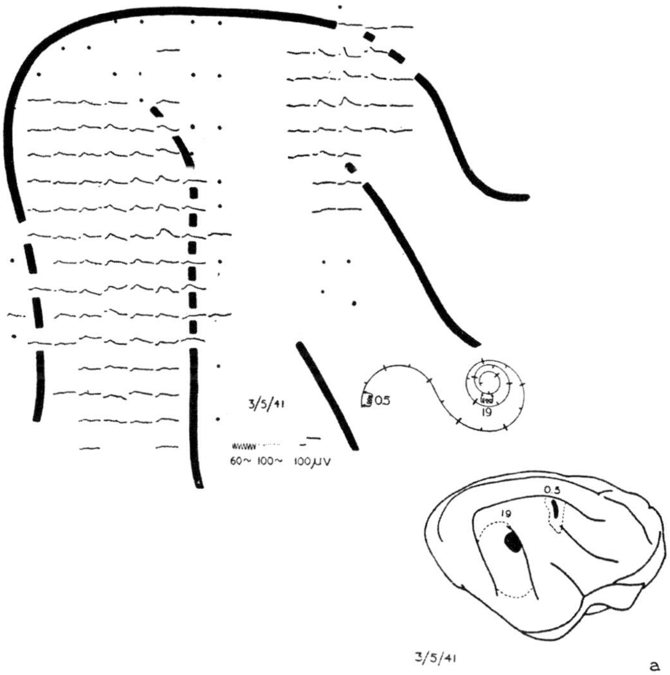

Fig. 7.2. Above: Experimental results on right cortex from stimulation of nerve fibers at 0.5 mm and 19 mm from basal end of cat left cochlea. Insert: Summary of response areas from same experiment. From C.N. Woolsey and Walzl, 1942, pp. 321, 329; Figures 3, 9a.

Delineation of Sensory and Motor Cortical Areas

With more sophisticated use of this method, Wade Marshall, Clinton Woolsey, and Philip Bard provided the first somatotopic localization of cutaneous sensibility in the cerebral cortex of cat and monkey (1937, 1941). Using focal light stimuli, Marshall and Samuel Armstrong Talbot (1903–1967), an electrical engineer, delineated the projection of the retina on the visual cortex and propounded a theory of color vision (Marshall & Talbot, 1942). Woolsey and Walzl (1942) similarly mapped the projection of tones from the cochlea on the auditory cortex (Fig. 7.2). As for the motor cortex, Woolsey ingeniously stimulated the bulbar pyramid and recorded the potentials conducted antidromically back along the corticospinal tract to their cells of origin in the motor area, the overall dimension of which turned out to be considerably larger than that determined by stimulating the cortex and observing motor responses.

Woolsey's Laboratory of Neurophysiology at Wisconsin

Clinton Nathan Woolsey (1904–1993; Fig. 7.3) was born in Brooklyn, attended Union College in Schenectady, New York, and gained his M.D. at

Fig. 7.3. Clinton N. Woolsey was the keystone in establishing the University of Wisconsin as one of the outstanding midwestern centers of neuroscience excellence. Photograph with kind permission of T.A. Woolsey.

Johns Hopkins Medical School in 1933. With aspirations to becoming a neurosurgeon, he worked with Marion Hines and Sarah Tower in neuroanatomy[1] while progressing from assistant to associate professor in physiology at Hopkins, from 1933 to 1948. In addition to teaching, Woolsey was deeply involved in the program of ablation studies with Philip Bard on hopping and placing reactions and mapping of evoked potential projections, as mentioned, with Wade Marshall. This was also the time of initiation of a fruitful collaboration with Jerzy Rose, an accomplished neuroanatomist émigré from Poland.

In 1948, Woolsey was appointed Schlicter Professor at the University of Wisconsin to develop basic research in neuroscience there. On the strength of a belief in the essential interdependence of anatomy and physiology, Jerzy Rose was brought to the laboratory which became a research unit in 1961 and in 1973 was made a full-fledged Department of Neurophysiology, still unique in American academia. In 1975, a Mental Retardation Center was added and Woolsey became its coordinator for biomedical research. Among many accomplishments of this program was the discovery, in rat to chimpanzee, of second cortical sensory areas adjacent to each of the somatic, auditory, and

[1] See Chapter 4.

visual fields; together with a supplemental motor area on the medial side of the hemisphere.

Proposed Cortical Suppressor Areas and "Spreading Depression"
In 1935, Warren S. McCulloch introduced electrophysiologic techniques to Dusser de Barenne's laboratory of neurophysiology at Yale (see Chapter 4), and began a collaborative study of cortical interconnections using Dusser's technique of placing strychnine on focal areas of cortex and recording evoked "strychnine spikes" at distant cortical areas to which they had been conducted. In the course of those studies, a strip of cortex along the anterior margin of area 4 was identified on which the application of strychnine arrested spiking and suppressed ongoing electrocortical activity elsewhere over the hemisphere. As noted (p. 115) concurrently, at Johns Hopkins, Marion Hines had found that ablation of this cortical strip was followed by spasticity in the opposite limbs. Dusser de Barenne then called this region "area 4-s," with s referring to "the strip of Marion Hines." Later, several of these strips were identified and described as 8-s, 2-s, and 19-s, with reference to adjacent cortical areas, s then connoting "suppressor." Interpretation of those areas as centers for spreading depression were shown to be erroneous: Wade Marshall (1950), then at the National Institutes of Health, and Norman Sloan and Herbert Jasper (1950), at the Montreal Neurological Institute, demonstrated that suppression was attributable to "Leão's spreading depression" provoked by drying or other physical debilitation of the exposed cortex.[2]

Unit-Recording with Implanted Microelectrodes in Behaving Animals
With the improvement of long-term (chronic) placement of microelectrodes in single cortical cells of alert animals (a technique widely used after its introduction by Hubel & Wiesel), Vernon B. Mountcastle at Johns Hopkins, in a series of studies with collaborators (1957 to 1978) analyzed the electrical activities of vertical columns of neurons in the monkey's somatic sensory cortex, evoked by natural excitation of peripheral receptors. Analogous vertical columnar organization had been identified on activation of the visual cortex by lighting the retina of the eye by David H. Hubel and Torsten N. Wiesel at Harvard (1968–1972). Sensory "barrels" were identified by T.A. Woolsey and Van der Loos (1970) in layer IV of the mouse S I region. Similar patterns of columnar organization of activity in the motor cortex have been reported by H. Asanuma and I. Rosen at Rockefeller University (1972).

[2] Aristides A.P. Leão from Rio de Janeiro discovered the phenomenon of spreading depression (a slow wave of reduced electrical activity passing over the cerebral cortex) while a graduate student of Walter B. Cannon's at Harvard. His career as a leading South American scientist included membership in the Brazilian Academy of Sciences and director of the Institute of Biophysics of the Federal University of Rio de Janeiro.

DEVELOPMENT OF ELECTROENCEPHALOGRAPHY

The claim of Adolf Beck in Cracow to have been the first to see spontaneous potentials from the exposed animal brain was counterclaimed by Richard Caton in Edinburgh, who could point to his publication in 1875 on "the electric current of the brain" (Brazier, 1961). The classic achievement of the reclusive Jena psychiatrist, Hans Berger, in solving the formidable problems of amplification of faint signals through the intact scalp was treated with neglect until Adrian and Matthews (1934) at Cambridge had repeated and extended Berger's 1929 finding of the alpha and beta waves.[3] After such notable validation, there was no hesitation in the rapid spread of electroencephalography (EEG) into American laboratories and clinics, as elsewhere.

American Competition to Rediscover and Develop the EEG

Perhaps in a more competitive spirit than abroad, American investigators and clinics were quick to apply the new technology to explore whatever electrical activity the intact brain produced, initially more from curiosity than with a predetermined experimental or diagnostic goal in mind. The earliest American papers on the electrical activity of the brain of laboratory animals were published from the Midwest (Lindsley, personal communication). At the University of Kansas, Bartley and Newman (1930) were recording evoked potentials with five-stage vacuum tube amplification and oscillograph from three "distinct" parts of canine cortex and found the action currents appeared similar to each other and to those seen in peripheral nerve. The implications, they concluded, were that "... specialization of function ... is a fact, but we do not know as yet the degree to which this specialization is true or its constancy over a period of time" (Ibid., p. 587). Simultaneously, and also looking for functional correlates, Travis and Herren (1930) at the University of Iowa were studying action currents on the cerebral cortex in rats (and one dog) during patellar and Achilles tendon reflexes. "These preliminary findings lend further support to the proposition that functionally the highest levels of the central nervous system are a part of the lowest levels" (Ibid., p. 693).

By 1935, a banner year, at least five pioneering laboratories were fortunate enough to command the expertise to design and construct the apparatus necessary for recordings through the human scalp. They were at Providence, Rhode Island; Harvard School of Medicine; Tuxedo Park, New Jersey; Western Reserve University, Cleveland, Ohio; and the University of Iowa. The basic elements of the rhythmic patterns and their distortion during

[3]The first full-length publication of human EEGs in an American journal was of Adrian's presentation at the 1934 meeting of the American Neurological Association in Atlantic City, New Jersey, alerting neurologists to a second form of brain imaging.

epileptic states were factors that contributed to the excitement of a new vista in neuroscience research. Some of that excitement permeates the details of James O'Leary's *Science and Epilepsy* (1976), written from the perspective of one of the first participants.

Early Contributions

Herbert Jasper and Leonard Carmichael, with the help of physicist and electronic engineer Howard Andrews, at Bradley Hospital and Brown University, Providence, Rhode Island, enterprisingly gained the first publication (1935) of "brain waves" recorded in this country. They confirmed the experimental results of Berger and of Adrian and Matthews and rashly proposed that the method might be useful in psychology and clinical neurology, even to revealing brain action as the electrocardiograph does for the heart. In 1938, Jasper left Brown to establish an EEG laboratory at Wilder Penfield's new Montreal Neurological Institute at McGill and assist in localization of cerebral epileptogenic tissue (summarized in Penfield & Jasper, 1943). Penfield's earlier studies had led to the concept of a "centrencephalon" in the upper brain stem (Penfield, 1936–1937; 1938), which he identified as responsible for consciousness in man, an idea that had little support and died.

Jasper was succeeded at Brown by Donald B. Lindsley who had been exposed to the EEG while a student with Hallowell Davis at Harvard. Four years before, Davis and Arthur J. (Bill) Derbyshire, had adapted the equipment built for their experimental work on auditory paths in the cat to present the first public demonstration of the EEG, with Lindsley as subject. Davis and his wife, Pauline, continued to explore the nature of the EEG, spending several summers at Alfred Loomis's private laboratory at Tuxedo Park, New Jersey, where a gigantic cylindrical rotating drum, supporting paper the size of a ping-pong table, was used to record the EEG of a night's sleep of human subjects (Loomis, Harvey, & Hobart, 1935; Fig. 7.4). In May, 1937, Loomis organized the first seminar on "brain waves," bringing together all then-active electro-encephalographers and some elder statesmen such as Alex Forbes. As Davis (1968, p. 6) stated, after reviewing the historical background from Caton through Berger,

> [t]he subsequent exchange of information and ideas was extremely helpful and the affair might be considered the first gleam in the eye leading to the establishment ultimately of first the Eastern and later the American EEG Society.

From Curiosity to Diagnostic Tool

The utility of the EEG as a diagnostic tool was soon apparent. In addition to its application in preoperative testings of Penfield's patients, Fred and Erna Gibbs and M.G. Lennox (1938) at the Massachusetts General Hospital (MGH) refined the diagnosis in epilepsy by relating symptoms and signs of

Fig. 7.4. An engineering feat, the Loomis chronograph constructed at the Tuxedo Park, New Jersey private laboratory made it possible to continuously record brain waves during a complete sleep cycle. Photograph from Loomis, Harvey, and Hobart, 1935, p. 397.

the disorder to specific abnormal rhythms in the electroencephalogram, which they suggested were related to the physiologic variables of blood sugar and carbon dioxide. Meanwhile, at the University of Iowa, Lee Travis and John Knott (1936) introduced the recording of brain potentials in normal speakers and stutterers (see Chapter 15), thus demonstrating the utility of the EEG as an experimental tool in psychology as well as in clinical neurophysiology. Remarkably, Herbert Jasper, Donald B. Lindsley, and John Knott all obtained their doctorates at the University of Iowa, under the mentorship of Lee Edward Travis (1896–1984; Fig. 7.5) as professor and chairman of the department of psychology. At Western Reserve University, in 1938–1939, Lindsley studied maturation of the EEG in children enrolled in a child study program, one of the many that sprang up in the United States in response to the aroused interest in child development. Later, he obtained the first "in utero" EEG, as well as subsequent serial studies of EEG maturation in the Lindsley children (Lindsley, 1938). That same year a three-part study of the EEG in normal infants and children by J.R. Smith, at Columbia University and Babies Hospital, New York City, added more significant data.

At Clark University in 1938, Hudson Hoagland, who had returned from a period with Adrian at Cambridge in 1930 with a Matthews oscilloscope, amplifier, and electrodes, was alerted to Berger's findings by Adrian and immediately reconfirmed them. Hoagland went on to collaborate with ongoing research on schizophrenia at nearby Worcester State Hospital, recording EEGs of some of the patients – who had trouble fitting into Hoagland's small screened cage which had been set up for evoked-potential studies of activity

Fig. 7.5. The department of psychology at the University of Iowa under Lee E. Travis remarkably influenced the clinical application of the EEG and American neuroscience through the people trained there (A.M. Grass, 1984). In addition, Travis remained focused on human communication throughout his career. Photograph from Ibid., p. 15, Figure 32.

in the lateral line organs of fish – with inconclusive results. Hoagland (1974) suggested, as had the group at the MGH, a relation between the alpha rhythm and blood carbon dioxide and glucose.

The (virtual) certification of the electroencephalograph as a diagnostic tool was conferred by the Second World War. It "was uniquely suited to the diagnostic needs of neuropsychiatric and neurosurgical services" (O'Leary & Goldring, 1976, p. 146): in addition to identification of epileptogenic centers, it was valuable in encephalopathy, penetrating wounds of the head, and various other wartime conditions. Wiring and assembling units in their basement,[4] Ellen and Albert Grass shipped about hundred 4-channel models to the armed services in the eastern and western theaters and to definitive-care hospitals at home. Fred and Erma Gibbs (1936) helped develop standards and manuals of the uses and limitations of the EEG; eventually the "flat" electroencephalogram became a criterion of brain death.

[4] See Volume 2 for the role of the Grass Instrument Company in furthering the progress of neuroscience in the United States.

In Experimental Laboratories

An early and unusual application of electroencephalography in humans was carried out for the U.S. Navy by neurologist Robert Cohn. His initial experience with EEG had been with Wade Marshall in 1935 in their extracurricular collaboration in construction of a machine for St. Elizabeths Hospital, the federal institution for the insane in the nation's capital. In addition to reading patients' records, Cohn was consultant on studies at the Washington Navy Yard on the effects of breathing various mixtures of compressed gas at simulated hyperbaric (diving) conditions. In 1943, he accompanied the installation of a unit at the Naval Hospital in Bethesda, Maryland and combined clinical and experimental studies there.

A second example of early experimental use of the EEG has already been mentioned (see p. 92). Briefly, in 1935, at Yale, with the aid of Warren S. McCulloch, Dusser de Barenne set up EEG recording to extend his strychnine studies of the monkey's sensory cortex. By midcentury, polygraphs were routine equipment in experimental laboratories as well as familiar in clinical settings. Together with the stereotaxic apparatus revived at Ranson's Institute of Neurology at Northwestern University Medical School, enabling precise placement of stimulating and recording electrodes, the cerebral subsurface had become accessible as never before. The axonologists of the 1930s were eclipsed by the "EEGers" of the '40s but in reality they were often the same individuals who, like Gerard, had been forced to plunge into "the great ocean of brain waves" (Gerard, 1975, p. 468).

Neurophysiologic Substrates of the EEG

Although many parameters of the electrical potentials at the surface of the cortex as an indicator of the net electrical activity of the brain were being examined, the challenging question was the underlying mechanisms. In what region of the brain were they generated? Those and other problems were addressed on several fronts.

Among serious investigations of underlying mechanisms, some of the most significant were undertaken by neuroscientists at Harvard. Robert Swain Morison (1906–1986; Fig. 7.6), anatomist Edward Wheeler Dempsey (1911–1975; Fig. 7.7), and others commenced studies of thalamocortical mechanisms responsible for the rhythmic spindle bursts and recruiting responses in the cortical EEG, inquiries that were fundamental to later recognition of the widespread influence of thalamus on cortex (Morison & Dempsey, 1942).

During the mid-40s, Donald B. Lindsley joined the department of psychology at Northwestern University, Evanston, Illinois, but found no facilities for his experimental program. On the invitation of Horace Magoun, he set up an EEG apparatus at the medical school in Chicago and subsequently

Fig. 7.6. Robert S. Morison, after carrying out studies on fundamental mechanisms in corticothalamic relations, attained high responsibilities in Rockefeller philanthropic activities. Photograph 1955 from the Rockefeller Archive Center.

Fig. 7.7. Edward W. Dempsey became interested in clinical medicine in spite of a brilliant start (with Morison) in neurophysiologic research. Photograph ca. 1964 from the National Library of Medicine.

collaborated with Magoun and others, including a talented young medical student, Tom Starzl, in studies of subcortical influences on the cortical EEG. In 1949, Giuseppe Moruzzi, a postdoctoral visitor from Italy, and Magoun identified in cats ascending reticular pathways in the core of the brain stem which evoked and maintained the EEG features and behavioral characteristics of wakefulness, arousal, and attention. With Lindsley and others, this brain-stem activating system was further investigated in studies that were replicated and extended to monkeys in the 1950s, when Magoun, and later, Lindsley moved to a new school of medicine at the University of California, Los Angeles,[5] in collaboration with neurosurgeon John D. French. Additionally, John Green, Ross Adey, and associates at UCLA determined that the relation of activity to behavioral state of the paleocerebral hippocampus was the inverse of that of the neocortex. During alert wakefulness, for example, the hippocampal EEG displayed a pronounced theta rhythm (trains of slow waves) whereas the neocortex recorded desynchronized activity (Green & Adey, 1956; Green, 1964).

Studies in Sleep/Waking

The electroencephalogram became the framework on which sleep studies progressed from the purely descriptive observations of "somnolence" in hypothalamus-damaged animals, as in Ranson's work, to the demonstrated relationship of behavioral states and specific patterns of cerebral electrical activity. As mentioned, observations by investigators both in the United States and abroad had shown that any type of afferent stimulus that arouses the subject to alertness also is capable of activating the EEG, "but the basic processes underlying it ... have remained obscure" (Moruzzi & Magoun, 1949, p. 455). Bearing on the problem was the chance finding that stimulation of the reticular core in the brain stem reproduced the EEG changes seen in physiologic arousal reactions: the slow-wave high-amplitude pattern characteristic of normal sleep changed to the low-amplitude fast activity of the alert subject. Moruzzi and Magoun, using the "encéphale isolé"[6] or chloralose-anesthetized cat, interpreted their results not as inhibition but as the projection of desynchronizing effects throughout the cerebral cortex (Fig. 7.8) and mediated, at

[5]The intense pressure for productive exploration of a novel and fascinating research landscape, coupled with unfinished construction of the new medical school in West Los Angeles, had prompted setting up temporary facilities at the Veterans Administration Hospital in Long Beach, 35 miles south of the university. For a decade, the research of Magoun's Department of Anatomy was carried out there until the Brain Research Institute became functional in its own building, as related in Volume 2 and in detail in *An American Contribution to Neuroscience: The Brain Research Institute, UCLA 1959–1984* (1984).

[6]In 1937, Frédéric Bremer at Brussels reported a waking EEG in the "encéphale isolé" cat (transection of the brain at the bulbospinal juncture), and a sleeping EEG in the "cerveau isolé" (following a midbrain transection) preparation.

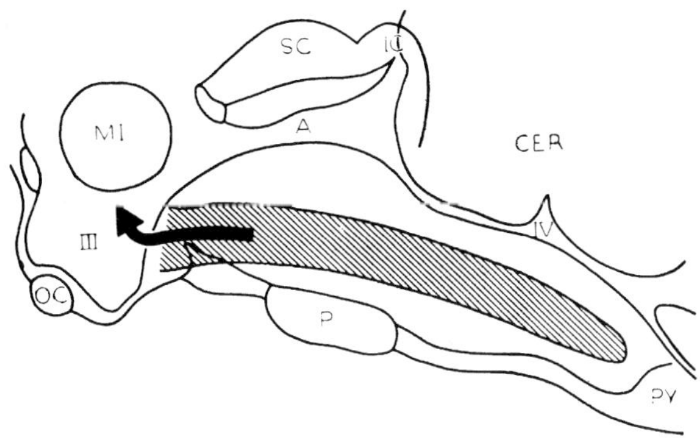

Fig. 3

Reconstruction of midsagittal plane of cat's brain stem upon which is projected, with cross-lining, the distribution of the ascending reticular activating system.
Abbreviations are as follows: A, aqueduct; CER, cerebellum; IC, interior colliculus; MI, massa intermedia; OC, optic chiasma; P, pons; PY, pyramidal crossing; SC, superior colliculus; III, third ventricle; IV, fourth ventricle.

Fig. 7.8. The initial diagram of the brain stem reticular formation was the forerunner of the logo of the Internet forum for neuroscience history <histneur-l@library.ucla.edu>. From Moruzzi and Magoun, 1949, p. 457, Figure 3.

least in part, by the diffuse thalamic projection system. They suggested the implications of the system in maintenance of wakefulness, thus shifting attention from an exclusive focus on sleep states.

The modern era of scientific study of natural human sleep commenced with a long-term research program carried out by the Russian-born psychologist, Nathaniel Kleitman (1895–1999; Fig. 7.9), at the University of Chicago. He and his students identified and characterized the stage of REM sleep, named for its associated rapid eye movements (Aserinsky & Kleitman, 1953; Dement, 1958).[7] Though the balance of the body musculature is flaccid in REM sleep, the accompanying neocortical EEG is low-voltage fast, and that of the hippocampus consists of theta waves – both features similar to the EEG pattern of alert wakefulness with eyes open. In humans, those findings are of possible relevance to the fact that this is the stage of sleep in which dreaming occurs. More recently (1959–1961), G. Moruzzi and his associates at Pisa, Italy, identified a bulbopontile synchronizing mechanism, electrical excitation of which

[7] Although Kleitman's role in initiating systematic sleep studies, including living in Mammoth Cave, Kentucky without external clues, was highly significant, his own estimate of his "really distinct contribution" to neuroscience was the discovery of the asymptotic nature of the acquisition of conditioned reflexes (personal communication, 1975).

Fig. 7.9. Intense interest in sleep research was inaugurated in the United States by the work of Nathaniel Kleitman, commencing in 1925 at the University of Chicago.

evoked large slow waves and spindle bursts in the cortical EEG, whereas inactivation was followed by the EEG arousal pattern. From its caudal position in the brain stem, this mechanism seemed to be phylogenetically old; possibly it serves to reduce visceral functions generally by diminishing the excitability of the brain.

By the 1960s, interest in sleep and waking research had become paramount in laboratories abroad and at home in which EEG recording made assessable the electrical activity of the whole brain but gave no clue to cellular-level activity. At the National Institutes of Health, Edward V. Evarts (1926–1985), was among the first to use the new technique of intracellular recording in chronic animals refined by Herbert Jasper and carried out a series of unit-recording experiments in cat and monkey visual and motor cortical cells; in a summary of his findings, Evarts (1967) suggested that comparisons of the behavioral and physiological components should be made before and after, rather than during a period of sleep. He interpreted his data as not supportive of the theory of sleep as restorative of depleted action-potential-generator substances. A continent westward, a center of sleep studies developed at the Brain Research Institute (BRI), University of California, Los Angeles, not surprisingly in view of the transplantation of H.W. Magoun's research program on the reticular arousal system from Chicago's northside to California in 1950. "Neurophysiologists ... gradually realized that the reticular formation was prominently important for the maintenance of the waking state" (Clemente, 1969, p. 80).

Neuronal excitation did not have the field to itself, however; evidence from acute and chronic experiments led C.D. Clemente and L.B. Sterman (1967) to postulate that sleep onset with suppression of behavior is correlated with a reduction in sensory stimuli to the reticular activating system that "sets the stage for ... active forebrain inhibition leading directly to the initial stages of sleep" (p. 87). Major programs in sleep research continue in the BRI setting: Ronald M. Harper's group is investigating the neural mechanisms underlying cardiovascular and respiratory activities during sleep and waking states. And in the laboratory of Michael H. Chase the control of sleep and waking is studied with molecular biology procedures, intracellular recording, microiontophoresis, and behavioral state analyses in the search for mechanisms accounting for the suspension of activity in motor neurons during slow-wave sleep.

Correlations

Inevitably, sleep research had become associated with biological rhythms and the search for the cerebral site of an endogenous clock in control of cycling. A basic rest-activity cycle in nervous system activity was readily shown in the EEGs of sleeping mammals, including the human, and less easily during the waking state (Kleitman, 1969). In 1972, two research teams reported evidence of the circadian center in the suprachiasmatic nucleus of the hypothalamus for drinking and motor activity (Stephan & Zucker, 1972) and for cortisone secretion (Moore & Eichler, 1972).

A major stimulus to the development of interest in sleep/waking research was the high incidence of information exchange opportunities, largely provided by federal funds in response to the information explosion. There were funds for symposia and publication of standardized terminology, techniques, and scoring of sleep stages (Rechtschaffen & Kales, 1968); and a bibliographic registry of the swelling current literature on the topic, titled *Sleep Research*, initiated by the Brain Information Service at the University of California, Los Angeles (see Volume 2). At Harvard Medical School, J. Allan Hobson initiated a system of managing the exploding literature which later resulted in publication of a valuable annotated bibliography (Hobson & McCarley, 1977) and choreographed a popular educational exhibit, one of the first in U.S. biomedicine, on sleep and dreaming. To inform physicians of the strides made in scientific progress, a symposium was organized on the physiology and pathology of sleep (Kales, 1969). Sleep research had become a subdiscipline in its own right and the Association for the Psychophysiological Study of Sleep was organized in 1971. The first international congress (1971) was held in Belgium with Bremer and Kleitman as honorary presidents and a program committee chaired by Michael H. Chase, who edited the proceedings.

With this concentration of interest in one of the universals of mammalian life, it would be a natural outcome to form an archives of sleep/waking neuroscience. That has not materialized, although the successful establishment in the field of another universal, the UCLA History of Pain Collection, is an instructive precedent.

AN INTEGRATIVE NEUROBIOLOGY

Major Foreign Antecedents

The anatomic roots nourishing the concept of a functional nervous system effecting an integrated behavior are found abroad in the "Old Country." Camillo Golgi, at the University of Pavia, Italy, discovered a chromate of silver method, "la reazione nera," for staining neurons and, applying it between 1870 and 1886, revolutionized earlier insights of the histology of the nervous system. Santiago Ramón y Cajal, at the University of Barcelona, Spain, first saw examples of Golgi's silver stain in 1887 and, modifying and applying it to the embryonic nervous system in which the nerve cells are relatively small and their fibers still unmyelinated, far exceeded Golgi's achievements. Cajal's results first received recognition at a demonstration before the German Anatomical Society in Berlin in 1889. Appointed to the University of Madrid in 1892, 2 years later he delivered the Croonian Lecture of the Royal Society in London, and was invited to lecture in the United States at the decennial of Clark University, in 1899. When Cajal and Golgi were jointly awarded the Nobel Prize in physiology or medicine (1906), in their ensuing lectures Cajal gave a clear exposition of the neuron doctrine, developed from his own studies; whereas Golgi attacked the concept and attempted to defend his own reticular hypothesis of a syncytial network of interconnected nerve fibers. In aggregate, those studies were the springboard from which rose all subsequent discoveries of neural function.

East Coast Tradition

A year after Johns Hopkins opened its medical school (1892), Franklin P. Mall, professor of anatomy, appointed an associate, Lewellys Franklin Barker (1867–1943), to teach histology. In 1895, at Mall's suggestion, Barker spent 6 months at Leipzig, where Paul Flechsig's lectures on the anatomy of the brain aroused his interest in the work of Golgi, Cajal, and Nissl. Barker became a champion of the neuron doctrine and, encouraged by Mall, prepared a profusely illustrated, 1,122-page account (Fig. 7.10) of *The Nervous System and Its Constituent Neurones* (1899).

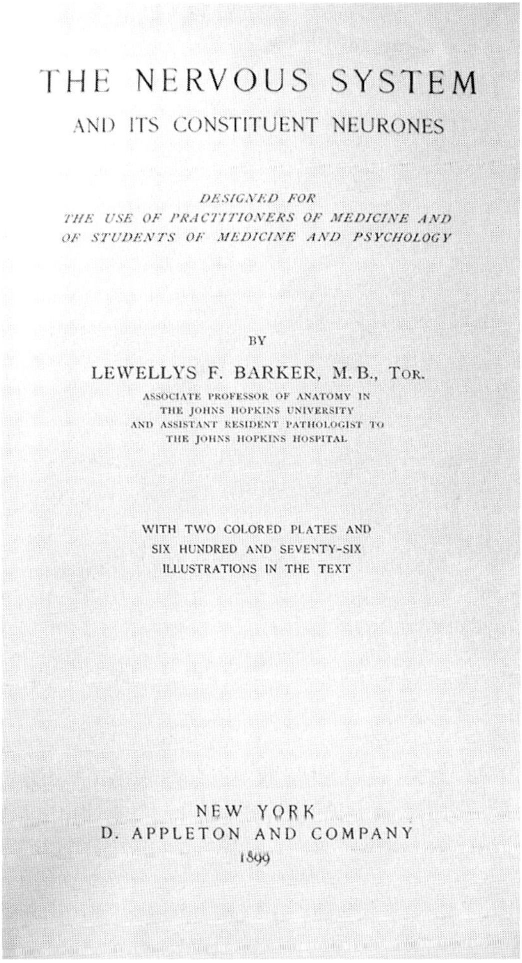

THE NERVOUS SYSTEM

AND ITS CONSTITUENT NEURONES

DESIGNED FOR
THE USE OF PRACTITIONERS OF MEDICINE AND
OF STUDENTS OF MEDICINE AND PSYCHOLOGY

BY

LEWELLYS F. BARKER, M. B., TOR.

ASSOCIATE PROFESSOR OF ANATOMY IN
THE JOHNS HOPKINS UNIVERSITY
AND ASSISTANT RESIDENT PATHOLOGIST TO
THE JOHNS HOPKINS HOSPITAL

WITH TWO COLORED PLATES AND
SIX HUNDRED AND SEVENTY-SIX
ILLUSTRATIONS IN THE TEXT

NEW YORK
D. APPLETON AND COMPANY
1899

Fig. 7.10. Title page of the first American textbook devoted entirely to the nervous system, written by Lewellys F. Barker at Johns Hopkins, who was especially interested in neuronal nutrition.

> I had thus the fine opportunity, Barker wrote, of entering the field of modern neurological histology in America, a field of description in which I am proud to count myself one of the pioneers .... [F]or the first time, the conduction paths of the central and peripheral nervous system were comprehensively and systematically described from the standpoint of the neurone doctrine (Barker, 1942, p. 61).[8]

[8] Barker's text was also the first to be written in America devoted solely to the nervous system.

Fig. 7.11. Extending his postdoctoral European exposure to embryology, Ross G. Harrison pioneered animal tissue culture and provided proof of the neuron's protoplasmic independence. Photographed in 1911.

Tissue Culture

In this nurturing milieu, in 1894, Ross Granville Harrison (1870–1959; Fig. 7.11) gained a Ph.D. in Biology at Johns Hopkins, under W.K. Brooks and, after a year in Germany, joined Mall's anatomy department in the medical school. In study at Bonn, Harrison had gained familiarity with the microsurgical and tissue-culture methods of experimental embryology. At Hopkins, he demonstrated that removal of the neural crest in amphibian embryos prevented development of spinal ganglia, dorsal roots, and sensory components of the spinal nerves. Reciprocally, removal of the ventral half of the neural tube eliminated all motor components and their innervation of muscles (1907). Next, he transferred bits of the amphibian medullary tube to sterile hanging drops of clotted frog lymph, and followed their continued growth and development in isolation, unobscured or influenced by surrounding tissues (1908). This, the first successful culture of animal tissue, demonstrating the outgrowth of unmistakable nerve fibers from neuroblasts, left no room for questioning of the neuron principle.

Subsequently, Harrison moved to Yale, where he revived a faltering effort in biology. One of the most notable of his doctoral students, Samuel R. Detwiler, became professor of anatomy at Columbia University and analyzed the influence of developing nerve fiber tracts on proliferation of neurons in the spinal cord. By excising or transplanting limb buds, he established

Fig. 7.12. Paul A. Weiss led the early movement in the United States to ensure adequate inter disciplinary training for neuroscientists as they developed from multidisciplines; Weiss's research focused on axoplasmic flow as a nutritive mechanism.

a conceptual foundation which later was elaborated in studies of limb transplants by Paul Weiss and neuronal regeneration relative to targets by Roger Sperry.

From the Midwest

Again reinforcing our argument (see Chapter 4) that during the decades of the 1930s and '40s, Chicago was the preeminent American center of neuroscience research, the consolidation of an integrated neurobiology was sustained by the work of two developmental psychobiologists at the University of Chicago.

Axoplasmic Transport

Paul Alfred Weiss (1898–1989; Fig. 7.12), a native of Austria, gained his Ph.D. in biology at the University of Vienna, and published his first papers on the functional and theoretical aspects of limb transplants in amphibia. Relocating in America, he continued those studies for 20 years at the University of Chicago and until 1964 at the Rockefeller Institute for Medical Research. His career achievements in developmental, cellular, and neural biology included experimental analyses and theoretical interpretations of growth control, differentiation, cell behavior, nerve regeneration, wound healing, and the coordination of nerve centers, accomplishments that were recognized in 1980 by the award of the National Medal of Science.

Weiss's observation of an accumulation of cytoplasm at the proximal end of a ligated peripheral nerve fiber led to the conclusion that new protoplasm was synthesized only at the nucleus of the nerve-cell and was continually moved down the axon at a rate of about 1 millimeter per day (Weiss & Hiscoe, 1948). From this first observation came the flood of subsequent research on axoplasmic transport more recently so elegantly investigated by his student, Bernice Grafstein (1971), and summarized by Sidney Ochs and R.M. Worth (1978).

The initial phase of Roger Sperry's research career, as a graduate student in Weiss's laboratory, established that the mammalian central nervous system has a basic circuitry for specific functions (Voneida, 1997). The experiments foreshadowed Sperry's later formulation, while at Harvard and Orange Park with Karl Lashley, of the chemoaffinity theory (1951) of nerve growth guided by chemical signals. Evidence supporting the neurotrophic factor hypothesis was reported by Frank Lillie's student at the University of Chicago, Marian Lydia Shorey.[9] By removing the limb bud from chick embryoes, she demonstrated a reduction in the number of neurons in the appropriate spinal cord center (Shorey, 1909). In a subsequent report Shorey (1911) suggested that a metabolic product of the target mediated the formation and maintenance of the connections between limb bud and spinal cord cells. Sperry's later work on duality of the cerebrum is described in Chapter 13.

In a somewhat similar pattern to that of Paul Weiss, Viktor Hamburger, a native of Germany, gained his doctorate under Hans Spemann at Freiburg, with a dissertation on the influence of innervation on amphibian limb development. Hamburger became well acquainted with Harrison during the latter's summer visits to Freiburg. In 1932, Hamburger received a Rockefeller fellowship to work with Frank Lillie, another friend of Spemann's. At Chicago, he applied Spemann's glass-needle technique to microsurgery on the chick embryo, a field in which he continued to experiment throughout his career. In 1935, Hamburger succeeded F.O. Schmitt as chairman of the department of zoology at Washington University, St. Louis, when the latter moved to the Massachusetts Institute of Technology. Hamburger's early publication on limb extirpation (1934) came to the attention of Rita Levi-Montalcini in Italy and under difficult conditions during the Second World War, she repeated this study with somewhat different conclusions. In 1947, at Hamburger's invitation, she came to St. Louis to resolve their inconsistencies; in the ensuing research, differential hyperplasia of ganglion cells, which showed growth

[9] We are indebted to Charles Howe at Mayo Clinic and Foundation for calling our attention to this work.

Fig. 7.13. The vast work on nerve growth factor commenced with the discovery by Elmer D. Bueker (1948) of a mouse tumor that was highly stimulatory to chick motor neurons.

potentials beyond their normal range, suggested the existence of growth-promoting agents.

Nerve Growth Factor

Beginning a search for such trophic agents, in 1949, they turned to the results of "a bold and ingenious experiment" of Elmer D. Bueker (1903–1996; Fig. 7.13), a former student of Hamburger's, "which was to change entirely the direction of my research (Levi-Montalcini, 1975, p. 249). Bueker (1948) had implanted tissue from a selected mouse tumor into the hind limb field of chick embryos to test whether or not nerve fibers would innervate those fast-growing tissues, which they did from nearby sensory and sympathetic ganglia. After an exchange of letters in which Bueker gave unrestricted consent, the experiment was repeated (with a more potent sample of sarcoma) and with dramatic results: a neurotization of the tumor by dense fiber masses from ganglia which were hyperplastic and showed the "halo effect" of sprouting neurites. To round out a classic example of internationalism in science, Levi-Montalcini traveled to Rio de Janeiro where the trophic element could be isolated in a tissue-culture laboratory of the Biophysics Institute directed by Carlos Chagas. After a quarter-century of subsequent study of its identity, the nerve growth factor "stands out today as one of the best known hormonal and growth factors, and its long-sought mechanism and site of action are slowly but steadily coming into focus" (Levi-Montalcini, 1975, p. 261). In 1986 she shared the Nobel Prize for physiology or medicine with Stanley Cohen.

Trophic Theory and Neural Connections

Only recently were the opposite views of Weiss and of Sperry regarding the controlling factor in neuronal organization fused into a middle ground. Those early studies and their interpretation, the functional control of connections (Weiss) and the rigid chemoaffinity of neurons and their target cells (Sperry), presented an ambiguity and constituted the prelude to a new proposal of a trophic mechanism. In trying to account for the fact that the number of neural connections accommodates bodily changes in size and form in ontogeny, let alone phylogeny, Dale Purves (1988) postulated a flexible regulation of axonal and dendritic branches of neurons engaged in a continuous balancing of feedbacks that maintains an equilibrium shifting in response to needs. In his words, "The trophic theory extends the general concept of trophic interactions from neuronal *survival* to neuronal *connectivity* . . . ." (p. 15). A recognition of the competitive importance of neuronal form (morphology) and the interactive role of cells and targets (dynamic physiology) and that those patterns continue into the maturity of the organism (development) contributed significantly to the concept of neuroscience as an integrator of many subdisciplines.

Although most of the evidence for the trophic theory of neural connections at present was derived from experiments on invertebrates and the peripheral nervous system (Purves, 1988, p. 95), including the autonomic, the mammalian central nervous system yielded a few important data (e.g., Henneman, Somjen, & Carpenter, 1965). The neuromuscular junction has been and still is the experimental locus of choice, as its accessibility and simplicity are conducive to "decisive analysis." The molecular basis for this malleability appears to be an intercellular signal such as the nerve growth factor (NGF).[10] In Purves's view, the evidence in the sympathetic nervous system suggesting NGF as a model may apply also to other systems. Like the differences found among neurotransmitter actions, the trophic agents may be as diverse in their effects.

The Pacific Escarpment[11]

The siren call of a climate that accommodated a continuous supply of marine animal subjects and the prospects of full-time research in a shore laboratory

[10] A lucid and well balanced account of NGF as a regulator of neuronal survival appeared in Purves, 1988, pp. 123–128.

[11] This section on Jacques Loeb is largely a condensation of "Jacques Loeb and his interests in neurophysiology" by H.W. Magoun, published in the *BRI Bulletin*, 11(1): 4–5; 10–11, 1987.

Fig. 7.14. The fruitful career of Jacques Loeb shown in his Laboratory of Experimental Biology at Pacific Grove about 1910, was marked by controversy, notoriety, and significant discoveries and propounded theories.

specifically constructed for him overcame the strong efforts to retain Jacques Loeb (1859–1924; Fig. 7.14) at the University of Chicago and in 1902 he moved to Pacific Grove, California. Deemed the "most visible exemplar of 'general physiology' in America, perhaps in the world" (Kohler, 1982, p. 111), Loeb had been born in Prussia at the end of a decade in which mechanistic conceptions of life had gained emphasis. His scientific descent was in direct line from that quadrumvirate of young physiologists – Helmholtz, Du Bois-Reymond, Brücke, and Ludwig – who advocated physiology as a chemicophysical foundation, as mentioned earlier. When he entered the University of Berlin Jacques's intent was to become a philosopher, but he developed a distaste for the subject and spent the next semester at Munich, where the lectures in psychology convinced him that understanding the mind and will should be approached by experimentation on the brain.

Loeb spent the next 5 years at the University of Strasbourg with Friedrich Goltz (1834–1902), a former student of Helmholtz, gaining his M.D. and formulating a materialistic conception of higher nervous activity based on "associative memory," determined by mechanisms found only in the vertebrate brain. From those beginnings he later developed ideas about consciousness (see Chapter 13). In his monograph, *Comparative Physiology of the Brain and Comparative Psychology* (1900), Loeb discarded a "psychic localization" in favor of a process of association in which the cerebral hemispheres act as a whole, not as a number of independent parts.

Tropisms

In 1887, Loeb moved to the University of Wurtzburg, where he became immersed in study of the behavior of much simpler animals. There, he was assistant to Adolf Fick, professor of physiology, a former student both of Carl Ludwig and Du Bois-Reymond, and a most enthusiastic adherent of the 1847 program of his mentors. Loeb also became an intimate friend of Julius Sachs, a plant physiologist who had applied the goals of the biophysical approach to botany and was then engaged in formulating the concept of plant tropisms.

Loeb conceived of founding a science of animal tropisms and in a paper (1905) on heliotropisms he described experiments with certain caterpillars that normally emerged in the spring, climbed to the tips of tree branches, and fed on the opening buds. Opposing the prevailing view that this conduct showed an instinct for self-preservation, Loeb demonstrated that if the only source of light was in a direction opposite to a supply of food, the animals kept their heads toward the light and starved to death. They were, he said, "photochemical machines and slaves to the light."

Loeb boosted the validity of this conception by studies with sessile annelid tube worms, which bent their heads toward light as did the plants of his mentor, Sachs. In 1909, Loeb addressed the Sixth International Congress of Psychology, in Geneva, on "The significance of tropisms in psychology," concluding with the statement "I believe indeed that tropisms ... will one day form the contents of a scientific psychology of lower forms." Three years earlier, however, Herbert Spencer Jennings, in a monograph on *Behavior of the Lower Organisms* (1906, p. 274), had concluded,

> [T]he local action theory of tropisms is of comparatively little value for interpreting behavior. This theory uses ... certain elements here and there observable in the behavior of some organisms .... It makes use of certain simple phenomena that actually exist, but elevates these into a general explanation of directed behavior, for which they are utterly inadequate.

In a prompt review, George H. Parker wrote (1907, p. 549),

> It would seem that Jennings in his enthusiasm for his own views has become blinded to the real strength of the tropism theory .... It is to be regretted that a book so excellent in so many particulars should be marred by so considerable, a deficit.

Over a number of years, the issue was debated further by associates of the two protagonists: H.B. Torrey, W.E. Garrey, and others for Loeb, and S.O. Mast for Jennings.

Another vista that had attracted Loeb in the 1880s was Wilhelm Roux's "developmental mechanics," involving experimental manipulation of the embryo. Although he did not study with Roux, Loeb spent time at the great seaside laboratory of marine biology at Naples, with its ready access to sea urchins and other classical organisms of invertebrate research. Here he had significant contacts with the enterprising circle of American embryologists and cytologists with whom he was to be associated for the remainder of his career. The most fortunate of those contacts was with Franklin P. Mall who was impressed with Loeb's brilliance, found an initial appointment for him in anatomy at Bryn Mawr College in 1891, and then persuaded Charles O. Whitman to appoint him in biology at the University of Chicago.

Parthenogenesis

The winter of 1898–1899 Loeb planned to spend in California at Stanford University's Hopkins Marine Station on Monterey Bay. Unable to get along with its director, he devoted the time to composing his *Comparative Physiology*, discussed above. In 1902, the University of California, which had established a small number of full-time chairs in the basic medical sciences, appointed Loeb professor of physiology. In addition to the provision of a laboratory at Berkeley, funds had been donated for an Institute of Experimental Biology constructed for Loeb's use on the ocean at Pacific Grove.

The significance of a phenomenon observed about 1900 by a former colleague at Bryn Mawr, Thomas H. Morgan – the development of a membrane by sea urchin eggs in slightly evaporated sea water as if they had been fertilized – was recognized by Loeb as an example of artificial parthenogenesis, to which he gave serious study. On a visit to Pacific Grove in 1906, the British physicist Rutherford described Loeb's progress:

> Spent the morning in Loeb's Lab., he is very busy there doing experiments in parthenogenesis and showed me all his methods of developing the sea urchin's eggs without fertilization. They were extraordinarily interesting. I first tried the normal method with the spermatozoon, and saw the egg develop in the course of a few hours. I then developed the beggar from the egg till he swam, by adding appropriate chemicals, without calling in the aid of the male. It took only about a minute to form the membrane, about three hours to get division of the cell and twenty-four hours to have them divided into sixty-four[-cell larvae] and swimming round. They are really wonderful experiments and he appears to me to be on the right track for great discoveries (quoted in Eve, 1939, p. 150).

Loeb's work in this field greatly enhanced his repute among scientists, of whom more than a 100 in 10 countries nominated him for the Nobel Prize, though he never received it. The newspapers made him a familiar name to the public at large. Maiden ladies were said to have given up sea bathing, humorists observed that vacations at the seashore were often remarkably fruitful, and barren couples earnestly wrote to Loeb for advice to help them start a family (Pauly, 1980). In the wake of this publicity at the beginning of the 20th century Loeb was one of the most effective spokesmen for the mechanistic conception of life (Pauly, 1987).

In 1910, Loeb left the University of California to become director of a division of general physiology at the Rockefeller Institute for Medical Research, New York, where he devoted the balance of his career largely to research in the fields of physical and protein chemistry. Historically, however, Loeb's initial interest was in the function of the brain and, in particular, its capacity for "associative memory". His early neurophysiological studies at Strasbourg and his research on the tropistic behavior of lower animals at Wurzburg, eminently qualified him as the first neuroscientist to hold an appointment in the University of California, and his seaside laboratory one of the first organized research units in the biomedical sciences on the West Coast.

Part 2
Behavioral Sciences

Chapter 8
Development of Experimental Psychology

A COMMON HERITAGE

Rise of Physiology and Psychophysics in Germany

Compared with his fellow countryman, Carl Ludwig, and the Leipzig school of physiology that developed during the last half of the 19th century, Wilhelm Max Wundt (1832–1920) had an even greater influence on American neuroscience. Wundt succeeded in defining a "new" field of experimental psychology, not only through his laboratory of physiological psychology and its students, but also by his voluminous writings. His career began as an assistant to Helmholtz in the physiological Institute at Heidelberg during 1857 to 1864, where he continued as an assistant professor for another decade. Parallel with Helmholtz's famous program of studies and discoveries in sound and visual perception, in 1862, Wundt began an annual course of lectures on "Psychology from the Standpoint of Natural Science," in which he incorporated Gustav Fechner's quantitative thinking for the measurement of sensations (Adler, 1977). The title of the lectures was later changed to "Physiological Psychology" and their substance was elaborated as *The Principles of Physiological Psychology*, the first half published in 1873, the second in 1874, both while Wundt was still at Heidelberg. The book went through six editions, the last in 1911, and in each was updated and enlarged, the last two editions to three volumes, reflecting Wundt's gradual metamorphosis from physiologist to psychologist and the beginnings of a new independent science.

Conjunction with Philosophy

After an interim year at Zurich, Wundt was appointed to a chair of philosophy at Leipzig where he developed in 1879 the first experimental laboratory of psychology, which less than a decade later was formally made an institute and "established this science on an experimental basis" (Hall, 1923, p. 155). It rapidly attracted students from all parts of the world, Americans and Russians predominating, and trained most of the first generation of experimental psychologists in America: among them, G. Stanley Hall, McKeen Cattell, Edward Scripture, Frank Angell, Edward Titchener, Fred Stratton, and Charles Judd. In 1881 Wundt founded a journal, *Philosophische Studien*, which by its content defined the initial subject matter of the field as predominantly sensation and perception, principally vision (Boring, 1929).

Contemporary accounts of Wundt, by American students studying in Germany in the latter part of the 19th century, began with the remark of William James. Writing home from Berlin in 1867, he commented,

> [P]erhaps the time has come for psychology to begin to be a science – some measurements have already been made in the region lying between the physical changes in the nerves and the appearance of consciousness .... Helmnoltz and a man named Wundt ... are working at it (quoted in Perry, 1954, p. 181).

Catalyst Wilhelm Wundt

Wundt's first American student, G. Stanley Hall, found that the theoretical work at Leipzig rested largely on "eye- and ear-made" experiments (1912, p. 416), and in assessing Wundt's *Physiological Psychology* that

> [t]he impulsion of [Wundt's] life has been to bring exact physiological methods to bear upon the study of psychic life .... But [his] book demonstrates the fact that an inordinate appetite for work, amounting to talent, can exist without a spark of genius .... (Ibid., p. 336).

Wundt's students noticed that he took copious notes during the seminars at which they reported the background and results of their research on the topics he had assigned to them. It seemed plausible that here was the source of the master's encyclopedic knowledge of the field.[1] Summing up his impressions, Hall concluded: "It would seem as if laboratory psychology in this country was now sufficiently developed so that it should be less dependent upon new departures made in Germany" (Ibid., p. 457).

Wundt's second American student, McKeen Cattell (see below), studied at Leipzig in 1881–1882 and 1883–1886, serving during his last year as Wundt's first assistant in psychology. Much of the work of the laboratory was then on reaction-time experiments, and Cattell was pleased that he was assigned a problem in that field because he had already done some work on sensorimotor processes. At his doctorate examination Cattell was grateful for the sympathy and kindness that Wundt showed and asked him things that he was sure to know. As Cattell (1921, p. 159) summarized:

> Wundt's leadership in laying the foundations of psychology ... are not here under consideration. The fact that his work for psychology was begun sixty

[1] *Editor's note:* A president of the National Academy of Sciences once commented that during his years as a department head he depended on his graduate students to keep him up to date on the current literature.

> years ago proclaims its fundamental character and accounts for its limitations. We advance over the temporary bridges built by men such as he, and they are more nearly works of genius than are the rubble and cement with which we later replace them.

A decade later, in 1894, Charles H. Judd studied in Wundt's laboratory, and later reported of his experience:

> There was in the *Institut* very little respect for the leaders in American psychology who had received their training elsewhere than in Leipzig. Especially was there a very pronounced antipathy to James.... [N]ot only had he criticized Wundt but... had allowed his criticism to take the form of witty sarcasm.... [H]e had indulged in that remark about patient laboratory work in a land where they did not know what it means to be bored. As a result, diplomatic relations were promptly and totally suspended (Judd, 1961, p. 215).

Judd was palpably awed by the care with which Wundt went over his thesis, prepared it for publication, and suggested changes backed by a wealth of material.

Hermann Lotze and Clarence Herrick

Significant as was Wundt's physiological psychology for transient students from the United States, another influence had an arguably greater effect on the movement of psychology away from the domain of philosophy. When the 23-year-old Clarence Luther Herrick (see p. 35) traveled to Leipzig to enlarge his horizons, he chanced across a slim treatise by Germany's leading philosopher of the time, Hermann Lotze (1817–1881).[2] Herrick's translation, titled *Outlines of Psychology* (1885), was put aside until later when, needing a text for his own teaching, he appended a section on brain structure to Lotze's two sections, "Individual Elements of the Inner Life" and "The Soul" [or Mind]. Herrick's publication of the three sections became widely used in college classrooms and represented a landmark in the transformation of psychology into a real science, at last separated from philosophy (Windle, 1979). In a final, helpful instruction, Herrick pointed out that Lotze was only an outline and for further enlightenment the reader should go to Wundt's *Lectures*. More recently, Lotze is credited with being the author of the first modern text on psychology, *Medicinsche Psychologie: oder, Physiologie der Seele* (1852).

DIVERGENCE: PSYCHOLOGY IN AMERICA AND ITS PIONEERS

When the 9th International Congress of Psychology was held at Yale in 1929 (the first on this side of the Atlantic), McKeen Cattell in his Presidential Address reviewed the rise of psychology in America. In describing the prior

[2]Lotze studied medicine in Leipzig, for 37 years lectured on philosophy at the University of Göttingen, and wrote a popular textbook on physiology (1851).

50 years, he designated William James and Stanley Hall as the only psychologists worth mentioning and, in unblushing hyperbole, deemed James's *The Principles of Psychology* (1899) an unequaled work of genius in any language or science.[3] Hall, at Johns Hopkins, had established the first American laboratory of psychology in 1883,[4] and George T. Ladd at Yale had published, in 1887, the first American text on the subject, *Outlines of Physiological Psychology*. Additionally, Cattell's appointment at the University of Pennsylvania had been the first in which the professor's work was confined solely to psychology. Cattell concluded his address with a recitation of the dozen chairs and laboratories that had been established from 1883 to 1900 and attributed the achievement to the growth of universities coupled with the introduction of elective courses which were conducive to the exploration of a "new" science.

William James (1842–1910)

In the development of the behavioral sciences in America, the careers of three early leaders were in various degrees analogous to that of Wundt in Germany; although their personalities were strikingly dissimilar, both from that of Wundt and from one another, William James provided the closest career match.

Born only a decade after Wundt, James (Fig. 8.1) obtained his M.D. at Harvard in 1868, started teaching physiology there in 1872, and introduced a course on "The Relations between Physiology and Psychology" in 1875. Promoted to professor of philosophy a decade later, his title was changed to psychology just in time for publication of his *Principles*, the composition of which had preoccupied him for a dozen years. In 1892, James recruited Hugo Münsterberg from Freiburg as director of the Harvard psychological laboratory and, thereafter, until his death in 1910, James's interests were increasingly in philosophy.

Appointment at Harvard

James's medical studies were twice interrupted, first by a collecting trip to the Amazon with Agassiz in 1865 to 1866, and then by a visit to Germany in 1867 to 1868; the latter was planned to improve his health, implement his interest in physiology, and perfect his knowledge of the language. From Berlin, he wrote Bowditch, his chief at Harvard, that Du Bois-Reymond's series of lectures were clear and brilliant and it was a great vexation that his health did not allow him to work in the laboratory. Those lectures confirmed James's growing belief that the way to approach psychology was through physiology (G.W. Allen, 1967), and he determined to go to Heidelberg to learn something

[3] For a discussion of the unusual nature and impact of *Principles of Psychology*, see Perry, 1954, p. 196.

[4] Unless James's work begun in 1876 at Harvard in physiology is counted.

Fig. 8.1. In the hermetic world of late-19th-century American neuroscience, the name of William James was, arguably, most often heard in conjunction with his writings on mind and consciousness, emotions and will.

of the senses from Helmholtz and Wundt. When he finally made the long-planned trip, he was chagrined to find the town without diversion and the university soon to shut down for vacation.

On his return to Harvard, James in 1876 set up an improvised research room, arguably the first for psychology in America, which was later developed into a laboratory for psychophysics. He had a metronome, a devise for whirling a frog, a horopter chart, and other bits of apparatus, according to his student, G. Stanley Hall (1923). By 1890 there were new quarters and James had introduced laboratory work into the regular undergraduate course in psychology. Graduate students sensed that their mentor, although recognizing the importance of developing psychological knowledge by experimentation, yet had a personal disinclination for laboratory work and it was for this reason that he had secured the appointment of Münsterberg for teaching experimental psychology.

An unusual student in the experimental course was invited by James to join his seminar in 1894 – Gertrude Stein. She published a report on motor automatisms (1898), a subject that tied in with James's interest in the relationship between the conscious and the unconscious mind and in a style that Stein applied later to the creation of a new type of fiction.

> The story of [Gertrude Stein's] final examination ... illustrates so well James's unprofessorial conduct as a professor. After reading the questions, Miss Stein wrote on her examination book: 'Dear Professor James, I am so sorry but really I do not feel like an examination paper today.' The next day James replied: 'Dear Miss Stein, I understand perfectly how you feel. I often feel like that myself!' Then he assigned her the highest mark of the class (G.W. Allen, 1967, p. 305).[5]

According to Allen, she was one of the most empathic students the professor ever had.

Challenged

The euphoria arising from progress in the development of experimental psychology and related laboratory study at a number of universities led to a certain rashness on the part of one of the most energetic participants, G. Stanley Hall. He was founding editor of *The American Journal of Psychology*, and in 1895 opened its seventh volume with an exuberant editorial listing the men who, at one time or another, had been associated with him at Johns Hopkins or Clark University. He then boastfully wrote that they were influential in founding departments and laboratories in the "higher institutions." James promptly sent Hall a blistering letter, in defense of the credit of his own university. He pointed out that it was he, James, who had inducted Hall into experimental psychology and praised him for what he had accomplished, but maintained that history should not be written in inverse order. Shortly thereafter, James, Ladd, Baldwin, and Cattell – representing the imputed universities of Harvard, Yale, Princeton, and Columbia, respectively – published a public contradiction of Hall's statements in *Science*. Ladd (1895, p. 627) said that, as an Alma Mater, Yale was proud of the 24 professors of psychology it had sent forth in the last 14 years, "but not less so, because she had imparted to them a distinctly different spirit and *morale* from that displayed by the writer of the editorial." Baldwin commented disparagingly on the ambitious claim to paternity of American psychology and Cattell wrote in a similar vein. Hall's rejoinder (1895, p. 734) not unexpectedly claimed his words had been misunderstood and hoped for future "hearty cooperation" and "virile competition" in psychology.

[5] The story also appears in Stein, 1933, pp. 97–98.

James's Stature and Contributions

As noted, William James's crowning contribution to the behavioral sciences was the publication of *The Principles of Psychology* (1899), in two hefty volumes, each exactly 689 pages long – a dimension appropriate to its prolonged gestation. It soon became the leading text in academia, and was translated into four languages. As for its assessment by James's peers, in 1890, Lincoln Steffens spent the winter at Leipzig and did some work in Wundt's laboratory. He related,

> One day ... [Wundt's] eye fell upon William James's great book on psychology, just out .... He picked up James,[6] ... and beginning at once to read, started off like a somnambulist for the door .... The next morning he came back, laid the book on my table and thanked me. 'You have read it?' I asked, astonished. 'All night long,' he answered, 'word for word, every word'.... As he was about to leave it with me, I stopped him with a question: 'What about it?'.... 'It is literature, it is beautiful,' he stammsered, 'but it is not psychology!' (quoted in Steffens, 1931, pp. 149–150).

With his customary expansiveness, G. Stanley Hall's review, in his own journal (1891), occupied 13 pages of exceedingly small type. It began with a lament that the book's volume and acumen made it difficult to review, then complained that the author seemed to be fascinated by the unsolvable problems, such as consciousness. Half-way from his start, Hall stepped back to gain a broader view and described James as an *impressionist* in psychology, with a portfolio of old and new sketches that stimulate or suggest. The favorite themes, selfhood and consciousness, struck the reviewer as a form of personal confession unequaled in the history of the subject. Hall praised the literary style of an able, mature man and hailed the work as the best in any language for teaching and study of the subject.

At this high watermark of James's career as a psychologist, it became obvious that the goals of his student period in Germany – to advance psychology through contributions to physiology – never reached fruition. Within a few years he was disillusioned and felt that the results of all the work in the laboratory seemed trivial and tedious. Of all James's conceptual contributions, his theory that emotions derive from organic sensations aroused by bodily expressions (derived incidently from his own introspection), first published in *Mind* in 1884, and reiterated in *Principles*, was most appropriate for physiological investigation. When these interactions were finally explained, however, initially by Walter B. Cannon in the physiological laboratory of Harvard Medical School, James's theory (combined with that of Georg Lange, as the James-Lange theory) was proven wrong.

[6] To psychology students of the period, the full two-volume set was called "James" and the abbreviated "shorter course," "Jimmy."

A major contribution to the future of scientific psychology was James's cordial and generous support of the research of two young men who initiated experimental studies of animal behavior at Harvard, in the decade bridging the turn of the century. In late 1896, his graduate student, Edward L. Thorndike, began experiments on chicks which he kept in his room until the landlady protested. When requests for university space were in vain, James accommodated the chicks in his own basement. The report of those studies, in Thorndike's doctoral dissertation, "Animal Intelligence," was a founding classic (1898) in the field of animal learning.

Two years later, in 1899, Münsterberg made space for students undertaking animal studies in the psychological laboratory at Harvard. In 1902, Robert M. Yerkes (another Harvard doctoral graduate) was appointed instructor in comparative psychology and in a few years a special laboratory for his research with a variety of experimental animals was constructed.

Succeeding generation of young investigators, applying a variety of novel techniques to experiments with animals, opened new fields of the advancing behavioral sciences in this country around the turn of the century. Further, in this same pattern, James's role in launching G. Stanley Hall's career in experimental psychology was highly contributory – again through Hall's protégé, Edmond C. Sanborn, and the latter's young doctoral students, Linus Kline and Willard Small, who were essentially contemporaries of Thorndike.

Granville Stanley Hall (1846–1924)

After obtaining the first doctorate in psychology in this country, under William James at Harvard, in 1878, Hall (Fig. 8.2) pursued further study at Leipzig with Carl Ludwig and Wilhelm Wundt. Appointed at Johns Hopkins as lecturer in 1882, he eventually developed the largest and most productive facilities for experimental psychology in the country and became "Professor of Psychology and Pedagogics" in 1884. Three years later, with $500 from a Philadelphian interested in spiritualism, he launched *The American Journal of Psychology*. Students in the department included John Dewey, James McKeen Cattell, Henry Donaldson, Joseph Jastrow, and later Edmund T. Sanford. The psychology he taught was almost entirely experimental and covered for the most part the material that Wundt had set forth in *Physiological Psychology* (1874). Hall's stay, which lasted most of the decade of the '80s, marked the zenith of Johns Hopkins's preeminence and leadership in psychology, in his own estimation (Hall, 1923). The student body was hardly less remarkable in quality than was the teaching force of young men fostering an intense intellectual activity. Hall believed, however, that H. Newell Martin felt that

Fig. 8.2. The steadfastness with which G. Stanley Hall devoted his energies to education made possible the survival of Clark University after the loss of most of its department of psychology.

experimental psychology, especially in dealing so largely with the senses, infringed upon his own development of physiology.[7]

Clark University

In June, 1888, Hall accepted the invitation to become initial president of Clark University, founded with an appropriation of 1 million dollar, by Jonas G. Clark, a retired businessman, at Worcester, Massachusetts. From educational leaders and scientists, Hall garnered suggestions and approval of the plan to launch a (solely) graduate institution, and decided with Clark to open the university with five related departments of natural science – psychology, biology, chemistry, physics, and mathematics – as a core curriculum from which the institution could grow. Hall managed to select a remarkably talented faculty; five of the ten were drawn from Hopkins.[8]

[7] A brilliant Englishman and former pupil of Ludwig, Martin established at Baltimore the second chair in physiology in this country (the first was Bowditch's at Harvard). Hall felt that Martin was never fully acclimated in this country nor quite happy in Baltimore.

[8] They were: Franz Boas (anthropology), Edmund C. Sanford, as well as Hall (psychology), Henry H. Donaldson (neurology), Warren P. Lombard (physiology), Franklin P. Mall (anatomy), Charles O. Whitman (biology), Arthur Michael (chemistry), Albert A. Michelson (physics), and William E. Story (mathematics).

The first 2 or 3 years of Clark University were a Golden Age which came closer to Hall's idea of a university than any other institution in the country ever did (Ross, 1972). Clark opened with 18 members of faculty grade and 34 students. Of this group, 15 had studied or taught at Hopkins, 19 had done graduate work at European universities, and 12 who already held the Ph.D. were pursuing postdoctoral study. As Hall was only 45, and most of the faculty in their early 30s, (and the students only a little younger), there was much informal contact and intellectual exchange among them.

Raided by Chicago
Shortly after the development of Clark University at Worcester, another such institution was conceived in the Midwest. In 1890 a successful move was made to establish a research university in Chicago together with the city's preparations for the World's Columbian Exposition. Marshall Field, the drygoods magnate, donated a tract of land adjacent to the exposition grounds and money to match a large contribution by John D. Rockefeller, enabling an organizing committee to proceed. The committee elected one of its own members, 34-year-old William Raney Harper, president of this new University of Chicago, and charged him with the task of securing a faculty in time to begin classes in the autumn of 1892. With the backing of such influential men, President Harper set out to build a university that he hoped would be second to none in America. In this important responsibility, Harper had more difficulties than did his counterpart, G. Stanley Hall, at Clark. Harper had to recruit the Faculties of Arts, Literature (together involving 16 departments) and Science (comprising 12 departments) at Chicago, a much more formidable task than Hall's. Harper's scholarly background, in Semitic languages and literature did not provide him much familiarity with the sciences. Hall had in 1890 sophisticatedly screened and recruited for Clark University the cream of talented young scientists in this country, both from Johns Hopkins and other institutions. It soon became apparent that the University of Chicago was having recruitment difficulties, particularly in the science area. No one but the Lord could have miraculously resolved this situation instantly but in an only slightly longer interval, William Raney Harper did so in a single morning!

At Clark University disturbing circumstances had developed. Following its initial year, Jonas Clark became disillusioned with his university and began to withdraw support. In trying to hide this deteriorating financial condition from the faculty, Hall developed arbitrary methods of effecting economies, displayed a general deviousness in his administrative activities, and in a variety of other ways increasingly antagonized and alienated his faculty. By April 1892 the possibility of improving relations appeared hopeless and Harper, with whom both Whitman (head of the biology department) and Mall (anatomy) had been in touch earlier, was notified of the situation and arrived in Worcester to meet personally with members of the Clark faculty at the home of Professor Whitman. By noon, the University of Chicago had acquired, for its opening in the Fall, most of the stellar science faculty that Hall had recruited for the opening of Clark 3 years before.

President Hall's account of this dismantlement of his institution evokes a feeling of woeful regret at the bludgeon-like features and devastating consequences of the blow. Seven years afterward, in his address at the university's decennial celebration in 1899, he referred to:

> The reductions of our force ... [were] sad to us almost beyond precedent, although helpful elsewhere .... [O]nly the Departments of Psychology and Mathematics remained nearly intact, [yet] we fortunately had left in every department young men as promising as any in the land. They needed simply to grow, and never has there been such an environment for a faculty to develop as in this 'paradise of young professors,' as a leading college president has called this University (Story, 1899, p. 51).

The Younger Generation

Among this outstanding young faculty at Clark was Clifton F. Hodge who was promoted to assistant professor of psychology and neurology in 1892 and initiated a number of research projects in neurobiology with his graduate students.

One of the most talented was Colin C. Stewart who, during 1895 to 1897, in a study of the effect of alcohol on motor activity, at Hodge's suggestion introduced use of the white rat in physiology. A second outstanding young member of the new faculty was Edmund C. Sanford, promoted to assistant professor of psychology also in 1892. Six years later, Sanford introduced laboratory practice with experimental animals into the psychology curriculum at Clark. One of his most enterprizing graduate students, Linus W. Kline, first used the white rat in experimental studies of animal learning, with puzzle boxes. Subsequently and at Sanford's suggestion, Willard S. Small tested white rats with a laboratory adaptation of the Hampton Court maze. These pioneering developments of Kline and Small at Clark, together with those of Thorndike at Harvard and Columbia, and of Watson at Chicago, marked the beginning not only of comparative psychology, but, additionally, of the behavioral revolution in this country. Within a decade or two, animal behavior had replaced the preoccupation of experimental psychology with introspective human studies, an American heritage from 19th-century Wundtian Germany.

Interest in Psychiatry

G. Stanley Hall had a career-long interest in the role of "scientific" psychology in relation to psychiatry in the study of mental abnormality. Psychopathology was included in his teaching at Hopkins, where he "took his students regularly to the Bay View Asylum, Baltimore's institution for the pauper insane" (Ross, 1972, p. 159). After moving to Clark University, Hall made similar arrangements for student visits to Worcester State Hospital. When Adolf Meyer moved to that hospital in 1895, he was additionally appointed docent in psychiatry at Clark and the following year initiated a course in psychiatry for its psychology students. When Meyer moved on to the New York State Hospital in 1902, Dr. Edward Cowles, superintendent of McLean Hospital, who had studied with Hall at Hopkins and was a trustee of Clark, continued this program.

As an extension of his interest in psychiatry, when Hall arranged a celebration to mark Clark University's 10th anniversary, in 1899, among the distinguished scientists invited from abroad he included Auguste Forel, professor of psychiatry at the University of Zurich, Switzerland, and director of the Burgholzi Asylum. Forel gave two lectures: one on ants, the study of which was his life-long interest, and the other on "Hypnotism and Cerebral Activity." discussing the work of Breuer and Freud in this field.

Freud's Visit

A decade later, at Clark's 20th anniversary celebration, Hall climactically brought Freud, Jung, and Ferenczi to America. Freud's lectures (one a day for 5 days) gave a general account of psychoanalysis, "in German, without any notes, in a serious conversational tone that made a deep impression" (E. Jones, 1961, p. 267). The university conferred an honorary doctorate on Freud who was reported to have been visibly moved during the first words of his speech: "This is the first official recognition of our endeavors" (quoted in Ibid.). Another contemporary witness, William James, was friendly and encouraging to Freud and his work.

Following the Clark program, Freud, Jung, and Ferenczi spent a recreational week at the "camp" of James Jackson Putnam (1846–1918), Professor of Diseases of the Nervous System at Harvard, in the Adirondack Mountains near Lake Placid, New York. The week passed pleasantly and, greatly to Freud's satisfaction, they sighted a wild porcupine. Before leaving Europe, he had said he was going to America to give some lectures and, as a diversionary goal, in the hope of catching sight of a wild porcupine. The phrase, "to find one's porcupine," became a recognized saying in his circle. Having achieved his double purpose, Freud was content to return home.

Another person at the Putnam camp that week was a 19-year-old Harvard sophomore, Alan Gregg, visiting Dr. Putnam's niece and nephew. In an unfinished autobiography, Gregg later wrote of some dominating events in his life which included reading William James and meeting Freud, whose ideas reinforced Gregg's notion that the tremendous force of the sex drive could accomplish great things. A quarter century later, in 1933, when Alan Gregg became medical director of the Rockefeller Foundation, he supported the development both of psychoanalysis and modern psychiatry which, in his generation, was still the "Cinderella of medicine." Beginning in the early 1930s, the foundation's Medical Sciences Division gave much attention to research and teaching in departments of psychiatry in leading universities around the country, and supported multidisciplinary studies through the National Research Council's Committee for Research on the Problems of Sex.[9]

James McKeen Cattell (1860–1944)

The third major pioneer in American psychology was McKeen Cattell, a graduate student at Hopkins working with H.H. Donaldson and others. When G. Stanley Hall was appointed at Hopkins, in 1882, Cattell (Fig. 8.3) assisted him in setting up a modest laboratory in a private house adjacent to the campus. The next year, Cattell was in Leipzig, brazenly applied as Wundt's initial assistant, and was the first American to gain a doctorate there, in 1886; his thesis dealt with individual differences in reaction time. For the next 2 years, his work was divided between Francis Galton's anthropometric laboratory in London, where the view prevailed that measurement of individual differences would be a promising field of the new psychology, and lecturing and starting a laboratory of psychology at the University of Pennsylvania. In 1888, he was appointed professor of psychology at Penn, as noted earlier, the first in this field as distinct from philosophy. Three years later he moved to Columbia University, where he spent the next quarter century. Dismissed in 1917 for "treason" because of his pacifist stand in the First World War, Cattell sued for libel and was later awarded $40,000 damages.

Demographic Research

At Columbia, Cattell extended his interest in individual differences to humans with estimations of the order of merit in subgroups of American scientists. By 1903, he had studied one thousand American men of science which he had selected from some 4,000 included in a biographical dictionary that he was compiling, the precursor of *American Men of Science*, later *and Women*. Within that group, Cattell found 200 (about 20 percent) who had advanced psychology, and from those he selected the ten leading to rank the remaining psychologists. The resultant statistical data were published in 1903 and the individual rankings not until 26 years later (Cattell, 1929).

Cattell was surprised at the large number of colleges from which the 200 psychologists had come and how they had gathered at relatively few universities for their specialized work, they having made selections on the basis of reputation and facilities, as well as fellowships and help in finding positions.

[9] See Volume 2 on the impact of the Rockefeller Foundation; also Robert Shaplen, *Toward the Well-being of Mankind, Fifty Years of the Rockefeller Foundation* (1964).

Fig. 8.3. McKeen Cattell's career-long search for individual differences was pivotal in shaping the acceptance of American psychology as a natural science.

At the turn of the century, William James ranked in the #1 position, head and shoulders above the next individual. With respect to the future development of psychology in this country, however, the key links in this demographic sequence – which Cattell was the first to analyze – were the Ph.D.-granting universities, which he described as being "few" and "small in number." Those descriptions were apt indeed for there were only 12 universities in the world which had awarded three or more doctorates to members of the group of 200 American psychologists that Cattell had assembled; nine universities were in the United States and three in Germany. Of those awarding more than five such doctorates, there were only seven universities in the world; six in the United States and one in Germany.

An Unpublished Autobiography

Although he wrote giftedly and frequently of the achievements of his many colleagues and associates, Cattell steadfastly refused to write anything autobiographical, and repeatedly declined invitations to contribute to Carl Murchison's three volumes of *A History of Psychology in Autobiography*

(1930, 1932, 1936). His excuse was that his pungent comments would lead to libel suits. The historian, Michael M. Sokal, in 1971 wrote his doctoral dissertation on Cattell, and attributed the latter's "autobiographical reticence" to bitterness at a perceived lack of recognition (Sokal, 1971). Cattell, when nearly 70, had been elected president of the 9th International Congress of Psychology held in 1929 at New Haven. His address, "Psychology in America," and its supplemental materials including individual names from the ranking of psychologists in 1903, became an important historical document (Cattell, 1929). Cattell must have been a difficult man to please if he clearly was bitter and disappointed at such striking recognition! Other honors included election (1901) to the National Academy of Sciences, the first in his field, and presidency (1924) of the American Association for the Advancement of Science. One finally wonders whether or not much of this attributed "bitterness" may not have been in the eye of the beholder.

In any case, the unpublished autobiography discovered by Sokal at the Library of Congress was written in 1936, when Cattell was 76 years old and is a record of the development of a psychologist in America during the late 19th century, and of his reaction to American psychology and the psychological community of the first third of the 20th century. With the permission of Cattell's two surviving children it was included in Sokal's published dissertation.

In this autobiography, the paragraph which Sokal identified as an example of the bitterness that greatly affected Cattell's last years, began:

> Certainly, I have no illusion that I am a figure in the world that will attract the interest or curiosity of a present or future public. My estimate, based on the attitude of others, is that I stand somewhere among the first hundred contemporary scientific men of the United States and also probably among the first hundred editors and promoters of science and education. This will put me among the first thousand in a Who's Who of the current sort, but this is not much .... (Cattell, in Sokal, 1971, p. 630).

This paragraph seems as readily interpretable as an instance of Cattell's jocularity, interrelating his penchant for the order of merit with his view of "a collective series of sketches as a study ... in the concealment of vanity" (Ibid., p. 629).

Cattell's Legacy

Among his contemporaries, Cattell's pronounced ideas and fearless aggression helped shape American psychology. As F.L. Wells, in one of the two memorials accompanying the double volumes of Cattell's selected papers (1947, v. 1, p. 5) wrote: "Few men so honest have had so good an understanding of the chicane. Nor has anyone in the field of psychology been master of such incisive wit .... He was among the principal influences in America that established this study ... as a discipline in its own right." That judgment was seconded by Robert S. Woodworth (1947) who pointed out that Cattell's extremely wide acquaintance among scientists and his varied services to psychology and American science in general contributed more than any other man to win recognition for this field among the natural sciences.

Laboratories for Psychology

During this formative period the establishment of a psychological laboratory became the hallmark of experimental psychology at successive educational institutions both in Germany and this country. Psychological experimentation, similar to Fechner's work on the psychophysics of perception, had been going on sporadically in physiological or physical laboratories for a long time (Boring, 1929). Beginning with Wundt and James, that same work in a psychological setting was introduced initially under the auspices of departments of philosophy before psychology was recognized. In the universities, typically one or two out-of-the-way rooms were provided, either under the stairs or in the attic, basement, or a run-down adjacent building. In these laboratories was assembled a variety of cumbersome "brass-instrument" apparatus for generating visual or auditory signals, recording time, or registering movements. For more than a half a century, almost all of the subjects were human, usually the professor's graduate students, and most of the experimental data consisted of introspective reports of sensation or perception, or motor responses to a perceived signal.

Growth in America

On their return to this country's universities, the American psychologists who had worked with Wundt at Leipzig established models of his laboratory.[10] The first was G.S. Hall at Hopkins (1883), Cattell was next at Penn (1887), following were H.K. Wolfe at Nebraska (1889), Frank Angell at Cornell (1891), E.W. Scripture at Yale (1892), George M. Stratton at California (1896), and Walter Dill Scott at Northwestern (1900). When members of this first generation moved from their initial institutions to others – as Hall from Hopkins to Clark, Cattell from Penn to Columbia, Frank Angell from Cornell to Stanford, and J. Mark Baldwin from Toronto to Princeton – they established second, frequently more elaborate, such laboratories, leaving behind hard-to-fill vacancies. Two desperate institutions imported foreign protégés of Wundt to their faculties: Hugo Münsterberg was recruited from Freiburg, Germany, to Harvard and Edward B. Titchener from England to Cornell, both in 1892. In Titchener's case, "the traditions of the Leipzig laboratory have been best maintained" (Cattell, 1928, p. 547).

Relatedly, doctoral graduate study in psychology was expanding at these same universities, where their laboratories provided the experimental resources

[10]The following data are from Garvey, 1929. See also Ben-David and Collins, 1966.

for dissertation research. As these American students completed their degrees, they went, off – in turn – like missionaries to more outlying heathen institutions, and established more such laboratories in the pattern of their alma maters'.[11] During the half-century from 1875 to the mid-1920s, this establishment of psychological laboratories at institution after institution became a talisman of the introduction and expansion of experimental psychology in America. From their beginnings at Harvard in 1875, and at Hopkins in 1893, the number of these laboratories increased to 16 in 1892, 27 in 1894, 49 in 1904, 74 in 1917, and 99 – just under a 100 – in 1925 (Garvey, 1929).

This last was established at Wittenberg College, Ohio, where in 1928, Martin L. Reymert, Professor of Psychology and the laboratory's director, arranged a week-long symposium on The Feelings and Emotions. The program attracted a remarkable number of notables to celebrate the completion of a new psychology building; with a special address on "Early Psychological Laboratories" by James McKeen Cattell at the dedication of its psychological laboratory. An undergraduate assistant at Wittenberg adding to the success of this remarkable symposium was Donald B. Lindsley, who was influenced by Professor Reymert to pursue subsequent graduate study with Lee Travis at Iowa and embark on his richly productive career in the neurobehavioral sciences.

We have, thus, by the first quarter of the 20th century passed through what Woodworth (1943) called psychology's adolescence, into a well established subdiscipline of neuroscience – experimental psychology – with laboratories, doctoral programs, and competent researchers and teachers forming a solid platform from which to move forward on the shoulders of succeeding generations of striving torch-bearers. In contrast to the older European schools, the American trend had been toward a functional and practical psychology. That "established historian of psychology" (Skinner, 1979, p. 72), Edwin G. Boring, would postulate that just as America had taken up Darwinism by virtue of the country's forward thrust and openness to new ideas, it was ready to do the same for the "new" psychology. Boring wrote (1950, p. 163) that the early psychologists such as William James "had a functional psychology" and they had only hastened a transition from introspection-as-research to research in animal behavior that had been underway since the last years of the previous century.

INTROSPECTION AS METHOD

Edwin Boring, to whom we owe many opinions on the history of psychology, labeled "introspection" as the chief method of physiological psychology and

[11] From their early identification, American psychologists have delighted in counting themselves, with a resulting plethora of data and their analyses, some of which are elaborated in Volume 2.

added that it was "essentially the study of consciousness" (1953, p. 170), a statement currently broadened to "the chief method of psychology" by the *New Encyclopedia Britannica* (1998, vol. 6, p. 360). Introspection was of primary importance to the pioneers of the "new" experimental psychology such as Wundt in Germany and E.B. Titchener, who had trained with Wundt before moving from England to the United States. The new physiological psychology had to be nourished by introspective data, but there was a problem: how were reliable data obtained when mental measurements were impossibly imprecise? This practice of "looking inward" to examine sensations and feelings and reporting them to an observer attempted to measure consciousness with all the philosophic and physiologic complexities of such studies in any era. The result of those arguments about data reliability and meaning led, in the 1920s especially in the United States, to questioning their functional usefulness and classical introspection was abandoned. The method itself as practiced scientifically in psychophysics, where the stimulus is the independent variable and the relative sensation the dependent variable, persists today as an indicator of experience of conscious events of a sensory nature. Significant studies were carried out with fragrances (Crocker, 1945) and pain (Hardy, Wolff, & Goodell, 1952).

The Titchener Era at Cornell

Early Years

The most famous American practitioner of the introspective methodology was a British expatriate. In 1892, Edward Bradford Titchener (1867–1927; Fig. 8.4), an Englishman from Oxford, came to Cornell as professor of psychology and director of a newly completed laboratory for research in physiological psychology, positions he held for the next 35 years. During 1916 to 1918, Cora L. Friedline studied and received her Ph.D. under Titchener. In her late career, Friedline was drawn to the history of psychology, and a number of her lectures were tape-recorded.[12] Three paragraphs from one of Cora's transcribed lectures convey a charming admixture of naivete and sophistication on the part of a female graduate student in psychology at a reputable Eastern university in this country in the mid-1920s:

> [D]uring my two terms at Cornell they were investigating organic sensitivity and the graduate students were the subjects. One swallowed a stomach tube very early in the morning and went around to lectures with that stomach tube in place all day long .... They went around to their lectures and reported periodically for an experimental period such as pouring hot water down through

[12] The tapes are on file in the Archives of the History of Psychology, at Akron University, Ohio. Quoted with permission.

Fig. 8.4. Edward B. Titchener's unbending promotion of introspection hastened the revolt of the experimental psychologists. Graduation photograph, 1904.

the stomach tube and the individuals would introspect as to the nature of that sensation. All that was the origin of a topic in Titchener's textbook called 'Organic Sensitivity.'

In addition to that, we all had little notebooks and, every time we went to the bathroom, we introspected. If you waited too long, that sensation of pressure and heat all over the body, restlessness and irritation, an irritated feeling and not being able to concentrate very well, and then a feeling of great relief and a cooling off after the elimination process. And this was differentiated for urination and for defecation.

Another topic which was investigated very extensively by the married individuals was intercourse. The research there was done by Boring and by Dallenbach. They had notes and their wives had notes on the nature of intercourse. And they had various bits of apparatus attached to them, which recorded the blood pressure and the change in heart beat and so on. This got around where I was living in the dormitory in Sage Hall and the laboratory was regarded as a very immoral place. Mrs. Barber who was the hostess would not allow any of the elementary students to go over to the Psychology Laboratory at night. Plus the fact that we were using condoms on the end of the stomach tube enhanced the nature of the immorality in the Psychology Laboratory. The tale over in Sage was that the lab was full of condoms and wasn't a safe place for anyone to go.[13]

Friedline's firecracker exploded in the third paragraph, which impugned two assistant professors in psychology at Cornell: Edwin G. Boring and Karl M. Dallenbach and their respective wives, who were "very extensively" making notes on "the nature of intercourse". Moreover, in the Wundtian tradition of Titchener's laboratory, they recorded the blood pressure, heart beat, "and so on."

Even in that era it was widely known that excitement and physical exercise raised the blood pressure and heart beat, but what did *and so on* connote? More rapid and deeper breathing? Or were records being made of the various organs and stages that have since been studied by others in the study of "intercourse"? The sequence of later American studies of human sexual activity (see Chapter 12) may have been introduced at Cornell in the 1920s, but if so, it is not retrievable from Dr. Friedline's quoted recollections. Both Boring and Dallenbach were married instructors in psychology at Cornell in 1916 to 1918 and, curiously, during that time, "[T]he Dallenbachs and the Borings occupied the two sides of a newly-built double house with a very thin partition between the halves. Several amusing anecdotes are based upon this unexpected intimacy" (Boring, 1958, p. 13). This brief sentence is the only such personal remark that we have been able to retrieve from either Boring's or Dallenbach's publications.

Margaret Floy Washburn (1871–1939)

As an earlier graduate student of psychology at Cornell, Washburn's experience of the Titchener era was less exotic than Friedline's and her training there had a far wider impact. Professor Washburn's achievements in theory development, experimental work, teaching, and professional service placed her as one of psychology's most influential early figures (Goodman, 1980) in the United States.

Drawn to science and philosophy as an undergraduate at Vassar College, Washburn (Fig. 8.5) determined to combine them in the new experimental psychology and in 1892 became Titchener's first graduate student at Cornell. On receiving her Ph.D. 3 years later, she remained close to Ithaca during her initial teaching appointment and continued to use the library and attend seminars to sustain her research. Publication of her dissertation, on the effect of visual imagery on tactile perception (1895), in Wundt's *Philosophische Studien* (which accepted almost exclusively work done at Leipzig) had been a signal honor and established her high reputation among her mostly male peers. That

[13] We are grateful to Professor Frederick B. Rowe, Department of Psychology, Randolph-Macon Woman's College (via J.V. McConnell) for providing this fascinating and historically important information.

Fig. 8.5. With a brilliant mind and unusual fortitude, Margaret F. Washburn developed her distinguished reputation for research in animal behavior in the midst of a predominantly male environment. She was photographed with M.L. Reymert, organizer of the Wittenberg College conference in October, 1927. From the Vassar College Archives, with kind permission.

year marked also Washburn's return to Vassar, where she remained committed to career and college, in spite of the absence of graduate students. This was also the beginning of a long association with *The American Journal of Psychology*. To minimize the lack of graduate students, Washburn instituted the practice of assigning simple research projects to her advanced undergraduates, the successful results of which were published in the *Journal* as "Minor Studies from the Vassar College Psychology Laboratory Communicated by M.F. Washburn."[14]

The greater part of Washburn's influence stemmed from publication of *The Animal Mind* (1908). It "met a critical need in the young field of comparative psychology" (Goodman, 1980, p. 75) for a textbook that presented systematically and cogently the results of scattered and inaccessible experimental studies, methodology, history of the movement, and analysis of the theoretical problems. Her interest lay in the evidence that animals have mind and consciousness and she propounded elements of learning theory later developed by Clark Hull. With six editions, the last in 1936, *The Animal Mind* became a classic in the field; she was the second woman[15] elected to the National Academy of Sciences, in 1931.

[14] *Editor's Note:* A short paper by Heywood and Vortriede (1905) was the first of Washburn's communications; the second author was my mother, and because she never mentioned it I know that she was not cognizant of its publication.
[15] Biologist Florence R. Sabin of Johns Hopkins was the first, in 1925.

Without a lineage of graduate students who went on to make their mark, Washburn's own significance has been overlooked. She did not hesitate to plunge into the mainstream of male-dominated psychology. In her autobiography, Washburn (1932, p. 143) made clear her logical stand: "I never followed Titchener when he developed his elaborate, highly refined introspective analysis .... It is worthwhile to describe conscious states, but not, in describing them, to turn them into something unrecognizable." Earlier, she had published her own ideas on introspection as an objective method (1922), deploring casting aside data that had been obtained with "as much care and patience" (p. 94) as in other sciences.

Perhaps the final word had been said a decade earlier by one of Wundt's peers. In a discussion of "the case against introspection" Knight Dunlap (1912) reviewed and quoted an array of contemporary experimental psychologists and concluded:

> There is, as a matter of fact, not the slightest evidence for the reality of 'introspection' as the observation of 'consciousness'.... We might keep the word to apply to observations of feelings, and of kinesthetic and cœnesthetic sensations ... but in view of the word's quite disreputable past it is probably better to banish it for the present from psychological usage (p. 412).

Organization

As the number of departments of psychology in American colleges and universities multiplied during the early decades, the urge to form organized aggregates rose, and Titchener accordingly initiated informal meetings of experimental psychologists. The unstructured character of the group allowed him to dominate the invitation list and confined the participants to those department heads of whom Titchener approved, creating dissatisfaction among some leading men such as John B. Watson. The "Titchener Club" was synonymous with generalized, human adult, normal experimental psychology based on introspective data. Without question the group's most significant meeting was in 1917, at Harvard, during the general excitement caused by the United States's declaration of war against Germany. Robert M. Yerkes, then president of the American Psychological Association, was made chairman of a committee to enquire how psychologists could help the military in forming a great army, and thus the deluge of Army personnel psychological testing was initiated. Soon after Titchener's death in 1927, a new society was planned that included animal psychology (assured by Yerkes's presence on the committee) and broke away from the conservatism of the "psychology of consciousness." In 1929 the committee organized a formal "Society of Experimental Psychology," in the pattern of an elite academy limited to no more than 50 members, with the intention that election should become a distinctive honor. The committee of 15 and 11 additional experimental psychologists were elected to membership, including two women – June Downey and Margaret Floy Washburn – the first to be admitted since 1901. In 1936, the society initiated the annual award of a Howard Crosby Warren medal (with the support of his widow), "for outstanding work in experimental psychology during the preceding five years." The first three awards were to: Ernest G. Wever and Charles W. Bray for studies of auditory function, Karl S. Lashley for work on the physiological basis of learning, and Elmer Culler for contributions to the physiological basis of the conditioned reflex. The achievements of all three rested on animal experimentation – Wever and Bray used cats in their studies; Lashley, white rats; and Culler, dogs – a far cry from the earlier Titchener Club.

Chapter 9
Animal Behavior and Comparative Psychology

A NINETEENTH-CENTURY STUDY OF ANIMAL BEHAVIOR

During the rise of psychology in 19th-century Germany, the focus was on understanding the human mind by experimental studies of sensation, reaction time, and other subjects involving introspection, under the leadership of Wilhelm Wundt. In England, psychologists were preoccupied with the study of animal behavior. The earlier methodologies were principally observational, or even anecdotal, with frequent anthropomorphic interpretations, as epitomized by George Romanes (1882). More than a decade later, the naturalist C. Lloyd Morgan, whose reputation was enhanced by his introduction of the law of parsimony in evaluating data, advocated a more cautious use of anthropomorphic analogies (1898). This early emphasis on animal behavior by English psychologists was in part attributable to the influence of Darwin's monograph *The Expression of the Emotions in Man and Animals* (1872). Some of the same influences found expression on the North American Continent.[1] Two major works in the European tradition of observation of instinctive behavior and other activities of animals roaming freely in their natural environments, exemplified by Konrad Lorenz and Nikolaas Tinbergen, were published in America at the turn of the century.

The American Beaver

It is little known that, early in the same period, Lewis Henry Morgan (1818–1881) (no relation to C. Lloyd), an American lawyer, mining company president, and legislator from Rochester, New York, pursued an avocational study of *The American Beaver and His Works* (1868). Lewis Morgan

[1] In the Spring of 1896, Lloyd Morgan gave the Lowell Lectures at Harvard, including reference to his incubation studies with chicks. Thorndike was a graduate student there at the time and it has been suggested that this influenced him to take up his experimental research using chicks in the Fall of that year (see p. 229).

Fig. 9.1. Engraving from Lewis H. Morgan's 1868 publication on the beaver, the first American contribution to animal-behavior studies. The striking, alert pose made the animal worthy of scientific attention, in Morgan's opinion.

exploited an opportunity created a quarter-century before when a railroad, for which Morgan was legal advisor, had been built through an uninhabited wilderness and passed through a beaver district, rendering it more accessible for minute investigation, perhaps, than any other of equal extent in North America. Morgan took up this subject for summer recreation and eventually accumulated material that "seemed worth arranging for publication" (Ibid., p. ix).

Sensory and Motor Organs

With the help of a physician in Rochester, Lewis Morgan also studied the comparative neuroanatomy of the beaver as well as its behavior. He found the eye to be disproportionately small, the optic nerve a mere thread, and its vision of short range, whereas its hearing and smell were very acute. The beaver brain, Morgan noted, was entirely smooth on the surface[2] and the cerebellum in its posterior development greater than in marsupials. Morgan argued from structural considerations that smell and hearing are the beaver's principal "informing senses," and because it frequently sits up on its hind legs to listen, he adopted that as the most suitable for its representation (Fig. 9.1).

[2]The beautiful work of Leuret on beaver and other brains was underway in France at this same period; Morgan quoted Leuret's figures for ratios of brain to body weights (Ibid., p. 77).

As cutting instruments (both for feeding and preparing their dams and lodges), the beaver is armed with powerful incisive teeth which are like chisels in form, structure, and efficiency. The other major motor organ of beavers is the tail, which assists the animal in diving and as a scull in swimming. It also is used to pack and compress mud and earth by repeated down – strokes and to give a signal of alarm to its mates.

Beaver Behavior

Moving now to the beaver's works, Morgan wrote of the marvels of the animals' dam constructions, their extent and great age accomplished by patient and long-continued labor. The resulting artificial ponds controlling the variable water level of the natural ponds and rivers Morgan interpreted as a sign that the beaver had moved from a natural to an artificial mode of life by means of dams and ponds of its own construction. In a concluding chapter on "Animal Psychology," Lewis Morgan expressed views that are more allied with those of his British counterpart, George Romanes, than with those of his more parsimonious name sake, C. Lloyd Morgan. In the "works of the beaver," Lewis illustrated the animal's intelligence and reasoning capacity, and suggested that the beaver is capable of reasoning processes indistinguishable from those performed by the human mind, but "inferior in degree."

A favorable review of Morgan's monograph was published by Jeffries Wyman in the *Atlantic Monthly* in 1868. Playing on the title, Wyman wrote, "The subject … is peculiarly an American one; for though beavers are found in the other hemisphere, they make no dams there, and a beaver without his dam is nobody" (p. 512). Culminatingly,

> [t]he posthumous career of Lewis Morgan reached a strange pinnacle when his book, *Ancient Society* (1877), the work of a railroad entrepreneur and a Republican, came to be viewed as a Socialist classic. Soon after reading it, Karl Marx had died in London, leaving instructions for Friedrich Engels to acquaint European Socialists with Morgan's discoveries …. In 1884, Engels published *The Origin of the Family, Private Property, and the State,* subtitled, 'In the Light of the Researches of Lewis Henry Morgan.' As a biographer put it, "Morgan had rushed in where Engels feared to tread" (Resek, 1960, pp. 160–161).

Primate Behavior

The protracted record of beaver behavior by Lewis Morgan was in contrast to a more sporadic series of observations that were even more adventuresome. The interests of Richard Lynch Garner (1848–1920) lay in the vocalization of nonhuman primates and in the course of recording phonographically and learning to imitate simian communication (see Chapter 14), he studied primate behavior in most of the zoological gardens of the eastern and midwestern

United States. He became so dedicated to his observations that he had a steel mesh cage constructed in demountable sections in which he lived and watched primates roaming free in their native habitat. In 1896 he published his first experiences in the French Congo, from which he concluded that apes have a sense of government, a sense of property, dignity, family life, justice, and that they "talk" (Garner, 1896).

DEVELOPMENT OF COMPARATIVE PSYCHOLOGY IN AMERICA

Morgan's (and Garner's) relatively unsystematic study of one animal species, impressive as it was, did not fulfil the requirements of a scientific study. Only when measured in a laboratory could animal behavior merit inclusion in the new subdiscipline of neuroscience – comparative psychology. And the animal that was of paramount importance here was another rodent, the Norway rat. Equally significant was the choice of the maze to test quantitatively the capability of this species (and later monkeys and humans) to remember and learn, two domains of knowledge that helped to carry psychology into neuroscience.

Albino Rats and the Maze

During the decade spanning the turn of the century, comparative psychology was initiated in this country by a small number of graduate students at four universities – Clark, Harvard, and Columbia in the Northeast; and Chicago in the Midwest. The program at Clark had the distinction, in the studies of Colin Stewart, Linus Kline, and Willard Small, of introducing the white rat to comparative psychology; as had that at the University of Chicago, shortly before, in introduction of this animal to comparative neurology. As the following arguments show, the initial source of the white rats used by psychologists and physiologists at the century's turn is undocumented and hence obscure. It may have been by a combination of circumstances, through importation by a single individual, and/or from general availability in the marketplace. The source is immaterial to our history, however, but the fact of the rat's introduction as an experimental subject is of utmost consequence in directing research to a manageable fusion of psychologic and physiologic concepts.

At the University of Chicago

At Chicago, as acknowledged later by Donaldson (1925), Adolf Meyer in 1893 introduced the white rat into the laboratory of his neurology course on the anatomy of the nervous system. After he moved to Worcester State Hospital with a docentship at nearby Clark University, he sent back a few dozen adult rats from which the Donaldson colony developed. Another version of the manner of Meyer's initial introduction of the white rat to the newly established University of Chicago in 1893 is somewhat more detailed. According to Curt P. Richter (1968), the Norway rat (of the albino variety), the first animal to be domesticated for strictly scientific purposes, was brought from Europe (from the department of zoology at the University of Geneva) to this country in 1893 for brain studies by Adolf Meyer, then a newly arrived Swiss neuropathologist at the University of Chicago. When Donaldson, professor of neurology at Chicago, first saw Meyer's rats he at once recognized their possibilities for scientific research (but see below) and was to establish the famous colony at the Wistar Institute in Philadelphia that for a time supplied rats to laboratories worldwide. More recently, Dr. Richter (letter, March 9, 1978) was unable to verify Meyer's having received or shipped white rats but confirmed that during his tenure at Hopkins Meyer championed the benefits and support of a large rat colony in Phipps from 1919 to 1977.

As for the individuals involved, the first of six early papers using the rat was authored by Donaldson in 1900; another single paper was by a graduate student, Alice Hamilton,[3] in 1901. The remaining four were by Donaldson's first Japanese student, Shinkishi Hatai, and were published in 1901 and 1902. Hatai continued as a research associate with Donaldson and throughout his career was a prolific contributor advancing research in comparative neurology with use of the white rat. In 1907, Hatai wrote of the many sources of procurement of laboratory rats, suggesting their much broader distribution in the United States at the beginning of this century than does the concept that they all sprang from those imported by Adolf Meyer and conveyed, first by Meyer, to the Worcester State Hospital and Clark University, in 1895, and later by Donaldson, to the reconstituted Wistar Institute in Philadelphia, in 1906.

At Clark University

Moving now to Clark University, Colin C. Stewart, a fellow in physiology during 1894 to 1897, was pursuing his doctoral research on the influence of alcohol on animal activity. The wild gray rat was used at first, but the difficulty of handling them forced a change to albino rats for Stewart's remaining 2 years; at the time, albino rats were on sale as pets in bird and animal stores in Worcester. Stewart's results were published in the initial volume of *The American Journal of Physiology* in 1898 as the first paper in which the albino rat was used as a laboratory animal in this country. The protocols of experiments were dated in his paper and this precision of timing increases only slightly the possibility that Stewart had obtained his white rats from Adolf Meyer. Meyer arrived at Worcester on November 16, 1895, and

[3] The American physician who became outstanding in the field of public health, specifically at the workplace.

left for an extended trip to Europe in April, 1896, leaving a period of little more than 4 months during which white rats might have been provided by Meyer for Stewart's use.

That possibility seems slim, however. In a survey of Adolf Meyer's early letters and other papers (Winters, 1950–1952), Eunice Winters (personal communication, January 14, 1979) has found no reference to his having imported white rats from Europe. Moreover, the publications of Donaldson's department of neurology lacked any animal-based papers during 1892–1897, but included five, all using the frog, during the succeeding 3 years. The first six rat papers mentioned were not published until 1900–1903; allowing 2 or 3 years of lag-time for undertaking research, writing it up, and getting it published, it seems clear – in retrospect – that Meyer's shipping Donaldson a large lot of white rats from Worcester, in 1897 or 1898, and urging him to make the rat, rather than the frog, the standard animal, formed the principal steps establishing the white rat in neurological research at the University of Chicago.

"Association" Studies

The second doctoral student so involved at Clark was Linus W. Kline, whose work with white rats was done probably with the lineal descendants of Stewart's animal stock. In late 1898 Kline began a study of "association" in rats by constructing wire mesh boxes in which food was placed and the small entrance door covered by paper or buried in sawdust. Hungry rats were first able to gain entrance to the box, and obtain food, by random exploration that tore away the paper or dug a channel through the sawdust. On repeated trials, the animals reduced the time spent in so securing food from more than an hour to less than a minute. In addition to his introduction of these "puzzle-boxes" for studies in rat learning, Kline was also involved in introducing maze-running studies of learning in animals.

The idea of using a maze came from Professor Edmund C. Sanford (in psychology at Clark) when Kline described to him the runways made by feral rats to their nests under the porch of an old cabin, which presented a veritable maze. Sanford suggested the possibility of using the pattern of the Hampton Court maze for constructing a "home-finding" apparatus. The maze as an apparatus for the systematic study of animal learning was constructed by Willard S. Small at Clark, who began an extended series of experiments with the white rat on maze learning, thus marking its first use in animal behavior. The published report (Small, 1900; 1900–1901) was a model of painstaking procedure and conservative interpretation of results; Small did not know of Thorndike's work until its publication after his own experiments had commenced and a thorough reading of the report convinced him that he was on the right track.

It is remarkable that so many of the early programs in comparative psychology in this country sought to extend the phylogenetic level of animals studied to include experiments on infrahuman primates, typically the monkey. The program at Clark culminated in this pattern, with A.J. Kinnaman, fellow in psychology, who published his observations in an article on the "Mental life of two *Macacus rhesus* monkeys in captivity" in 1902. Kinnaman's paper began, "To Dr. Edward L. Thorndike, of Columbia, belongs the honor of having first taken the monkey into the psychological laboratory. He experimented for several months with three *Cebus* monkeys from South America." In a variety of experiments involving learning to manipulate complex latches and locks to open boxes and obtain food, together with tests of color discrimination, Kinnaman undertook his magnum studium – the repetition of Willard Small's maze-running experiments – with monkeys. His chicken-wire maze had a shortest path from the entrance to the food in the center of 105 feet with 27 corners to be turned and was regarded as learned whenever the animal succeeded in passing through it ten times consecutively without error. On average, male monkeys took almost twice as many trips as females to reach that standard. Kinnaman concluded that the maze tests did not reveal new insights on the general problem of the monkey's intelligence or method of learning and were an interesting point of connection with Small's work on rats.

Thorndike at Harvard and Columbia

Studies on Chicks

Concomitant with, but independent of, the research developments of Stewart, Kline, Small, and Kinnaman at Clark University in 1894 to 1901, analogously pioneering studies of animal behavior were initiated by the young graduate student already mentioned, Edward Lee Thorndike (1874–1949; Fig. 9.2), at Harvard and continued at Columbia University. Thorndike, for whom psychology had been terra incognita, in college was assigned chapters from James's *Principles*, and found them immensely stimulating. During 2 years of graduate study at Harvard, 1895 to 1897, he registered for the course under James, and is said (Skinner, 1979, p. 87) to have attended lectures in Boston by Lloyd Morgan of University College, Bristol. In his lecture on the acquisition of habits, Morgan pointed out that young chicks learned rapidly to gobble up small white butterflies but avoided without fail the acrid-tasting caterpillars of the cinnabar moth, and asked what were the physiological concomitants of the behavior. Presumably extrapolating to man, he answered that as actions afford satisfaction or dissatisfaction, they are respectively enforced or suppressed. Soon thereafter, Thorndike began his own study of chick behavior but with experimental methods. A chick, 2 to 8 days old, was placed in a "pen" formed by books stacked to make an enclosure with two to four

Fig. 9.2. Edward L. Thorndike (ca. 1897) succeeded in developing a theory of learning from the early data on chicks and rats, thus bridging psychology and physiology.

exits, one leading out to the other chicks and food, and the other(s) leading to another pen with no exit. These pens formed rudimentary mazes which the hungry chick had to traverse correctly to locate the exit.

Thorndike can be credited with introducing a simple model of the maze at Harvard a year or two before the concept of using the more complex Hampton Court maze for experimentation was proposed by Sanford at Clark University, in the Spring of 1898; and 3 years before Willard S. Small actually had constructed a rectangular model of the Hampton Court maze and employed it in studies of rat learning. The remarkable coincidence of these essentially concomitant, independent developments, by young doctoral graduate students at Harvard and Clark Universities, had the additionally curious feature that Thorndike developed his rudimentary maze first, and only later devised his puzzle-boxes, an inversion of the sequence at Clark, where Kline, Small, and Kinnamon undertook their puzzle-box studies first, and Small and Kinnaman later pursued maze studies. Moreover, in his puzzle-box studies, Thorndike put the animal inside the box, from which it had to escape to obtain food; whereas at Clark the animal was placed outside and had to learn how to gain entrance to the food.

As to the animals involved, for this was the very beginning of comparative psychology in America, the Clark University program, from Colin Stewart through Willard Small, was indubitably the pioneer in introducing the white rat to the general field of psychology, and all of its subsequent areas of investigative specialization. As Kinnaman (1902; see p. 229) so gracefully acknowledged, however, Thorndike, by then at Columbia, was the first to introduce the monkey to psychology in this country. His three *Cebus* monkeys, from South America, were only a year and one animal ahead of Kinnaman's two macaques, from India, at Clark. In terms of introducing standard animals, Clark was again in the lead, for the common primate in most fields of investigative use in this country today is the Macaque and not *Cebus*.

Having to support himself by time-consuming tutoring at Harvard, Thorndike applied for and received a fellowship to continue his studies at Columbia, where Professor Cattell approved his work on the mental life of animals for his dissertation. Among his fellow graduate students were C. Judson Herrick, Robert S. Woodworth, and Shepherd Ivory Franz. At Columbia, Thorndike used cats and dogs and an assortment of self-carpentered puzzle-boxes, for experiments with a common design: Hungry cats or dogs were put into enclosures from which they could escape, by some act within their capability, to gain adjacent food. The numbers of trials and times required to escape were recorded until they became stabilized at a constant, brief, "learned" level. Motivation and its reward were critical factors in those experiments and, as Thorndike (1911, p. 54) remarked, "Never will you get a better psychological subject than a hungry cat."

The Laws of Learning

It was chiefly through those insightful analyses and generalizations of "cause and effect" that Thorndike exceeded his contemporaries at Clark University. Kline had designed and constructed more elegant puzzle-boxes, and Willard Small's elaborate maze was beyond Thorndike's technical resources and possibly his capabilities. From essentially the same experiments and results, however, Thorndike was able to conceptualize the two major "Laws of Learning" which still form foundation blocks in this field today. In Thorndike's words:

> The Law of Effect is that: Of several responses made to the same situation, those which are accomplished or closely followed by satisfaction to the animal will ... be more firmly connected with the situation, so that, when it recurs, they will be more likely to recur; those which are accompanied or closely followed by discomfort to the animal will ... have their connections with that situation weakened ... .
>
> The Law of Exercise is that: Any response to a situation will ... be more strongly connected with the situation in proportion to the number of times it has been connected with that situation and to the average vigor and duration of the connections (Thorndike, 1911, p. 244).

Thorndike also recognized that attention strengthened and facilitated the process that it accompanied. In contemporary terms, these laws emphasized the importance of affective reinforcement, as well as repetition in learning.

Peer Evaluations

Thorndike's doctoral thesis, titled "Animal Intelligence: An Experimental Study of Associative Processes in Animals," was published in 1898. It received some negative judgments, as did his 1911 publication (see Waters, 1934) and also the highest praise. In his Silliman Lectures at Yale, Charles (later Sir Charles) Sherrington asserted that the psychological study of behavior had great potential for physiology,

> New methods of promise seem to me those lately followed by Franz, Thorndyke [sic], Yerkes, and others .... By combining methods of comparative psychology ... with the methods of experimental physiology, investigation may be expected ere long to furnish new data of importance toward the knowledge of movement as an outcome of the working of the brain (Sherrington, 1906, p. 307).

In his *Lectures on Conditioned Reflexes*, translated into English by W. Horsley Gantt in 1928, Ivan P. Pavlov acknowledged that American psychologists had been carrying out experiments similar to his, Thorndike's having preceded his by several years. He considered Thorndike's book, *Animal Intelligence*, a classic, both for its bold outlook and for its accuracy.

Much later, in *A History of Psychology in Autobiography*, B.F. Skinner apologized for not having acknowledged enough of Thorndike's work. "It has always been obvious that I was merely carrying on your puzzle-box experiments but it never occurred to me to remind my readers of the fact." Thorndike replied, disarmingly, "I am better satisfied to have been of service to workers like yourself than if I had founded a 'school' " (Skinner, 1967, p. 410). Other reactions to his "Law of Effect" were so contentious that Thorndike complained that it was disfavored by psychologists and neglected by physiologists. Among a variety of alternative suggestions, H.L. Hollingworth (1931) made the witty proposal "that it should more properly be called the Law of Affect." In 1934, a survey of its status as a "Principle of Learning" by R.H. Waters included 73 references and identified "feedback" of the effect as the criticism most frequently mentioned. The next major review, by Leo Postman in 1947, included 332 references and identified the problem of reinforcement as still a subject of heated controversy. In addition, by then a second concern had developed around integration of the law with theories of learning. In Clark Hull's (1942) drive-reduction theory, the law occupied a central position. In contrast, it was attacked by Edward Tolman (1949) who opposed all reinforcement types of learning theory.

THE LAW OF EFFECT: FROM THORNDIKE TO SKINNER, OLDS TO MILLER

Among later investigators of the ideas of learning and feedback, one was exceptionally successful in imprinting his name and concept on the field.

Burrhus Frederic Skinner (1904–1990)

A major contributor to the learning issue, B.F. Skinner (Fig. 9.3) during 1927 to 1931 was the second talented young psychology student in this sequence at Harvard. Skinner's precocious brilliance was displayed by a skit prepared for a research course at Harvard. As Skinner recalled,

> I had not been impressed by the ape experiments of Yerkes and Köhler and wrote a short scenario of a demonstration film poking fun at them. A 'scientist' in white coat is seen ... pointing to the essentials of the experiment – a basket hanging from a high branch of a tree on a long rope, some boxes to be piled up by the ape to reach the basket, and a banana .... The scientist picks up the banana, climbs a ladder against the tree, and reaches for the basket. He slips, grasps the basket, and finds himself swinging from the rope. He begs the ape to pile some boxes under him so that he can get down, but the ape refuses until the scientist first throws him the banana (Skinner, 1979, p. 31).

Early Work

Skinner had earlier acquainted himself with Pavlov's conditioned reflex studies and in a physiology course given by Hudson Hoagland had covered the

Fig. 9.3. The sublime self-confidence of B.F. Skinner sustained his progressive refinement of apparatus and operant conditioning.

postural and spinal reflex research of Sherrington and Magnus. Moreover, he had collaborated with physiologists (Lambert, Skinner, & Forbes, 1933) in the medical school in studies of the reflex control of movement. By the end of the year, however, he was analyzing the behavior of a rat as it traversed a suspended runway to obtain food at its end. The amplified vibrations from the rat's passage, recorded on a kymograph smoked drum, looked like those of a torsion-wire myograph, but to Skinner they reported the conditioned behavior of the organism as a whole (Skinner, 1956).

Focusing on the rate at which the rat sought food, he dispensed with the runway and, in a series of further reductions, simplified the apparatus until everything transpired within a small soundproof box with a horizontal spring-loaded lever on one wall. When a hungry rat depressed the lever, a pellet of food was released into a tray just below, and a cumulative recorder was triggered. Skinner initially thought this resembled Thorndike's famous puzzle-box and he first called his set-up a "problem box." Reputedly (Skinner, 1979, p. 205), it was first called a "Skinner Box" by Clark Hull, who introduced it at Yale. With this apparatus, Skinner studied increase, or other change, in the rate of responding, later called a "schedule of reinforcement."

Skinner's interest in conditioned reflexes had been substantially reinforced, in the second year of his graduate study in 1929 when Pavlov himself was enthusiastically received in Boston as honorary president of the first International Congress of Physiological Sciences to be held in this country. As a cherished souvenir of the occasion, Skinner succeeded in obtaining the original of the signature used on Pavlov's official photograph. Much later, in 1966, when the Pavlovian Society of America met at Cambridge, Skinner gave the dinner address on these and other recollections, under the appealing title, "Some Responses to the Stimulus 'Pavlov.' "

Operant Conditioning

Following his doctorate in 1931, Skinner continued at Harvard in postdoctoral study of the distinctions between patterns of conditioning. His paper on "Two types of conditioned reflex and a pseudo type" differentiated between Pavlovian and what Skinner had by then identified and would later call *operant conditioning*, modestly designating this as Type I and relegating Pavlov's historic model to Type II. "The essence of Pavlov's *respondent conditioning*," Skinner explained, "is the substitution of one stimulus or signal for another. In *operant conditioning* there is no substitution of signals … ." (Skinner, 1935, p. 75). In a review of Skinner's *The Behavior of Organisms* (1938), Ernest Hilgard (1939, p. 122) pointed out that "[t]he distinction between respondent and operant has been implicit in Thorndike's work all along, but it did not become explicit … . This clear distinction is perhaps Skinner's most significant conceptual contribution."

In 1946, Skinner began a series of summer conferences on "the Experimental Analysis of Behavior" which eventually met at the same time as the American Psychological Association (APA) and later formed a separate Division 25 of the APA. Considering the wide range of Skinner's investigative interests and the variety of applications of his findings, it is curious that, except for his limited early collaboration with Alexander Forbes, throughout his long career Skinner maintained an isolation from the central neural substrates and processes underlying the behavior with which he was engaged. In a summation of his accomplishments, he wrote: "I rejected Sherrington's physiology not because, like Jacques Loeb, I 'resented the nervous system,' but because I wanted a science of *behavior*." And again, "In my thesis [on "The Concept of the Reflex in the Description of Behavior," Harvard, 1931] I had redefined the properties of Sherrington's synapse as laws of behavior rather than as properties of the nervous system" (Skinner, 1979, pp. 68; 166).

Skinner had strikingly implemented those remarks in 1956 by highlighting the significance for comparative psychology of three tracings that recorded operant behavior to successive reinforcements by a pigeon, a rat, and a monkey which showed astonishingly similar properties (Fig. 9.4). In his enchantment with these sparkling schedules of reinforcement, Skinner seems not to have recognized that he had unwittingly contributed more to the cerebral localization of operant conditioning – in those three records – than had his contemporary, Karl Lashley, with myriads of rats running mazes after surgical ablation of their cerebral cortex, in the latter's prolonged but unsuccessful "search for the engram." As in Thorndike's earlier studies with chicks and

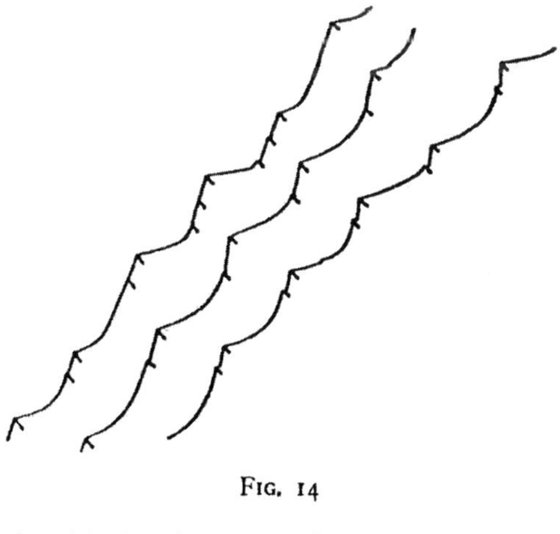

FIG. 14

Fig. 9.4. The similarity of the learning curves of three phylogenetically dissimilar species – pigeon, rat, and monkey – supported B.F. Skinner's contention that subcortical structures were responsible for operant conditioning. From Skinner, 1959, p. 374, Figure 14.

those of Yerkes with crows, Skinner demonstrated the identical performances of a pigeon, rat, and monkey. The perfect accomplishment of the pigeon, a species which had been completely decorticate from birth, as had all its ancestors for millennia, suggested that the cortex was unnecessary for operant conditioning. The cerebral hemisphere of the brain of birds consists entirely of a greatly hypertrophied striatum, totally lacking the cerebral neocortex of the mammalian brain. The subjacent limbic forebrain and brain stem of birds, however, closely resemble or are identical with those of mammals. It is in those ancient subcortical neural structures that the seat of operant conditioning should be sought, as will now be made clear.

James Olds (1922–1976)

The third talented young doctoral student at Harvard in this program, James Olds (Fig. 9.5), devoted a part of his thesis on "The Acquisition of Motives" (Harvard, 1952) to "A Neural Model for Sign-Gestalt Theory" (1954),

Fig. 9.5. James Olds's work was a brilliant exemplar of the power of neuroscience to uncover new concepts: his combination of anatomic, physiologic, and psychologic techniques led to in-depth investigations of motivation. Photographed about 1970.

interrelating aspects of Edward Tolman's cognitive theory of learning with Donald Hebb's theoretical construct of cell assemblies in the brain. In retrospect, its greatest significance lay in motivating Olds to spend the next 2 years in Hebb's psychological laboratory at McGill University. Because of Hebb's preoccupation with writing a book, his assistant, Peter Milner, cheerfully taught Olds methods of electrode implantation and electrical stimulation and recording in the brain. When Olds discussed his projects with Hebb, including recording from and stimulating rats during their behavior in a "Hebb-Williams maze," Hebb became interested and suggested that testing for motivational effects be done first as they seemed easy to carry out. The list of other things he might do took shape in pencil scrawls on the back of an old envelope, including whether or not brain stimulation might be aversive, as had been reported by Neal Miller at Yale in his presidential address to Division 3 at the American Psychological Association meeting in September, 1953. Shortly afterward, Olds was engaged in a practice experiment, the first of many, involving an albino rat with an electrode that had been accidentally implanted in the septal region of the brain. Lead wires suspended from the ceiling connected the animal to an electric stimulator, which Olds controlled by hand switch. Each time the animal entered the test corner of the open field it received a one-half-second stimulation. By the third stimulation, it remained in the corner with "an excited happy look."[4] In successive experiments, to gain a brain-stimulation reward rats ran faster and faster, or ignored a food reward, or endured painful foot shocks. To delineate the critical region and basic characteristics of his surprising discovery, this recent doctorate from Harvard naturally chose a Skinner box.

Bar Pressing for Pleasure

With Peter Milner, Olds (1954) constructed a small box with a large lever at one end. It had already been determined that brain stimulation equaled or exceeded a food reward, so Skinner's elaborate food-dispenser was omitted. Instead – in an inspired step – the lever was connected with a microswitch that when depressed briefly completed the stimulation circuit to the implanted electrode. For the first time in the history of neuroscience, this arrangement enabled an animal to push a lever and deliver a stimulus – in this case, obviously reinforcing and, presumably, pleasurable – to the interior of its own brain. The later popular term for this behavior – "self-stimulation" – was introduced by the Rioch group at Walter Reed Army Institute of Research (Porter, Brady, Conrad, Mason, Galambos, & Rioch, 1958).

[4] Olds continued: "You may wonder how I know, but I have 'gone among them and learned their language' " (Olds, 1972, p. 267).

The distribution of responsive sites was totally subcortical; the most effective region, as measured by the rate of self-stimulation, was the forebrain bundle in the lateral hypothalamus, in immediate proximity to neural centers for the basic drives of feeding, drinking, sex, and sleep. High scores of stimulation in the anterior commissure region started at about 3,000 per hour, and worked up to 6,000; the region which consistently yielded the highest scores, about 8,000 responses per hour, was proximal to the mammillary body where the medial forebrain bundle crosses from the hypothalamus into the tegmentum.

The ingenuity and novelty of this coup by a young postdoctoral psychologist and his senior associate immediately captivated the attention and interest of persons in a variety of fields. It is difficult, almost a half-century later, to recreate the excitement evoked by their major breakthrough interrelating the behavioral and neural sciences. From 1955, Olds's presentations enlivened and highlighted most major conferences and symposia in this general area, many of them international in scope. Comments by his peers convey something of the flavor of appreciation and regard with which his contributions were received.

An example was the 1955 Nebraska symposium on motivation, where David McClelland commented: "The research results reported by Olds are the stuff of which dreams ... are made" (p. 139) and went on to stress their importance for motivation theory. In an article on "History of Brain Stimulation" in 1973, Elliot Valenstein, who is not often given to hyperbole, assessed Olds's contribution as possibly containing the key to understanding the physiologic basis of motivation and reinforcement.

After his "practice operation" that settled his future research direction, Olds spent a short further period of postdoctoral research at the nascent Brain Research Institute at the University of California at Los Angeles. The years 1957 to 1967 were at the University of Michigan, where he and his wife, Marianne, studied many of the anatomical, pharmacological, and behavioral aspects of brain-stimulation reward. Returning to California, at the California Institute of Technology the Oldses undertook single-cell studies of reinforcement in the reward and learning systems, in an attempt to find an integrated theory of learning. Unfortunately, James Olds died at the age of 54 just when he felt they were close to that achievement (see Peter Milner, 1977).

The Accolades Continued

In *The Biology of Reinforcement* (1980), Aryeh Routtenberg wrote that the discovery in 1954 of brain self-stimulation was a scientific explosion, fusing the behavioristic, physiologic, anatomic, and neuropsychologic domains in a single event. In the same publication, dedicated to Olds, Neal Miller emphasized Olds's key contribution to the development of interdisciplinary work that produced new information on the brain and learned behavior, and for having "the wit to recognize the possible significance of this completely unexpected behavior" (Miller, 1980, p. 3). And again, in Peter Milner's perceptive memorial (1977, p. 140), he wrote:

> Olds was a great psychologist whose impact in the field of learning was comparable to that of Tolman and Lashley.... Olds will also be remembered for a change that he, more than anyone else helped to bring about, a change in the attitude of psychologists to those parts of the brain that lie below the cortex. Until the 1950's, it was generally assumed that the hypothalamus and limbic systems were concerned only with instinctive and emotional behavior, and could safely be ignored by those interested in learning or cognition. Olds dramatically put an end to that era.

Attention to Memory

The new era of research on mechanisms of learning and memory which had been ushered in by self-stimulation was just as dramatic as the discovery. As noted, Olds had been motivated to explore further the aversive effects of brain stimulation reported by Neal E. Miller and collaborators in cats (Delgado, Roberts, & Miller, 1954). The work was a combination of Delgado's investigations of brain structures susceptible to nociceptive stimuli and Miller's behavioral techniques to analyze learning. They found three brain regions in which a pain-fear-like reaction to electrical stimulation had the drive characteristics of a true emotion in reinforcing trial-and-error learning and seemed congruent with the drive-reduction hypothesis.

Neal Elgar Miller grew up in the State of Washington and after attending college there he completed his graduate work in psychology with Clark Hull at Yale's Institute of Human Relations. Whereas Skinner had "honed Thorndike's law of effect into a precision tool for the control of behavior" (Jonas, 1972, part II, p. 34), by placing emphasis on the rate of response, Hull's adaptation was a theory to explain all behavior as based on motivation/drive to maintain bodily homeostasis with appropriate responses to internal and external stimuli. Miller in turn set himself and his students the daunting task of trying to account for all reinforcement by a "single physiological mechanism" (Ibid., p. 42). At the 1963 convening of the famous series of conferences called the Nebraska Symposium on Motivation, he suggested yet another alternative to the drive-reduction hypothesis of reinforcement: a "go-mechanism" based on extending the traditional definitions of stimulus and response and making six assumptions about them. The parsimony of the theory was appealing and characteristically Miller immediately devised experiments to refute it.

Operant Visceral Conditioning

By 1967, Miller and Carmona had shown operant learning of a visceral response in dogs (salivation in response to a water reward) and soon had taught rats to change vital functions such as respiratory and heartbeat rates, blood pressure, and vasodilatation. Those demonstrations of visceral learning proved that "biofeedback" could be utilized experimentally and clinically to control certain normal and pathologic states in humans and the technique became part of the psychosomatic armamentarium. As a leader of the school of psychology that believes all behavior follows a single law not yet thoroughly understood, Miller has devoted his career to exploring the relationship of brain, behavior, and the environment. His commitment to dissemination of this broad approach to neuroscience is seen in his energetic participation in the organization of the Society for Neuroscience in 1969, as described in Volume 2. Relatedly, when interviewed in 1988, Miller stated that he believed his most significant contribution to progress in neuroscience was the conference and resulting brochure on explaining brain research to journalists timed to coincide with the ascendency of the discipline.[5]

After following the unfolding of the trail of Thorndike's laws of learning to the final years of the 20th century, we now refocus on the earlier decades to discuss one of the most important movements in the behavioral sciences.

SPREAD OF BEHAVIORISM

As a deviation to the left of American functionalism, objective behaviorism had a strident, short history, lasting about a quarter-century (Harrell & Harrison, 1938). By stepping around the self-consciousness of introspection, the behaviorists – personified by J.B. Watson – observed and manipulated animal behavior in the field initiated by Small, Kline, and Thorndike. The University of Chicago served as behaviorism's breeding-ground with President Angell and John Dewey presiding over the movement to bring psychology closer to biology.

John Broadus Watson (1878–1958) at Chicago and Hopkins

As a graduate of Furman University in South Carolina, Watson's (Fig. 9.6) interests in pursuing advanced study at the University of Chicago were stimulated both by one of his professors having spent a sabbatical there, and by the leniency of its classical language requirements compared with those at Princeton, where his mother had expected him to go. Starting in 1901, Watson spent three "frightfully busy years" taking his Ph.D. (Watson, 1961, p. 273).

[5] Interview, Series CON Code MIL, Oral History Project, Neuroscience History Archives, UCLA. Brochure, "The Scientist's Responsibility for Public Information: A guide to effective communication with the media" (N.E. Miller, 1979). Miller's goals were echoed two decades later by an eloquent "new" scientist, Stephen Jay Gould (1999) as the endeavor to narrow the gap between C.P. Snow's "two cultures" continues.

Fig. 9.6. John B. Watson met two challenges in his career – the defense of objective behaviorism and the rationale of studying sex in human subjects

During two of these, he earned his living by "dust[ing] Mr. Angell's desk and clean[ing] apparatus" and, in the third, by "keep[ing Professor Donaldson's] white rats for him." As for Watson's studies, his research, jointly under both men, was on the relation between the increasing complexity of behavior in the developing white rat and the growth of myelination of its central nervous system. In his dissertation, published in 1903 with $350 lent by Donaldson, the rat's development was studied to determine the earliest manifestation of the associate process. A nest with a mother and litter of young rats was kept at one end of an elongate box throughout the experiment. Vertical slats could then be inserted across the remaining space, dividing it into three consecutive compartments. Gaps in the slats, large enough for baby rats to pass, but too small for the mother, were then opened or closed, converting the route from these compartments back to the nest from a straight to a criss-crossed one. Starting at the age of 10 days, young rats were taken from the nest, put in the nearest compartment, and allowed to make their way back to the mother. As soon as they were able to return, the pathway was made longer and more complex. By 16 days of age, the young rats were weaned and their testing was extended to food-box, latched-door, and labyrinth (maze) experiments. As Watson summarized:

> From the twelfth to the twenty-third day, there is a graduate … increase in the complexity of the memory processes …. At approximately twenty-four days

> we felt justified in saying that the rat is psychically mature… [able] to learn anything that a rat at maturity can learn (Watson, 1903, pp. 86; 119).

Because the process of myelination is incomplete at those ages, Watson could answer two main neurophysiologic questions: medullated fibers in cortex are not the sine qua non of rat's forming and retaining definite association and the complexity of psychic life increases much more rapidly than does the medullation process in the cortex.

On obtaining his Ph.D. in 1903, Watson continued at Chicago but in 1904 spent the summer at Johns Hopkins, under Professor William Howell in physiology, working up operative techniques on animals. Back in Chicago and with Harvey Carr, Watson first determined the average time required for trained adult white rats to traverse Willard Small's Hampton Court maze, and then applied his newly gained surgical techniques to determine the sensory factors involved in this learned behavior. He found that turns were made solely on the basis of proprioceptive impulses in a kinesthetic reflex system. In addition, Watson started work on animal vision, using monkeys as subjects, and on terns in the Dry Tortugas, carried out at the Marine Biological Station of the Carnegie Institution of Washington.

Appeal for an Experiment Station

In 1906 Watson had written eloquently on the need of an experimental station for the study of animal behavior, pointing out the benefits of investigations requiring the continuity of many generations of a single species, or the continuous observation of animals from birth to old age. Additionally, he proclaimed that knowledge of the behavior of insects and birds would remain in a backward state until a natural environment with adequate space for experimentation was ensured. Watson's participation in the Carnegie Institution's Marine Biological Station represented an effort to implement this aspect of his concern with bird behavior. His later recruitment of a Hopkins graduate student, Karl S. Lashley, in studies of the behavior of terns at this Caribbean station first introduced Lashley to the advantages of such research stations, in one of which, the Yerkes Primate Research Center at Orange Park, Florida, Lashley would conclude his career.

Interaction with Yerkes

Returning to maze studies, Watson's activities had paralleled those of Robert M. Yerkes. Commencing in 1899 at Harvard and Woods Hole, Yerkes had employed rudimentary mazes in comparative studies of learning in crab, crayfish, frog, and turtle. In 1906, he began using a reduced Hampton Court maze in studies of dancing mice and became interested in analysis of the animal's course through the maze such as the distribution of its errors. In the same year, a young neuropsychiatrist, Gilbert V. Hamilton, in training at nearby McLean Hospital during 1905 to 1907, spent a valuable year's work under Yerkes at Harvard, in which he developed a method for getting at this issue. He numbered each of the 24 cul-de-sacs in the Hampton Court maze, from 1 in the center to 24 at the exit (i.e., backwards). In each animal trial, the errors made were marked by their appropriate numbers on a diagram of the maze, so as to indicate proximity to entrance, midcourse, or finish with food reward. Similarly, any preference for errors to the right or left of the proper course was displayed, the direction of the 24 cul-de-sacs being L-11 and R-13.

In relation to those studies, in May 1906, Yerkes opened a correspondence with Watson at Chicago concerned with the design of an improved, standard maze which he and William T. Porter were planning to construct at the latter's Harvard Apparatus Company, just established in 1904.[6] Watson was in favor of a standard maze and liked the sketch (from Hamilton) that Yerkes sent to him. That same month they considered collaborating on building one and testing it with rats, but nothing came of this practical suggestion. In the fall of 1908, Watson accepted an appointment at Johns Hopkins, as professor of experimental and comparative psychology and director of the Psychological Laboratory. Happy with the turn his work without supervision had taken, he and Robert Yerkes (in Baltimore for a visit ostensibly to perfect himself in animal surgical techniques under Harvey Cushing) began to collaborate on visual apparatus and finally published a monograph together, *Methods of Studying Vision in Animals* (Yerkes & Watson, 1911).

Seemingly with the goal of preparing a second collaborative monograph, after a 5-year hiatus, Yerkes reopened the earlier labyrinth correspondence with a letter to Watson (10/6/1911), which began, "Here goes for the labyrinth method … ." and proposed writing it up after systematic review of their voluminous material, with first a history "from Adam on to Watson," second, a critical discussion of values and requirements, and third, a description of plan and method. Yerkes claimed doing the history, and his share of the construction and criticism, while leaving the "devising of a perfect labyrinth mostly" to Watson. A year later (10/17/1912), Watson replied: "We need different kinds of mazes for different kinds of work" and that he favored the circular maze "for common work in all laboratories." In another letter (1/20/1913) he made the first reference to Yerkes's proposed graphic recording on a printed form of the maze, possibly experimenting with a camera lucida image coinciding with the form. Two months later, the pace had quickened and because each man favored his own method, wisely, they were written up separately and published one after another

[6]The role of the Harvard Apparatus Company in the development of laboratory physiology, and hence, neurophysiology, is discussed in Volume 2.

in *The Journal of Animal Behavior* (Yerkes & Kellogg, 1914; Watson, 1914a), with Yerkes given full credit for the original idea of a graphic record.

The publication of these two short papers concluded this Yerkes-Watson correspondence on maze improvement which was joined briefly by Madison Bentley at Cornell. The outcome was commemorated by Yerkes, however, in the subsequent assignment of an entire room of his laboratory of animal psychology at Harvard to maze studies. He described its features:

> Room 42 is equipped with the Watson circular maze and the Yerkes and Kellogg graphic record device. The latter enables an observer to obtain accurate records of distance and errors, in addition to those of time, in all maze experiments. Thus the value of the maze-method is trebled (Yerkes, 1914b, pp. 177–178).[7]

Emergence of Objective Behaviorism

In one of Watson's letters to Yerkes on the maze, he wrote (3/12/1913) of the course of 12 lectures he was engaged in at Columbia on Cattell's invention, the introduction titled "Psychology as the Behaviorist Views It" (Watson, 1913). Watson's manifesto began:

> Psychology as the behaviorist views it is a purely objective experimental branch of natural science. Its theoretical goal is the prediction and control of behavior. Introspection forms no essential part of its methods, nor is the scientific value of its data dependent upon the readiness with which they lend themselves to interpretation in terms of consciousness.

In the same vein, he concluded, "The position is taken here that the behavior of man and the behavior of animals must be considered on the same plane" (Watson, 1913, pp. 158, 176). An addendum to this paper, "Image and Affection [in current terms, "Affect"] in Behavior," followed and, culminatingly, an expanded, 400-page book, *Behavior: An Introduction to Comparative Psychology* (1914b). For Watson, psychology had obviously become a "science of behavior" rather than the traditional "science of conscious experience." Clearly the majority of his fellow American psychologists, reacting against the earlier concept of psychology as a study of mind, were receptive to these views: The following year at the age of 37, Watson was elected president of the American Psychological Association. The title of his presidential address, "The Place of the Conditioned Reflex in Psychology," marked an extension of his advocacy for objective experimental research from lower animal to human subjects.

[7] For a history of the maze as method, see Volume 2.

Conditioned Reflex Studies

Later, Watson paid tribute to the influence that Lashley had had on him when the latter was his postdoctoral student. For Watson's conditioned reflex studies, Lashley first designed a suction cup, fitting over the orifice of the parotid duct in the mouth, for recording salivary flow in human subjects, following the paradigm of Pavlov's studies in dogs. It was awkward to use, however, and of dubious accuracy. He and Watson then shifted to Bechterev's mode of conditioning, which had been developed for human studies from the start, and employed the easily recordable reflex responses of skeletal muscles, such as the knee jerk. With the elegance of design that characterized most of Watson's instrumentation, arrangements were developed for conditioning the flick of a single finger – a great contrast to Bekhterev's ponderous mechanical contraptions.

Following the First World War, Watson returned to Hopkins where Adolf Meyer provided him with a psychological laboratory for animal experimentation on the third floor of the Phipps Clinic and arranged through other departments and the hospital for his study of infants. In 1919, a talented young graduate student, Curt Richter, joined Watson as a laboratory assistant and in his dissertation research initiated career-long studies with the albino rat. Although assigned a problem on learning in rats, Richter instead was interested in their activity cycles, thus opening the door to his concern with periodic phenomena and "biologic clocks."[8] Based on 65 years of studies comparing domesticated and wild forms of the Norway rat, Richter deduced the changes in anatomy, physiology, and behavior of civilized humans in "Rats, man, and the welfare state" (1956) with pessimistic foreboding.

Studies with Human Subjects

With the novel opportunities afforded him in the medical school and hospital, the main thrust of Watson's work during 1919 to 1920 was with human subjects. This began with the study of innate reflexes in newborn babies and extended to experimentation with conditioned emotional responses in infants. Other studies with adult subjects explored the effect of alcohol on human motor performance. Additionally, there are recent indications that during that period Watson initiated the study of sexual responses in the human female, some 50 years before Masters and Johnson (see Chapter 12).

An account of Watson's study of instincts and early habits was prefaced by this disarming statement: "The human infant in general is sturdy and well able to stand all of the simple tests we need to apply." He compared the birth

[8] Announcement of the current availability of Richter's enormous database, "Biological Clocks in Medicine and Psychiatry," on the World Wide Web was made in 1988 by Nancy McCall and Lisa Mix, archivists at Johns Hopkins University.

process as being more stressful and described the precautions taken when testing the "grasp reflex" in the human newborn when any small object is placed in the palm. He and assistants found that normal infants can suspend themselves for appreciable intervals (as long as one minute), and the instinct disappears permanently at 4 months of age. Of equivalent interest were observations on the Babinski reflex and defensive responses, the latter persisting throughout life.

Conditioned Emotional Responses in Infants

Moving from these studies of innate reflex responses, Watson and his graduate student, Rosalie Rayner, became engaged in the study of conditioned emotional responses in an 11-month-old boy, "Little Albert." A normal, healthy child, he had been frightened by a series of loud noises 2 months previously. The same feared noise paired with presentation of an animal, e.g., a white rat, conditioned Albert to produce a fear reaction to any furry animal, showing that the emotional reaction had been retained over time and was subject to transfer.

From those studies with human subjects, Watson's role as a pioneer in a number of applied areas is gradually being recognized. Behavior modification, a current approach in clinical psychology and psychiatry, is such an area; "aversive conditioning" involved in behavior modification is a simple repetition of Watson's conditioning of Little Albert. In the introduction to a collection of *Readings in Modern Methods of Treatment Derived from Learning Theory*, H.J. Eysenck (1960) paid tribute to the work of Watson and Rayner, and of Mary Cover Jones, as having been the theoretical foundations of "behavior therapy" and the start of a new advance in the scientific treatment of neurosis.

Observations on Adult Humans

Watson's move from rats to the study of human subjects was by no means limited to infants, as correspondence on his "alcohol experiment," in the Spring of 1920, has come to light. In a letter[9] to Edward L. Thorndike, concerning standard forms of mental tests, Watson described a controlled experiment on the effects of alcohol on 14 hours of dart-throwing: there were no appreciable changes in accuracy. He wanted to repeat the experiment using mental tests, asked if Thorndike thought it "worthwhile," and proposed collaboration. The merit of the experiment was not clear to the Hopkins administration, however, (it was during Prohibition), and it was never carried out.

[9] Watson to Thorndike, 5/13/1920, Johns Hopkins Medical School.

The last area of adult studies in which Watson had seemingly begun activity in 1919 to 1920 was the field of human sexual behavior. The initial published acknowledgment that Watson "was one of the first Americans to investigate the physiological aspects of the [human] sexual response" appeared in James B. McConnell's lively textbook, *Understanding Human Behavior* (1977). "Since the medical sciences had studiously ignored the subject, Watson set out to investigate the matter himself – at first hand ... by connecting his own body (and that of his female partner) to various scientific instruments while they made love. He fathered what were probably the very first reliable data on the human sexual response ... ." (p. 273). Unfortunately, there was a notorious divorce and the boxes of carefully annotated scientific records were confiscated and Watson's academic career ruined. He joined a large advertising agency and stayed with it the remainder of his professional life, while continuing to write authoritative articles for popular magazines.

Recent supportive evidence of undocumented observations (Magoun, 1981b) has come to light in a cigar box containing Watson's apparatus for study of sexual responses, which was retrieved in 1945 by psychologist Reginald Bromiley, from a storeroom at Hopkins. Included were a vaginal speculum, a tambour sensor for recording changes in intravaginal pressure, and a pair of contacts designed to count contractions. Although Watson never published any explicit results of study of human sexual behavior, a number of his other publications indicate a familiarity with and interest in this field. An example is the introduction he was invited to write for a book by G.V. Hamilton and Kenneth Macgowan, *What is Wrong with Marriage* (1929). Watson began with a description of his experience when asked to help plan a research program for the study of sex. When the committee met, Watson asked if was to be a "real study of sex" or rats, rabbits, and guinea pigs again? When assured it was the latter,

> I tucked my plan into my pocket and went back to the office lamenting the fact that, while we were in the twentieth century in science, we were still in the fourteenth century in folk ways ... .
>
> The study of sex is still fraught with danger. It can be openly studied only by individuals who are not connected with universities (Ibid., p. xiii).

Cultivation of Respect

This depressing episode was recounted some 60 years afterward by a newly minted Hopkins historian, P.J. Pauly (1979), placing it in the context of an administration striving to mitigate the suspicious regard in which the university was held by the social leaders of Baltimore, a southern, conservative city. Although Watson's world-class reputation had brought psychology at Hopkins to a high point, the scent of scandal was perceived by faculty luminaries like Adolf Meyer as threatening to the university's honor and respect.

After Watson's dismissal, Knight Dunlap was advanced to head the department. He was one of many anticipators of Watsonian ideas (Herrnstein, 1969) and while increasing the production of well trained professional psychologists he kept the department on an even keel without the intellectual ferment of the former years. But the university's financial condition was unfavorable for support of a relatively young department and it was discontinued after Dunlap left for a position in the University of California at Los Angeles. Of greater significance, Watson's dismissal gutted the leadership of the behaviorist movement. Without a university base nor professional outlet for his research, the influence of the former leader was rendered impotent and his ideas unchallenged[10] and undefended, in spite of their partial support from a few contemporary figures including Robert Yerkes.

An evaluation of behaviorism by a respected contemporary psychologist merits notice. In 1931, Robert S. Woodworth wrote on the significance of behaviorism, which he described as having first appeared as a reform of the method of introspection, not as its elimination. Behaviorism removed consciousness from psychologic concepts and hastened their severance from philosophy. It also went to the other extreme and extended into sociology, Woodworth wrote, and "won the public ear," so that interest in child development became focused to the point where many looked on it as a new hope. "It is a religion to take the place of religion" (Woodworth, 1931, p. 92).

Knight Dunlap (1875–1949)

A native Californian, Dunlap was born in Diamond Springs in Gold Rush country and began his undergraduate studies at the University of California in 1895. In his junior year, he took a course in experimental psychology given by Professor George M. Stratton, who had just returned with a doctorate from Wundt's institute at Leipzig and had promptly installed a new laboratory on the Berkeley campus. "Here I found what I wanted," Dunlap wrote (1930/1961, p. 38), and his career in psychology had begun. For graduate study, he was drawn to Harvard. The outstanding faculty there included William James, Hugo Münsterberg, John Dewey, and in philosophy, Josiah Royce (another Californian, born in Grass Valley).

Completing his doctorate under Münsterberg, Dunlap returned to Berkeley and then followed Stratton to Hopkins. The effect of John B. Watson's 1908 recruitment to Hopkins from the University of Chicago was mixed. "As a laboratory director, as well as in our personal relations, Watson was admirable," Dunlap wrote (Dunlap, 1930/1961, p. 44), but, "I have always felt that

[10] William Sahakian has suggested that Titchener's attacks, for example, "dignified behaviorism by centering the attention of the entire community of psychologists on this budding school of thought, instead of permitting it to go unnoticed" (1975, p. 369).

behaviorism held back the development of scientific psychology" (Ibid., p. 46). After the First World War, the department was split with Dunlap heading the Homewood division and Watson the division at the Phipps Psychiatric Clinic, directed by Adolf Meyer, with Curt Richter and Horsley Gantt his research colleagues. The latter three, Meyer in particular, were self-identified psychobiologists, but it was Dunlap who, in 1917, established a *Journal of Psychobiology*.[11]

During 1927–1928, when Dunlap was chair of the Division of Anthropology and Psychology at the National Research Council, one of the conferences the division sponsored was on experimental psychology, held at Carlisle, Pennsylvania and attended by representatives of the principal psychologic laboratories in the United States. The group recommended establishment of a National Institute of Psychology, preferably in Washington, D.C., or immediately adjacent thereto as well as further institutes connected with universities. This National Institute of Psychology, actually incorporated in the District of Columbia in June 1929, was legally empowered to advance scientific work in psychology by all appropriate means.

A small board of directors, with Dunlap as president, was elected in September 1929 and adopted by-laws providing for an active membership of fifty qualified by research in experimental psychology. The board proposed, subject to membership approval, the establishment of a national laboratory with a permanent staff and adequate equipment for research in human and animal psychology to undertake problems too large to be handled by university laboratories and also to foster the development of a real comparative psychology (Dunlap, 1930). In his autobiography, Dunlap (1930/1961, p. 56) reported, "the Institute is a healthy youngster, needing now for the members to take charge, and large funds for research." Neither of those goals was met and the project was abandoned.

Some precedence for research institutes in psychology existed at that time. With support of the Laura Spelman Rockefeller Memorial in 1924, two of Dunlap's fellow students at Harvard, R.P. Angier and R.M. Yerkes, had been instrumental in the establishment of an Institute of Psychology at Yale. In 1929, Yale's president, James R. Angell, also a psychologist, expanded it into an Institute of Human Relations, with major support from the Rockefeller Foundation and the related General Education Board.

Reaching for a National Institute of Psychology

An informative example of the importance of timing, a decade earlier other such proposals had been unsuccessful. Shortly before the First World War,

[11] This journal later merged with Yerkes's *Journal of Animal Behavior* as *The Journal of Comparative Psychology,* and is called since 1947 *The Journal of Comparative and Physiological Psychology.*

E.E. Southard, Bullard Professor of Neuropathology at Harvard, had vainly attempted to establish an institute around the research program in nervous and mental diseases at the Boston Psychopathic Hospital. After the war, Harvey Cushing returned home with plans for founding a national institute of neurology to aid the continuing care of the servicemen with disorders and injuries of the nervous system and also to provide laboratories for experimental work. Cushing would be in charge of neurosurgery, and Lewis H. Weed (then professor of anatomy and later dean of the Johns Hopkins Medical School) had agreed to become director of laboratories. The cost of $10 million was prohibitive to a reactionary Congress, and the plan remained a dream. In the words of a biographer: "To Cushing it was more than a personal disappointment, it was a public tragedy" (Thomson, 1950, p. 203).

Thirty years after Southard's and Cushing's proposals and 20 years after Dunlap's feckless efforts, the National Institute of Mental Health in 1949, and a National Institute of Neurological and Communicative Disorders and Stroke in 1950, were established "immediately adjacent" to Washington, D.C., in Bethesda, Maryland. These institutions fulfilled the goals of those earlier proponents beyond their wildest dreams.

Turning Homeward

In 1936, Knight Dunlap returned to the University of California to chair the department of psychology on the Los Angeles campus. As befitted a man educated by two of Wilhelm Wundt's doctoral graduates, Münsterberg and Stratton, Dunlap maintained a career interest in experimental, or as he often called it, "scientific" psychology. In addition, like Wundt before him, he had the breadth of vision to advocate research institutes in this field and, as noted, had actually attempted to establish one on a national scale.

By the end of the first half of the 20th century, American psychology had become a solid science. It had shed the vagueness of philosophy and the subjective methods of introspection and psychoanalysis. Behaviorism had been assimilated. An infrastructure of departments, laboratories, and students was in place and the discipline had proven its usefulness in two world wars. The behavioral sciences had evolved to focus on the individual over masses of group data and inevitably would join the neural sciences.

Chapter 10
Prefrontal Cortex of Infrahuman Primates

The historic example of the effects of a dysfunctional prefrontal cortex – the accidental frontal lobectomy of Phineas Gage in 1848, his miraculous survival, and the change in his behavior – are well known (see p. xvi). Gage's physician noted that: "The equilibrium ... between his intellectual faculties and animal propensities, seems to have been destroyed" (Harlow, 1869, p. 13). Abroad, David Ferrier (1873), initiating frontal lobe ablations in monkeys at Brown Institution, London, wrote that on seeing his operated animals little effect might be perceptible, but that a decided change was produced in their behavior: they appeared to have lost the faculty of intelligent and attentive observation. Similar descriptions of frontally lobectomized monkeys were published by the Italian, Leonardo Bianchi (1895), in his monograph, *The Functions of the Frontal Lobes*. It opened with a chapter on the "History and Evolution of the Frontal Lobes,"[1] followed by a summary of his many observations of animals and humans with frontal lobe injuries. John Fulton (1951, p. 23) perceptively remarked,

> Both Ferrier and Bianchi were good observers, but the behavioral changes which they described were difficult to appraise because at the time they worked objective methods had not yet been developed for analyzing behavioral disturbances in quantitative terms.

EXPERIMENTAL RIGOR

Pioneering Experiments by Shepherd Ivory Franz (1874–1933)

A beginning in formulation of quantitative measurements of behavior which Fulton found lacking at the turn of the century was made by a member of the American school of psychologists, who advocated experimental studies under

[1] Curiously, in this chapter, Franz is referred to as "Ivory Shepherd."

Fig. 10.1. Shepherd Ivory Franz instilled systematic testing and rigorous standards into measurements of animal behavior carried out before and after making experimental lesions, thus improving the reliability of the data.

controlled conditions, in contrast to "naturalistic" observations. Not only were rigid protocols followed (as in the learning experiments with mazes described in the preceding chapter) but the animals of choice were phylogenetically close to the human and were studied as individuals or in small groups. An American pioneer in those investigations, Shepherd Ivory Franz, may not have received due recognition for his role in promoting those changes.

Education

As an undergraduate student at Columbia in the 1890s, young Franz (Fig. 10.1) was attracted to physiological psychology by reading Wilhelm Wundt's *Psychologie* and assisting McKeen Cattell, who had begun his own career as assistant and doctoral student of Wundt at Leipzig in 1886. At Columbia, "The laboratory, as in many other universities where it was not found in a basement, was located in an attic" (Franz, 1961, p. 91). Among his fellow-students were C. Judson Herrick, Edward L. Thorndike, and Robert S. Woodworth. After gaining his Ph.D. in 1899, Franz spent the next 2 years assisting Henry P. Bowditch and William T. Porter in physiology at Harvard Medical School. At that time Walter B. Cannon had begun his observations on the effect of emotions on gastric motility, using x-rays, in Harvard's physiological laboratory. Shortly before, in the basement of Professor William James's home in nearby Cambridge, Edward Thorndike had completed his experiments with chicks which demonstrated the importance of reinforcement in learning (see the previous chapter).

Reading of Bianchi's studies on frontal lobectomy in dogs and monkeys, Franz conceived of combining them with those of Thorndike. As he recalled, "it was a simple step, from [Thorndike's] results to try the combination of animal training and [cortical excision] as a method whereby some additional cerebral problems might be solved" (Franz, 1933, p. 97). Beginning in 1900, Franz pioneered in relating experimental psychology with the physiological study of brain in animals by evaluating the effects of extirpation of brain regions upon the maintenance, loss, or recovery of learned performance. He found, first in cats and then in monkeys (1907), that complete frontal lobectomy led to the loss of recent, but not of earlier acquired habits; and those habits lost could later be relearned. This finding and the analogous recovery from initially impaired movement following ablation of the motor cortex, suggested that many functional deficits after brain lesions might be recovered, depending on the plasticity of the residual tissue.

Franz's published findings attracted the attention of Professor Sharpey-Schäfer at Edinburgh and Sherrington at Liverpool, both of whom commended him by letter. Fifteen years later, Franz introduced Karl Lashley to this approach, which provided a basis for much of Lashley's later research. In addition, those early studies led Franz to rehabilitative therapy in clinical psychology. Moving to Dartmouth to teach physiology at its medical school, Franz's interest in abnormal psychology was aroused by attending a course of lectures given by Edward Cowles, superintendent of McLean Hospital for the Insane at Waverly, Massachusetts. Joining the staff of the McLean Hospital in 1904, Franz established a laboratory for physiological psychology and began to apply the results of his research to brain-damaged patients. Later he wrote: "Everything tended to show that there are not the definite and exact functions for parts of the cerebrum which are posited, but … rather a possibility of substitution" (Franz, 1907, p. 101). Among other activities, he initiated a training program for aphasics, to overcome their speech handicaps, with a success that encouraged the rehabilitation of patients with other cortical lesions.

Leaving McLean in 1907 for what appeared to be a greater opportunity, Franz became psychologist and later scientific director at the Government Hospital for the Insane (St. Elizabeths) in Washington, D.C. A new psychology laboratory was equipped with an extraordinary collection of apparatus acquired and loaned by the Federal Bureau of Education. In his annual report for 1907, the hospital's superintendent, William A. White, pointed out that the laboratory was a step toward creation of "a strictly scientific department for the study of mental diseases, necessary because the development of the psychoses has a 'psychogenic element' and must be studied in the clinic." He predicted that the psychological laboratory would become as necessary to a mental hospital as the pathology laboratory.

Franz's early laboratory was a forerunner of later productive research units at St. Elizabeths, among them the Blackburn Laboratory, home to an outstanding collection of brains accumulated from autopsies of the large population of documented cases of dementia. An example of the significance of pathological research in a mental hospital may be mentioned. In 210 cases selected from the large population of documented autopsies and clinical reports at St. Elizabeths, Meta A. Neumann (1896–1996) and Robert Cohn showed in 1953 the silver-staining placques and neurofibrillary characteristics of Alzheimer's disease to be more widely distributed than anticipated, and thus helped distinguish the disease from senile dementia accompanied by vascular changes.

Collaboration with Lashley

A decade after moving to St. Elizabeths, Franz began a collaboration with Karl S. Lashley, who had been engaged in studies of instinctive behavior with H.S. Jennings and John B. Watson at Johns Hopkins. Their findings of the effect of cortical ablations on "habit-formation" in the albino rat again resembled those of Franz in the cat and monkey: Lesions of the frontal lobes did not prevent or abolish learned behavior unless they were extensive, long-established learned behavior persisted in any case, and behavior abolished by extensive cortical lesions could subsequently be relearned. From Franz Lashley acquired the surgical and histologic skills for his career-long effort to comprehend the neural basis for learning. After the First World War, Lashley took over this problem and devoted the balance of his career, as he put it, *In Search of the Engram* (1950). Throughout this long hegira, Lashley continued to utilize the interrelated techniques of learned behavior and cortical ablation, developed by Franz in his postdoctoral research at Harvard at the turn of the century.

By contrast, during and following the First World War, as noted, Franz had become increasingly involved in clinical psychological studies and their application to therapy of patients with war-related brain injuries or mental illness. His publication in *Psychological Monographs* (1915), on "Symptomatological Differences Associated with Similar Cerebral Lesions in the Insane," heralded his interests in this direction. His concomitant concern for improved methods of therapy was elaborated in the postwar monograph, *Nervous and Mental Re-Education* (1923). Franz's term "re-education" of more than half a century ago is today called "rehabilitation" but many of the same problems remain. In a number of instances Franz anticipated present insights. For example, he pointed to the value of singing as a subsidiary method in speech therapy.

"[S]ome motor aphasics can sing connected words at times when they are unable to carry on a conversation .... Some sensory aphasics also are able to associate words more readily with music than when the words are used in conversational speech" (Franz, 1923, p. 200). Current studies of cerebral dominance have referred musical capabilities to the recessive hemisphere, supporting Franz's use of singing in the recovery of speech in aphasic patients.

At the beginning of the '20s, there was a "volcanic explosion" of interest in psychoanalysis to which William A. White was a major contributor and the prospects at St. Elizabeths no longer appeared attractive. In 1924 Franz became the pioneer neurobehavioral scientist at the just developing southern branch of the University of California at Los Angeles by accepting an appointment as the founding professor of psychology as well as chief of the Psychological and Educational Clinic at Children's Hospital, Hollywood. Two years later, he was selected to deliver the second, prestigious Faculty Research Lecture at UCLA, in which he presented an elegant survey and synthesis of his investigative contributions to "How the Brain Works." In his remaining years Franz was engaged, with a number of advanced students, in a series of *Studies in Cerebral Function. I-IX*, published in 1933, the year of his death. He was succeeded as chairman of psychology at UCLA, in 1936, by Knight Dunlap, from Johns Hopkins (see p. 248). One of Dunlap's major responsibilities was to plan and design the first psychology building on the campus in Westwood, described at the time as probably superior to all in the country. Most appropriately, it was named Franz Hall.

Karl Spencer Lashley (1890–1958)

The association of Karl Lashley (Fig. 10.2) with Franz brought together two men with very different backgrounds but similar aims – to make experimental psychology more rigorous. A chance enrollment in a course in zoology as a graduate student at the University of West Virginia was Lashley's introduction to his life's work – the professor was the outstanding neurobiologist, John B. Johnston. When Johnston departed at the end of the spring term, his replacement appointed Lashley as departmental assistant. Receiving his master's degree in 1911, Lashley spent the summer at the Cold Spring Harbor Laboratory, where his work attracted the attention of H.S. Jennings who offered him a fellowship at Johns Hopkins and the opportunity for his Ph.D. in genetics, awarded in 1914. His association with James Watson in field work with terns at the Dry Tortugas and with Franz at St. Elizabeths in learning surgical and histologic techniques to test the behavioral effects of brain lesions (described earlier) solidly launched Lashley's research career. Lashley's innate care and precision in experimental procedures were compatible with Franz's laboratory philosophy and carried even higher the aim of experimental psychologists for reliability in measurements of behavior.

Fig. 10.2. Experimental psychology to Karl. S. Lashley meant precision and careful detail in his long series of studies on vision and on learning and memory.

Years at Minnesota

Lashley was recruited to psychology at the University of Minnesota by Robert Yerkes, who had agreed to chair and develop the department with new blood but then had remained in Washington in government activity after the First World War. During his 9 years at Minnesota, Lashley became the major figure in the technique of decorticating rats and then testing their capacity to solve problems in a variety of activities as he sought to determine the neural basis of learning and sensory discrimination, first in rats and later with primates. He departed Minnesota in 1926 for a research associateship with the Institute of Juvenile Research in Chicago and then the University of Chicago as professor of psychology until 1935, when he was appointed at Harvard. He was named director of the Yerkes Laboratory of Primate Biology in 1942 and retired from the laboratory and Harvard 13 years later. Through a pattern of proposing hypotheses which he carefully tested and often disproved, Lashley propounded two "laws." Borrowing from chemistry and embryology, the Law of Mass Action stated that the more cortical tissue available the more rapid and accurate the learning. And learning depended on the amount of cortical tissue available but not on what particular tissue – the Law of Equipotentiality.

The history of behavioral sciences has multiple references to Karl Lashley, not only because of the large number of his students and colleagues but also due to his prodigious and consistent canon of work as manifested by his bibliography (Beach, 1961). In 1920, Lashley published the first report of a series he called "Studies of Cerebral Function in Learning" and the 13th paper in the series in 1944; an even denser group commenced in 1930 on "The Mechanism of Vision" and ended 18 years later with the 18th of that series. Before Lashley's retirement four former students and colleagues planned an appropriate tribute which turned out to be a memorial volume of his selected papers (Beach, Hebb, C.T. Morgan, & Nissen, 1960). A generation later, a younger group of his admirers (Orbach, 1982) honored the publication of Lashley's *Brain Mechanisms and Intelligence* (1929) to signal their indebtedness to the insights of this "independent, self-determining and fairly individualistic creative scholar, shunning administrative opportunities and demands and giving little heed to recognition and personal preferment" (Yerkes, quoted in Orbach, 1982, p. 93). With greater perspective, we can say that the intellectual challenges to the advancement of psychobiology in the first half of the 20th century were probed more deeply by Karl Lashley than by other American psychologists and with more lasting influence.

PRIMATOLOGY AT YALE UNIVERSITY (1930–1950)

During the decade 1895 to 1905, the Universities of Harvard, Clark, Chicago, and Columbia constituted the major American centers contributing to progress in comparative psychology (Yerkes, 1943a). A quarter-century later, Yale University joined that prestigious group following events that stemmed from the overlapping of the careers of three men: James R. Angell, John Fulton, and Robert Yerkes. Through their synergistic influence, research in psychobiology moved as close to the human animal as was then possible. Angell was an experimental psychologist before becoming president of Yale and his recruitment in 1930 of John Fulton to the department of physiology secured for his university one of the youngest departmental chairs as well as a promising young neurophysiologist. Robert Yerkes had been installed since 1924 as professor in the Institute of Psychology, and at Yale from the mid-20s, it was "no more rats, no more cats; nothing but scrapes with monkeys and apes."

Cortical Neurophysiology and Primate Behavioral Studies

Ancillary Events

In physiology at Yale during the 1930s, the primate interests of both J.G. Dusser de Barenne and J.F. Fulton reflected the British pattern developed during the preceding decade with Charles Sherrington at Oxford. Dusser (as noted in Chapter 4), from Utrecht, had extended his strychnine studies of cortical sensory areas from cats to monkeys at Sherrington's laboratory and during the '20s, Fulton had spent a total of 6 years in Sherrington's program, principally in studies of muscular contraction and spinal reflexes. Additionally, early in his career, Harvey Cushing, who spent his last years at Yale, had visited Sherrington's laboratory, then at the University of Liverpool, in the summer of 1901. Sherrington had acquired a chimpanzee, an orang-utan, and a gorilla, and spent the fortnight before leaving for his vacation in intensive study of the excitable motor cortex of those apes, with Cushing – already oriented toward neurosurgery – helping with the craniotomies and experiments. In a belated report[2] (Grünbaum & Sherrington, 1917), the authors thanked Cushing of Boston, A.W. Campbell of Sydney, and Alfred Fröhlich of Vienna for their assistance in many of the experiments. In the meantime, in 1904, Sherrington had delivered the inaugural Silliman Lectures in the life sciences at Yale, published only 2 years later as *The Integrative Action of the Nervous System,* one of the classic contributions to the development of contemporary neuroscience.

By the 1930s and '40s, a major development of cortical neurophysiological and behavioral studies utilizing monkeys and chimpanzees was in full swing in three separate programs at Yale University. In the Laboratory of Neurophysiology, Dusser de Barenne, Warren McCulloch, and Leslie Nims concentrated on the application of strychnine neuronograpny to delineation of the sensory and suppressor areas of the cortex. At the Yale Laboratories of Comparative Psychobiology, and of Primate Biology, Robert M. Yerkes and others in the Institute of Psychology were preoccupied with chimpanzee studies in the former carriage house of an estate willed to the University. In the school of medicine, John F. Fulton, professor and chairman of physiology, sought in every feasible way to interrelate the neural and behavioral aspects of research in this primate studies field.

An example of Fulton's endeavors was his support of preparation and publication of a seminal volume in the field – *Bibliographica Primatologica. Part I* (Ruch, 1941). In his introduction to this comprehensive compendium, prepared by a postdoctoral student, Theodore Cedric Ruch (1906–1983), Fulton pointed out that the "gifted" Linnaeus in 1735 had the courage to group mankind with the apes and monkeys, as anthropomorphia. In the 10th edition of *Systema naturae* (1758), Linnaeus used the Latin term, "primate," for the new order. Ruch used the "new and useful derivative," "primatology," in the title and to avoid a judgment about "higher" and "lower" forms, the phrase "Primates other than man." Fulton also declared that primates are much more than "research

[2] Because of an altercation with Victor Horsley, leading to an exchange of letters in *Lancet* in 1894, Sherrington published only brief abstracts of these experiments and delayed their full presentation until 1917, 2 years after Horsley's death.

animals: ... they are research material of special quality" (Ruch, 1941, p. xii). As he continued: "To compile and analyse the literature of primates ... seemed a logical step in strengthening the structure of primate biology ... ." Having collected the information *on paper*, Fulton proposed to collect it *in fact*, with the well known result – a world-class medical library.

The bibliography first listed historical citations to 1800, then proceeded through the anatomy and physiology of the body systems. The final section, Psychobiology, filled only 25 pages in the slightly more than 200 of the total, a measure of the newness of the concept of brain and behavior. For his "objective annotation in compact, codified form," Ruch examined more than 4000 items and determined for each the extent to which it dealt with primates, the type (lecture, abstract, review, etc.), frequency of plates and tables, and, most difficult of all, the genera. As Fulton took care to note, the index of about 2000 authors and less than 5000 entries were minuscule compared with similar compendia on fishes or birds, for example, but the literature on primates had taken a spurt in the 1920s, perhaps because of the sudden interest in them as experimental medical subjects.

The Laboratory of Physiology

In his analysis of the substrates of spasticity which typically accompanied paralysis following internal capsular lesions in man, Fulton gave prior attention to the evolution of cortical dominance in primates, first noted in print in his monograph with Allen D. Keller titled *The Sign of Babinski: a Study of the Evolution of Cortical Dominance* (1932). The topic was also of central interest in a presentation with Margaret A. Kennard (see Chapter 4) which included observations on the nature and mechanism of hemiplegic spasticity (Fulton & Kennard, 1934). They concluded with a ringing statement: "Painstaking anatomical studies of the connections of the premotor area are urgently needed" (p. 160).

Frontal Lobe Function

Evidencing the remarkable range of his interests, Fulton additionally instituted collaborative research exploring the role of the prefrontal cortical area, in front of the motor and premotor regions, first in monkeys and then in chimpanzees. His principal collaborator was Carlyle Ferdinand Jacobsen (1902–1974) who in 1926 was one of Karl Lashley's last doctoral graduates at the University of Minnesota, with a dissertation on "A Study of Cerebral Functioning in Learning: The Frontal Lobes" (1931). Lashley, after studies with Franz at St. Elizabeths Hospital had moved to the University of Minnesota in 1917 and a decade later left for Chicago. No more serendipitous conjunction of initially disparate programs of research could be conceived than the sequence – during some 30 years – joining Franz's initial frontal lobe studies at Harvard, his collaboration in this field with Lashley at St. Elizabeths, Lashley's induction of Carlyle

Jacobsen into this subject at the University of Minnesota and, culminatingly, Jacobsen joining the Fulton program of frontal lobe studies at Yale. Although appointed in Yerkes's laboratory, Jacobsen worked exclusively with Fulton and after 6 years of productive collaboration received "an attractive call to Cornell" which Yale did not meet, despite Fulton's strenuous efforts. Fulton's diary (June 14th–20th, 1937) recorded his feelings: "I am inclined to look upon Jacobsen's work as the most important that has taken place in recent years.... Yale is letting go one of the most distinguished younger men it has ever had on its faculty...." In his foreword to *The Functions of the Association Areas in Monkeys* (1936), Jacobsen explained that monkeys were used in this first extensive series of behavioral experiments on cerebral function in primates to define specific problems so that the more expensive chimpanzees might subsequently be used to better advantage. He attributed the experiments' inception largely to Lashley's recognition that theories of cortical function should be tested in an animal more closely resembling man than the rat, and their success to the facilities of the Laboratory of Physiology, but also to Fulton's participation in the physiological and surgical aspects. A fuller collaboration is difficult to imagine. Among the findings, bilateral extirpation of the frontal association area resulted in total and permanent loss of capacity for delayed responses. Almost paradoxically, solving the execution of visual discrimination tests and simple puzzle boxes was not significantly altered, suggesting that behavioral adaptations which utilized immediate memory depended on those association areas. Lashley (1929) perhaps had provided an answer, having demonstrated that memory may not be a unitary process, but was mediated through different neurological mechanisms. More generally, the operated animal was more easily distracted than the normal subject – the mnemonic influence of immediately past experience was lacking. A repetition of the experiments (Malmo, 1942) showed that loss of the frontal association areas resulted in memory impairment because the animals had become more susceptible to "retroactive inhibition."

Lobotomy for Mental Illness

The postoperative personality changes (a general calming effect) of two chimpanzees with prefrontal lobotomies, named Becky and Lucy and studied by Carlyle Jacobsen, John Fulton, and their associates, were presented by Fulton at the International Neurological Congress in London in 1935. Egas Moniz, professor of neurology at the University of Lisbon, Portugal, was in the audience and in the discussion asked whether or not this procedure could be applied to patients with mental disorders. A month later, Moniz with his neurosurgical colleague, Almeida Lima, had performed the first operation which introduced "psychosurgery," his own term, to the armamentarium of psychiatric therapy.

The chief figures in the United States to become associated with frontal lobotomy were Walter Freeman and James W. Watts, manifested by their publication *Psychosurgery in the Treatment of Mental Disorders and Intractable Pain* (1942). Freeman introduced orbital lobotomy, the technique utilized by Walter B. Cannon for decortication of cats in 1925. An accumulation of disturbing reports from the major U.S. programs in psychosurgery included those edited by Milton Greenblatt, R. Arnot, and Harry C. Solomon, *Studies in Lobotomy* (1950) and *Frontal Lobes and Schizophrenia* (1953); and Fred A. Mettler, *Selective Partial Ablation of the Frontal Cortex* (1949). Especially critical evaluation came from the University of Chicago psychologist, Ward C. Halstead, author of *Brain and Intelligence* (1947). The initial apparent success of the procedure had a short life as evidence of the long-term deleterious effects became apparent and it was abandoned as quickly as it was taken up. In the laboratory, meanwhile, interest in specific regions of the frontal lobes had been aroused and studies of the orbital surface of the frontal lobes (Livingston & Fulton, 1948) were undertaken at Yale and elsewhere, of the cingulate gyrus and personality (Ward, 1948) at the Illinois Neuropsychiatric Institute, Chicago, and were extended to the role of amygdala in violent and aggressive behavior. A quarter-century later, there was an especially significant recognition of the complex interdisciplinary nature of the basic context of aggression in a symposium held at the University of Texas Graduate School of Biological Sciences in Houston. The agenda took not only the expected biological approach but also was sensitive to anthropomorphic, social, and ethical concerns (Fields & Sweet, 1975). Further understanding was shown in the entry (Barratt, 1999) for aggression in the *Encyclopedia of Neuroscience*.

Robert Mearns Yerkes (1876–1956)

The third and most significant figure in Yale University's 4th-decade rise to eminent status as an active center of American research in neuroscience was Robert Yerkes (Fig. 10.3). He entered Harvard as a premedical student in 1897, but shifted to graduate study first in biology and then psychology – eliding the two through the balance of his career, as "psychobiology." Years later, Yerkes (1943a) wrote of the prevailing *Zeitgeist* at the turn of the century – from tales of Thorndike's chicks to the chilly relations between the presidents of Yale and Clark – which limited any interchange of information about the comparative work of Small, Kline, or Kinnaman to reading their published reports.

Early Investigations

During his graduate study at Harvard, Yerkes spent the summer of 1899 as Thorndike's assistant at the Marine Biological Laboratory at Woods Hole, where he pursued an experimental study of the associative processes of the common

Fig. 10.3. The eminence of Robert M. Yerkes in American comparative psychology rested securely on the ranging variety of species he studied – invertebrate to subhuman primate – and on his administrative abilities.

speckled turtle, to determine to what extent and with what rapidity turtles could learn. Timing of a direct route through a labyrinth to the nest was chosen as the method. Additionally, at Woods Hole and assisted by George H. Parker, Yerkes studied the behavior of a small jellyfish and its modifications following surgical ablation of peripheral receptors and portions of the central nervous system. His dissertation results, published as "A contribution to the physiology of the nervous system of the medusa .... (Yerkes, 1902) represented an early achievement in the developing field of invertebrate neurophysiology. After gaining his doctorate with Münsterberg in 1902, Yerkes continued his studies, with graduate student help, on a large array of species.

Those investigative accomplishments in comparative psychobiology which utilized the resources of the Marine Biological Laboratory impressed on Yerkes the importance of institutional factors in advancing research, a conception which greatly influenced his subsequent career. As he described it (1943b, p. 289), it was an idea "which became a daydream, then a plan of action, and finally a program of scientific inquiry, with its institutional embodiment." At that time, there was nowhere in the world any such research institution, with laboratories for "structural and varied sorts of physiological inquiry." Soon another element was taking shape, the belief that monkeys and apes, then almost unused as subjects in experimental research and seldom thought of as laboratory animals, almost certainly would prove peculiarly useful for the study of behavior, if they could be made conveniently available for research.

For the next 15 years, however, Yerkes was fully occupied with the comparative study of infraprimate animals, and progress was made toward establishing a "special research institute for comparative psychobiology," in the pattern of a summer field station. Between 1899 and 1912, he published work on habit-formation in the phylogenetic order of the earthworm, green crab, and crawfish, among the invertebrates; and in the green frog, speckled turtle, and dancing mouse in the vertebrates; in all of these the maze or labyrinth were utilized. In addition, in 1915, there came studies on the crow and the pig, using a multiple choice method which he had developed in 1911, as an improvement over that devised by G.V. Hamilton.

Laboratory resources at Harvard had increased gradually during that period. In 1899, Hugo Münsterburg made two rooms available for animal studies in the Harvard Psychological Laboratory. In 1906, a newly planned laboratory of five rooms and space in an unfinished attic were provided. Seven years later, this attic was developed into a 12-room laboratory of animal psychology and was extended by the formal acquisition of a summer field station in Franklin, New Hampshire, where "naturalistic studies" could be pursued.

Learning a Pattern

In his maze studies, Yerkes utilized a simple pattern for invertebrates which were placed in a compartment with two exiting passages, right and left, one of which led to escape or to food and the other yielded an electrical shock or was blocked by a glass plate. In the case of the earthworm, after some 30 trials, the animal learned to turn right to escape, taking an average of 66 seconds in the process. The cephalic five segments of the worm containing the cerebral ganglia or "brain," were then amputated with a sharp pair of scissors. The wound healed without incident, but learned behavior was severely impaired postoperatively. After 2 weeks, the worm regained some choice, but its performance was markedly slowed, averaging 250 seconds. In the 3rd to 5th weeks, improvement to an average of 130 seconds was attributed to new learning by the regenerated nervous system. Increasingly challenging mazes were presented as Yerkes tested his way to the mammalian series. His first mammalian study using the maze was a monograph, *The Dancing Mouse* (1907) in the preface of which (p. viii), Yerkes wrote, "It is my conviction that the scientific study of animal behavior and of animal mind[3] can be furthered more … by intensive special investigations than by extensive general books. Methods of research in this field are few and surprisingly crude, for the majority of investigators have been more deeply interested in getting results than in perfecting methods." True to his word, Yerkes tested the dancers in four increasingly complex

[3] Note the title of Margaret Floy Washburn's survey of comparative psychology, *The Animal Mind*, first published in 1908.

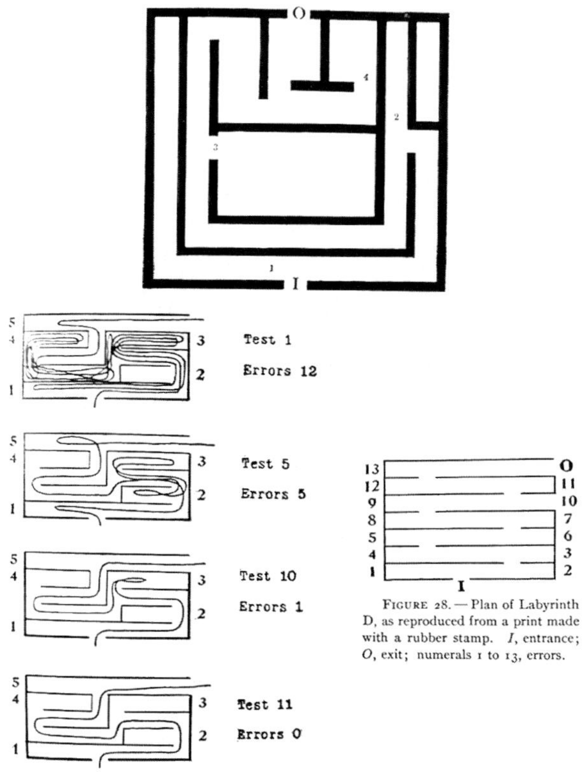

Fig. 10.4. Diagrams of three of the four labyrinths used by Robert M. Yerkes in testing the ability of the dancing mouse to learn. Top: modified Hampton Court maze with 4 errors. I, entrance; O, exit; 1–4, blind alleys. Left: a record sheet of labyrinth C with 5 blind alleys showing paths of successive trials. Right: "regular" labyrinth D with 13 blind alleys to test learning of repetitive, "machine-like" choices. From Yerkes, 1907, p. 211, Figure 25; p. 219, Figure 27; p. 222, Figure 28.

labyrinths, three of which are shown in Fig. 10.4. The upper labyrinth represented Yerkes's first, much-simplified, adaptation of the Hampton Court maze, which by then had become a standard fixture at Clark and The University of Chicago.

Multiple Choice

At a symposium on the "Study of Human Behavior," at Cold Spring Harbor, Long Island in 1913, Yerkes described briefly,

> two methods of analyzing behavior that have recently been devised. These methods, unlike those in general use by students of animal behavior, are applicable alike to man and to other mammals, even to birds as well.... The two... are the quadruple choice method of Hamilton and the multiple choice method of Yerkes (1914a, pp. 630–631).

Fig. 10.5. The career of Henry H. Goddard furthered reliable testing of intelligence among American recruits and among children in the child care movement of the early 20th century. Photograph ca. 1940 from Zenderland, 1998, p. 336, with kind permission.

The latter method consisted in presenting a bank of keys (like those of a piano keyboard) to the subject which is given to understand, verbally or through actual experience, that pressing one of the keys will yield a desired result, such as food. As noted, Yerkes continued his phylogenetic studies of the behaviors of crow and pig, with Charles A. Coburn in 1914 to 1915 at the Franklin Field Station. The outcome of the pig series of tests with multiple choice led Yerkes and Coburn (1915, p. 225) to state:

> While hesitating to claim that we have demonstrated the presence of ideas, we are convinced that the pig closely approaches, if he does not actually attain, to simple ideational behavior .... Our results indicate for the pig an approach to free ideas which we had not anticipated.

With the entrance of the United States in the First World War, Yerkes was appointed chairman of the Emergency Psychologists Committee, National Research Council, and served 1917 to 1919 as a psychological officer in the Surgeon General's Office. In that capacity, he was responsible for intelligence testing, the psychological examination of recruits; in other words, to identify with simple and reliable tests those men who were incompetent. He was ably assisted by the appointment to his committee of Herbert Henry Goddard (1966–1957; Fig. 10.5), "the country's leading expert on feeblemindedness" (Zenderland, 1998, p. 282). To develop the tests, Yerkes needed space and funds and when these were not forthcoming from the Surgeon General, his plan was rescued by the invitation of Goddard to the training school for feeble-minded at Vineland, New Jersey. Quaker-educated at Haverford College and Clark University, Goddard was infused with G. Stanley Hall's belief in education as an intellectually independent exercise, freed from rote learning. The tests that Yerkes's committee devised and implemented "were the common sense judgements most prized by Progressive America ... that emphasized utility, efficiency, and pragmatism." (Zenderland, 1998, p. 286). Goddard's career as a superintendent of the training school at Vineland was influential in the child-study movement and became manifest in publication of *A Psychology of the Normal and Subnormal* (1919). The wartime effort also helped resolve the turf dispute with psychiatrists over the educational and clinical superiority of psychologists in measuring minds, leaving mental disease to the psychiatrists.

Chimpanzees at New Haven and Havana

Yerkes stayed in Washington as chairman of the Research Information Service, NRC, but yearned to return to research, if possible to the study of anthropoids. The break came in the summer of 1923 when Yerkes purchased two chimpanzees, "courageously" spending two-thirds of his savings. The male chimpanzee, named Chim and the female Panzee, had a great time at the Franklin Field Station with the Yerkes's children, David and Roberta, and on return to Washington in the Fall, they were given the third-story back bedroom and sleeping porch of the Yerkes's home overlooking Rock Creek Park. In the Spring of 1924, Yerkes accepted a professorship in the new Institute of Psychology of Yale University where he could devote himself to research in comparative psychobiology and the promotion of facilities for the scientific utilization of anthropoid subjects. To prepare for the responsibility, he went that summer of 1924 to Havana, Cuba, where he observed a large colony of primates, thanks to the generosity of Senora Rosalia Abreu and the Carnegie Institution of Washington.

In a later account of the development of his plans for an anthropoid station, Yerkes (1943b, pp. 292–293) wrote:

> Early in the century, Mrs. Rosalia Abreu ... began to keep primates as pets. Wealth enabled her to indulge this expensive hobby freely, and in a few years she had built up a world-famous collection of New and Old World monkeys and of anthropoid apes. Of the latter, only the gorilla was lacking. She was the first to breed chimpanzees successfully in this hemisphere. Mrs. Abreu was an intelligent, strong-willed, fearless person, who held firmly to her ideas, but she lacked scientific training or special knowledge of biology, and although she earnestly desired that her animals might be used to promote human welfare, her collection remained to the end an assemblage of pets in an extraordinarily beautiful tropical environment. I am deeply grateful for her generous and friendly

help.... Her practical experience and wisdom proved of the greatest value to me.[4]

On returning from Cuba, Yerkes took up residence in New Haven, and commenced to bask in the favorable light cast on his work by Yale's president, who was now James Rowland Angell, with whom he had discussed at length (while they were associated in Washington) how psychology could best be advanced in America. Their plan was that by establishing in each of a few geographically distributed universities a strong department, with adequate provisions for training students, those special centers would become magnets for advanced students who wished to make psychological science a lifework. At Yale, Angell was able to secure funds for an Institute of Psychology with Yerkes as professor to work with advanced students at his own research. With a few 100 dollars, a university-owned brick barn was converted into suitable quarters for the animals with laboratory and office space, which served as "The Primate Laboratory" for 5 years. The chimpanzee was chosen as the most favorable animal with which to undertake a psychobiological study, and a pair was purchased in 1925, the male named Bill for William Jennings Bryan and the female called Dwina for obvious reason. A second pair was bought and named Pan and Wendy, and serious work could begin in New Haven at the end of the summer. Before the year was complete, however, Yerkes accepted the offer to study a young female gorilla (named Congo) in Jacksonville, Florida for about 2 months in each of three successive winters, 1926 to 1928. She proved to be a fascinating subject.[5]

An Anthropoid Breeding Station in Orange Park, Florida

The next 4 years demonstrated the feasibility of using chimpanzees in New Haven for experimental purposes, but no means of breeding them, the logical next step. At President Angell's request Yerkes prepared a brief description of his plan for an anthropoid breeding and experiment station, with an estimate of the cost of continuation and maintenance over 10 years. The Rockefeller Foundation granted Yale's request for an appropriation of $500,000 and the way was opened to select a location and erect buildings. A site about 15 miles from Jacksonville was found and in June, 1930 a "primate Eden in Florida"

[4] A different impression of Signora Abreu was recorded by the dancer, Isadora Duncan. Vacationing in Havana to restore her health, she and her entourage visited a member of "one of the oldest families who ... receiv[ed] her guests with a monkey on her shoulder and holding a gorilla [!] by the hand" (Duncan, 1927, p. 330). Commenting on Abreu's plan to bequeath her animals to the Pasteur Institute for experimental use, Duncan wrote: it "seemed to me a peculiar form of showing post-mortem love" (Ibid.).

[5] Yerkes repeated from a paper by Rothmann some single-word descriptions: "I have found the chimpanzee to be sanguine, the orang-utan melancholy, and the gorilla reserved" (Yerkes, 1927, p. 179).

was completed. The four New Haven chimpanzees became the nucleus of the Orange Park colony and were soon joined by Henry Nissen with 16 wild chimpanzees presented to Yale by the Pasteur Institute in Paris. An additional 15 were donated by Senora Abreu's children after her death in Havana; the most valuable were the adults, because the chimpanzee requires 8 to 10 years to mature sexually so that the gift saved at least 5 years in the development of a breeding colony. Before the end of the first year at Orange Park, some 35 chimpanzees as experimental animals had been assembled; best of all was one home-bred infant, for Dwina had borne a daughter, appropriately named Alpha. The research diligently carried out at Orange Park included all possible characteristics and problems of behavior, yet to Yerkes the work seemed only the beginning. To ensure a broader context, Yerkes had the primary role in establishing *The Journal of Animal Behavior* in 1948, an event of major importance in the evolution of a subdiscipline of neuroscience.

On his retirement in 1941, the Orange Park Station was renamed the Yerkes Laboratories of Primate Biology and was operated jointly by Yale and Harvard Universities. Karl S. Lashley, research professor of neuropsychology at Harvard, succeeded as director and Henry W. Nissen, associate professor of psychobiology at Yale, continued as assistant director. The business management became the responsibility of each university, and control of research and educational activities were delegated to Lashley and a board of scientific directors. The colony continues today, but in Atlanta as the Yerkes Regional Primate Research Center, under the aegis of Emory University, the only one of the six regional and one national NIH primate centers to be named for an individual.

THE PRIMATE CENTER DREAM

An Initial Proposal

Toward the end of his presentation in 1913 at Cold Spring Harbor, Yerkes elaborated the advantages of using lower animals in comparative studies in psychology, because of their availability and controllability, so that preliminary exploratory and problem-defining observations on their behavior could be made. When the experimental subject is sufficiently oriented and reasonably skilled in technique, then the problems of human behavior may "with better effect" be attacked. In conclusion, Yerkes announced to the broad spectrum of assembled scientists his plan for implementing that goal. To apply knowledge of animal behavior to that of man, it seemed to him, a "systematic, thoroughgoing study" of the behavior of the anthropoid apes and higher monkeys was called for. He proposed a permanent, tropical station where apes could be bred, reared, and observed for long-term experiments and behavioral development in their semi-wild state.

At least two incentives may be identified for this dramatic proposal. The first, which Yerkes later acknowledged, was the paper of Max Rothmann on "The Establishment of a Station for Research on the Psychology and Brain Physiology of Apes" in 1912.[6] The second was that Yerkes's plans to visit the Prussian Academy's research station at Tenerife in the Canary Islands had to be abandoned because of the outbreak of war and as the opportunity for future cooperative studies of anthropoids with the Germans had been lost, an American initiative was in order. Yerkes's broad vision was evident:

> This plea for special and unique facilities for the systematic study of the apes is presented to you because upon students of genetics, eugenic investigators, and sociologists, quite as heavily as upon behaviorists and psychologists, must rest the responsibility of carrying out any such proposal. Moreover, I can urge the plan upon your consideration with enthusiasm because I fully believe that this apparently round-about way to knowledge of the laws of our own behavior is in reality the most direct and desirable way. Certain it is that if we neglect our present opportunities to study the anthropoids, our children's children will condemn us for neglecting invaluable opportunities. Today, the chimpanzee, the orangutan, the gibbon, as well as many species of monkey, are at hand for observation. A generation or two hence, many of the primates may be extinct (Yerkes, 1914a, pp. 632–633).

First American Primate Laboratory

It is also likely that Yerkes had been influenced during earlier years by his former student, Gilbert van Tassel Hamilton (1877–1943), who in 1909 had joined the psychiatric staff caring for Stanley McCormick on a large estate, Riven Rock, at Montecito, California – a wealthy resort community, described by Hamilton as "the Newport of the West." With support from the family, Hamilton had constructed a laboratory on the estate with facilities for housing a monkey colony. He wrote Yerkes (25 December 1913; see Footnote 8) "Mrs. [Katherine Dexter] McCormick is here, and much interested in the laboratory. She is a graduate of Tech [MIT], and knows her biology pretty well." By that time, the colony contained 16 monkeys, 3 baboons, and an orang-utan, but Hamilton's only publications on his studies of those animals, both in Yerkes's *Journal of Animal Behavior*, were on the multiple choice experiments (1911) and the animals' "sexual tendencies" (1914). In correspondence with Yerkes (10 letters during 1909 to 1913), Hamilton invariably discussed some aspect of his interest in these primates. There was considerable background, therefore, for the outcome of Yerkes's interrupted plan to spend 1915 in Tenerife and he gladly

[6] We are most grateful for a copy of a translation in English, apparently by Yerkes, of the original German publication kindly provided by the archivist, Mrs. Nellie O. Johns, at the Yerkes Primate Center, Atlanta.

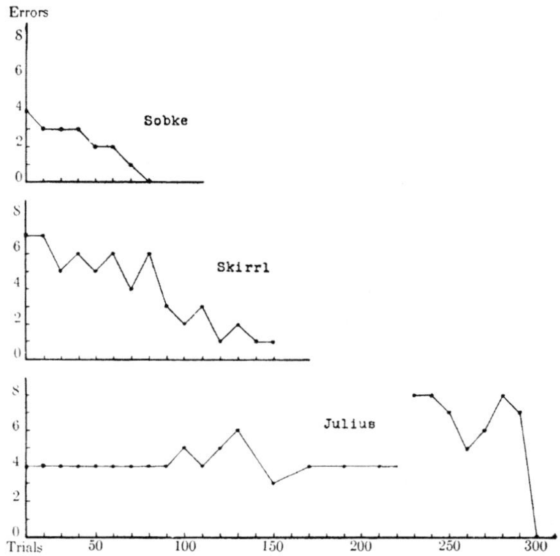

Fig. 10.6. The first learning curve of an anthropoid ape published in America was of "Julius", studied by R.M. Yerkes at the primate colony promoted by Catherine McCormick in Santa Barbara, California. From Yerkes, 1916a, p. 27, Figure 18.

accepted Hamilton's invitation to spend 6 months in Montecito where he luxuriated in the facilities offered.

At this first primate research laboratory constructed in the United States, and with a young orang-utan named "Julius," Yerkes initiated the long series of anthropoid studies which occupied the balance of his career. His multiple choice apparatus, first tried with human subjects at the Psychopathic Hospital, Boston, and then adapted for animals at the Fairfield Station, was installed and with its use, Yerkes wrote, "I have the satisfaction of being able to present … the first systematic experimental study of any feature of the behavior of an anthropoid ape" (Yerkes, 1916a, p. 128). And of recording the first curve of learning of an anthropoid ape (Fig. 10.6), which Yerkes interpreted as indicative of a high order of ideation compared with the curves for various learning processes exhibited by other mammals. "But," he concluded, "contrasted with that of man, the ideational life of the orang utan seems poverty stricken. Certainly in this respect Julius was not above the level of the normal three-year-old child" (Ibid., p. 132).

Further Detailed Arguments

In the last chapter of his monograph on studies at Montecito, *The Mental Life of Monkeys and Apes: A Study of Ideational Behavior* (1916a), Yerkes extended his advocacy of "Provision for the Study of the Primates, and Especially the Monkeys and Anthropoid Apes."

> There should be provided, he reiterated (p. 137), in a suitable locality a station or research institute which should offer adequate facilities (1) for the maintenance of various types of primate …; (2) for the[ir] successful breeding and rearing …; (3) for systematic … observation under reasonably natural conditions; (4) for experimental investigations from every significant biological point of view; and (5) for profitable cooperation with existing biological institutes or departments of research throughout the world.

As to location, Yerkes favored Southern California over Florida, and on American soil rather than abroad and envisioned six "highly trained and experienced" biologists with necessary support staff and an endowment of about $1 million. Yerkes's shorter appeal published in *Science* concluded with a stirring sentence: "For my part, I am so entirely convinced of the scientific importance and human value of this kind of research that I am willing to devote my life wholly to it" (1916b, p. 234). And so he did, achieving one of the most significant contributions to the advancement of neuroscience in the century: a solid foundation in the scientific knowledge and husbandry[7] of mankind's closest living species.

Dashed Hopes for a Center at Montecito

Yerkes's preference for locating the primate center he envisioned in Southern California was not due simply to nostalgia for his 6-months visit to Montecito the year before. From a dozen letters recently uncovered in the remarkable Yerkes archives at Yale University, it appears that he and Hamilton had developed plans for expanding the existing Montecito laboratory and bringing Yerkes from Harvard, both as its key investigator and also to engage in behavioral studies of Stanley McCormick and possibly contribute to his therapy. Yerkes, it may be recalled, had since 1911 been a half-time psychologist at the Boston Psychopathic Hospital under E.E. Southard.

Any plan had first to be cleared with Adolf Meyer, professor of psychiatry at Johns Hopkins, who had overall supervision of Stanley McCormick's treatment, and through whom the plan would have to be presented to the conservators of the estate. No actual document has been found, but a variety of selected references to it, some guarded and others more explicit, may be presented seriatim.[8]

An Outspoken Thought

Three months after his return to Harvard, Yerkes wrote Cattell (10/18/1915),

> I had to terminate my work in California in the midst of most interesting findings, but I am living in the hope that the time is not far distant when I shall be

[7] What Yerkes called "housekeeping" – animal care and management.
[8] The correspondence is in the Robert Mearns Yerkes Collection of the Cushing-Whitney Library of Yale University, to which we are extremely indebted for permission to quote.

> able to continue it indefinitely under excellent conditions. Just now I am busily engaged in preparing a report on what I was able to accomplish during my few months at Montecito.

Six months later (4/10/1916), Yerkes received a letter from Adolf Meyer at Hopkins: "I am glad to see you are East again, but afraid that the untimely death of Dr. Favill [the chief psychiatrist at Riven Rock] may have interfered with some of the cherished plans."

And in Baltimore

By October of that year, Yerkes had become acutely restless at Harvard and wrote (10/16/1916) to John B. Watson:

> I am quite as eager to escape from it, and ... have no regret for having chosen to gamble on the anthropoid scheme instead of choosing a big academic opportunity.
>
> At any moment, you may have opportunity to help toward the starting of work with anthropoids. Hamilton is even now in Baltimore with a proposition [that] may mean everything to us during the next few years. It may or may not come directly to you through him, or through Meyer.

In the immediate reply (10/18/1916) Watson informed Yerkes,

> I have had one or two glimpses of Hamilton, but as yet no sustained conversation with him. He is coming to have a little talk with me on Monday and I believe Meyer is going to have the two of us to dinner. Neither Meyer nor Hamilton has hinted anything about the primate station, but I am ready to lend my helping hand if there is anything that I can do ....

Within a week the anticipated conversation had taken place and Watson wrote Yerkes (10/24/1916):

> Hamilton, last night coming home from a dinner at Meyer's, broached the subject which I know is very near his heart and yours too. I sincerely hope that this may be effected and anything I can do in the matter shall be done.

Four days after the dinner Hamilton (still in Baltimore) wrote Yerkes (10/27/1916),

> I have been hoping that Meyer would define his attitude toward the prospective appointment more clearly than he has done thus far, hence this delay in writing. He is very cautious, and I do not believe that he will commit himself until he can sound the Conservators. I do not want to push him, and I do not feel altogether satisfied with his sparring for time.

On receiving Hamilton's note, Yerkes apparently could not stand the uncertainty any longer and wrote directly to Meyer (10/30/1916):

> [Hamilton] has presented to you a plan in which I am concerned, that naturally interests me quite as much as it interests him, and which must, I fear, present

> curiosity many baffling features to those of you who are responsible for a difficult administrative and medical situation.

Reasserting his "friendly terms" and after an expression of deep respect for Meyer's judgment, Yerkes continued:

> Ordinarily, I should not for a moment consider such a proposition, but as it happens, I am at present not only willing but eager to have opportunity to consider it. The reasons, briefly, are these. First, I am extremely anxious for early opportunity to proceed with the anthropoid research.

As Yerkes explained, the Montecito opportunity would demonstrate his plan to the Carnegie Institution of Washington which might in a few years be interested in it. He continued:

> In the second place, my interests are in the medical sciences, and despite the fact that I am not trained as a physician, I am perfectly willing to undertake certain approaches to medical work. I have definitely chosen to direct all of my scientific and practical work toward medicine instead of toward education .... I do not for one moment pretend that I can take the place of a regularly trained physician, but I do believe that I could be of substantial aid to Doctor Hamilton, else I should not consider the proposition.

And finally:

> As to the personal aspect of the matter, I told Doctor Hamilton at once that I should wish to discuss it fully with you and with Mrs. McCormick. Indeed, without the approval of both of you, I should hesitate to undertake the work, and I certainly could not undertake it happily without the enthusiastic urgency of both of you.

Yerkes concluded his plaintive, self-justifying letter by repeating that he counted on Meyer to decide what was best for patient and family.

The Obstacle

In his prompt reply (11/2/1916), Meyer admitted that he had envisioned a physician with practical experience. The idea of the possibility of scientific advances impressed him, but he felt he needed evidence of the relation of investigation "to the problem at hand" and requested a statement from Yerkes. Then Meyer wrote about his beliefs in psychology's direction.

> I am delighted to see from your letter that you recognize the so-called medical direction of psychology as one fully as worthy as the educational one. I think myself that no matter how much educational psychology takes to heart Rousseau's valuation of the nature of the child, there will always be dogmatic and didactic purposes dominating the studies; whereas in what we call medicine the human problems assert themselves in such a way that even the doctor forgets his habit of wanting to help first even if he should not know what he is trying to help. The situations are so fascinating that they appeal most strongly to our scientific curiosity.

In convoluted phrasing, Meyer asked for "any suggestion" and added, "Please consider this as a strictly personal feeler. I am anxious to get my views clear before I bring the matter up to the Conservators again."

Yerkes responded to Meyer (11/6/1916) in evident relief that he could do so without Hamilton as a go-between. Without a definite statement as to what was expected of the new appointee, Yerkes admitted that his own anthropoid research would be only indirectly related to the practical psychiatric problem, but he was confident that when he knew the particulars of the case, he would be able to contribute "analytic studies of the behavior of the patient," bringing his research, training, and experience to bear upon "this particular human problem," and "to be able to add something valuable to the situation." In exchange, Yerkes summarized, he would have the opportunity to try to breed anthropoid apes in Southern California.

A procrastinating response from Meyer on 11/8/1916 brought the prospect of early development of a major primate station at Montecito to a grinding halt. With Machiavellian skill, Meyer confounded patient needs, research plans, and conservator decisions. "It is, however, quite plain that we shall first have to canvass the situation whether a man psychiatrically trained would be available," and wait until the problem has been straightened out.

In a next to final letter (11/29/1916) in this series, James Watson wrote Yerkes from Hopkins,

> I should certainly be tremendously interested in seeing a joint station for the study of primates, infrahuman primates, and the young human established. It is one of my dreams now to have an experimental nursery. I have literally cartloads of problems that could be worked out. It would be fine to be associated with you in this work. Maybe these dreams won't come true in our lifetime ... but here's hoping.[9]

A concluding letter from Hamilton to Yerkes (7/23/1917) put a "Montecito Primate Center" to its final rest: "I have definitely severed my connections with Riven Rock, have moved family and goods to Ohio, where we have been since 24 March."

Monkeys at Cayo Santiago, Puerto Rico

Some 40 years later, during 1960 to 1963, in a program that exceeded Yerkes's wildest dreams, the National Institutes of Health not only moved the renovated

[9]In 1940, Yerkes published a short note in *Science* which began, "Quarters have recently been completed at Yale Laboratories of Primate Biology, Orange Park, Florida, for an *experimental nursery* [italics added]. Infants separated from their mothers at birth are to be reared under controlled conditions and used as subjects of a special program of research, of which Dr. Henry W. Nissen, assistant director, will be in charge." Watson's dream of an "experimental nursery" did, therefore, come true in his lifetime, but with infant chimps.

Yerkes Anthropoid Station from Florida to Atlanta, Georgia, but also established other primate research centers in various parts of the United States. The prodromata of that undertaking, initiated in the later 1930s, overlapped Yerkes's activities at Orange Park and although he played no formal role, his influence on its development was significant. The institutions involved were Columbia University and the University of Puerto Rico, rather than Yale; and its participating faculty were C. Ray Carpenter, primatologist; Philip Smith and Earl Engles, endocrinologists; and George Bachman, director of the School of Tropical Medicine of the University of Puerto Rico, under the auspices of Columbia University.

Clarence Ray Carpenter (1906–1975)

In 1931, on completing his doctorate at Stanford in psychology, with a dissertation under Stone and Terman on the reproductive behavior of pigeons, C. Ray Carpenter (Fig. 10.7) obtained a National Research Council postdoctoral fellowship for the sociological studies of primates at Yale's Laboratories of Comparative Psychobiology, under Yerkes. He was promptly assigned to investigate the familial and sexual behavior of Howler monkeys in the tropical forest of Barro Colorado Island and its biological laboratory in the Panama Canal Zone. Carpenter's research was outstandingly productive and comprehensive, making the first complete analytical and geographical census of a total primate population, of extraordinary interest to the psychologist and sociologist alike (Price, 1968).

Moving to Bard College of Columbia University in 1935, Carpenter next directed the Behavior Research Division of an Asiatic Primate Expedition to Indo-China, undertaken jointly with members of Harvard's Peabody Museum in 1937. During 4 months in Siam, he studied the behavior and social relations of some 90 gibbons, with supplementary motion pictures and sound recordings, new research tools for that time. After this rather long period of intensive and successful field work, he went on for another 2 months of preliminary study of orang-utans in Sumatra before returning home in September 1937 with seven gibbons. Bachman stepped forward to provide quarters, feed, and care for the animals in San Juan and urged his colleagues to plan additions to the colony, of gibbons as well as other primates. In particular, he stressed macaque monkeys from India, the number of which used in research in this country had by then reached more than a 1000 per month. Growing apprehension of another world conflict with curtailment of imports from abroad pointed to the need for initiating breeding colonies within U.S. territory. Bachman had explored a number of small uninhabited islands off Puerto Rico[10] suitable for

[10]Yerkes had included "Porto (sic) Rico" among several other semitropical sites appropriate for establishing primate breeding colonies (1916b).

Fig. 10.7. The extraordinary success of C.R. Carpenter's field work and humane collecting of primates was a contributing factor to the explosion of interest in comparative animal studies in the first half of the 20th century. Photograph at age 63; drawings by Mort Künstler. From Price, 1968, p. 83.

colonization and selected Cayo Santiago, half a mile off the mainland, near San Juan. On 10/1/1937, back in the United States after collecting a large number of macaques, Carpenter wrote (see Footnote 8) to Yerkes that a primate building had been constructed in connection with the medical school and he "had the extraordinary good fortune of having 100% of my animals arrive safely." In a month, 11/8/1937, Carpenter wrote again, stressing the progress toward preparation "of a simple field laboratory" for a visit to the island from Yerkes, whose advice he was anxious to have. Two months later, the trip was made and Yerkes immediately described (1/11/1938) to Carpenter his very good impression and within a few months had recommended granting of support by the John and Mary Markle Foundation of New York for Carpenter to make a collecting trip to Indo-China and Sumatra. The major problem, during the 2 years Carpenter was primarily responsible for the colony, was the fact that collecting and housekeeping demands left no time for research.

The contrast was extreme between the carefully planned, adequately funded, orderly developed, and excellently administered primate center at Orange Park, Florida and that – in every respect the opposite – at Cayo Santiago, Puerto Rico. Nonetheless, when the monkeys had been established on Cayo Santiago, several people tried to make use of them. One of these was James Watt, later director of the NIH National Heart Institute, who became acquainted with the colony in January, 1940 and moved his Shigella research program from Bethesda to San Juan for 2 years to study an epidemic of diarrhea among the monkeys.

The U.S. Public Health Service Assists

Puerto Rico was isolated during the Second World War and getting enough food for the people, let alone the monkeys, became very difficult. The authorities decided to sell the monkeys and there was a period after the war when the NIH Heart Institute supplied the University of Puerto Rico with a grant of $5,000 a year to help support the remaining colony. But nobody did any research with them and, after a few years, support was withdrawn. In 1953, William F. Windle joined the National Institute of Neurological Disease and Blindness, as chief of the Laboratory of Neuroanatomical Sciences, and resumed his earlier studies on neonatal asphyxia, using monkeys kept at NIH. In 1955, he was invited to transfer his research to Caye Santiago and arrangements were made for NIH to purchase the animals with the understanding that the money would be spent in constructing a laboratory for Windle's program. The U.S. Public Health Service Quarantine Station at San Juan made additional space available and by 1960, there were at least 400 breeding female monkeys and a budget of nearly three-quarters of a million dollars a year, supporting a full-fledged Laboratory of Perinatal Physiology, of which Windle was "chief". In 1963, a new director of the institute was appointed and Windle resigned.

Perspectives

In a recent paper on "The Cayo Santiago Primate Colony: Its Relationship to Establishment of Regional Primate Centers in the U.S," Windle (1980) reviewed James Watt's early association with the Cayo Santiago monkey colony and expressed his belief that the colony's most significant outcome was its contribution to the birth of an idea. Doctor Watt, who had by then become director of the National Heart Institute, had visited the much larger primate colony at Sukhumi, U.S.S.R., where his observations led him to recognize that the size of the investment would influence the ability to get support in the U.S. Congress. Coupling that idea with the Puerto Rican experience and Bachman's dream of a "proper laboratory monkey," the institute got behind the movement to create national primate colonies; thus the Puerto Rican colony had been a catalyst and to Windle it seemed regrettable that it did not become one of the Regional Primate Centers, the development of which are described in Volume 2 of this publication.

An essay on the history of apes and monkeys as objects of enquiry by the English-South African anatomist, Sir Solly Zuckerman, documented eloquently the indebtedness of human knowledge to the anthropoids, with which "man has a special zoological affinity" (1970, p. 205). Both Yerkes and Fulton were singled out as having in parallel carried out or inspired experimental work on primates that moved the science of brain and behavior forward during the first half of the century. The essay concluded with a caution against excessive exploitation of primates in scientific discovery and disease alleviation to the point of their extinction. That warning is more relevant three decades later with the added deprivations by native and foreign hunters.

Chapter 11

"That Whale Among the Fishes – the Theory of Emotions"[1]

One of the areas in which neural and behavioral studies were closely associated almost from the turn of the century, was that of feelings and emotions. As Sherrington expressed this in 1906: "Of points where physiology and psychology touch, the place of one lies at 'emotions' " (p. 257). A little more than a decade after Charles Darwin's *Expression of Emotion in Man and Animals* (1872),[2] William James, professor of philosophy at Harvard, and Carl Georg Lange, professor of medicine in Copenhagen, in independent papers, presented the first conceptual explanation of this relationship in what was promptly called the James-Lange theory of emotions.

THE JAMES-LANGE THEORY

Limiting himself to the emotions with obvious bodily expression, James (1884/1967, p. 13) pointed out

> Our natural way of thinking about these standard emotions is that the mental perception of some fact excites the mental affection called the emotion, and this later state of mind gives rise to the bodily expression. My thesis on the contrary is that *the bodily changes follow directly the PERCEPTION of the exciting fact, and that our feeling of the changes as they occur IS the emotion* (James, 1967, p. 13).

[1] Title of the essay by M.F. Meyer, 1933.

[2] And a century later, Oppenheim (1978) pointed out that its publication marked an important milestone in comparative psychology's inevitable embrace of neuroscience. Behavioral psychologists began to consider the fruits of research in embryology, ontogeny, physiology, and anatomy as relevant to their own studies and to pursue lines of inquiry opened by Coghill, Windle, and the Herricks (see Chapter 2).

And he logically concluded, "Without the bodily states following on the perception, the latter would be purely cognitive in form, pale, colourless, destitute of emotional warmth". Indicating that the arguments had come from introspective observations and their relation to cortical centers in the brain, James found a scheme "perfectly capable of representing the process of the emotions" (James, 1967, p. 28).

Carl Georg Lange presented his view more succinctly:

> Whatever the causes may be that arouse affections [emotions], the effects on the nervous system are identical in one point; in the effect upon the vasomotor center, that group of nerve cells which regulates the innervation of the blood-vessels. The stimulation of these cells, which lie chiefly in the [bulbar region] between the brain and the spinal cord, is the root of the causes of the affections .... (Lange, 1967, p. 73).

Interpretations

The James-Lange theory was by no means out of the woods, however, and the succeeding attacks – on neurophysiologic grounds – ultimately established contemporary insights in this field. Curiously, the first of these attacks utilized an approach that James had himself proposed, but in support of his view that blocking the backflow of impulses from the viscera would eliminate all prospects of emotion. As Walter B. Cannon (1927, p. 109) later summarized,

> James attributed the chief part of the felt emotion to sensations from the viscera, Lange attributed it wholly to sensations from the circulatory system. Both affirmed that if these organic sensations are removed *imaginatively* from an emotional experience nothing is left. Sherrington and [the Harvard group] varied this procedure by removing the sensations *surgically*. In their animals all visceral disturbances through sympathetic channels – the channels for nervous discharge in great excitement – were abolished. The possibility of return impulses by these channels, and in Sherrington's animals by vagus channels as well, were likewise abolished .... The animals *acted*, however, insofar as nervous connections permitted, with no lessening of the intensity of emotional display. In other words, operations which, in terms of the theory, [should] largely or completely destroy emotional feeling, nevertheless leave the animals behaving as angrily, as joyfully, as fearfully as ever.

Neurophysiological Dissent

Of the three great physiologists of the early part of this century – Sherrington in England, Pavlov in Russia, and Cannon in this country – all were involved in the neurobehavioral sciences. Two were in gastric physiology (Pavlov and Cannon), and the latter was also in neuroendocrinology, as noted (Chapter 5). As part of his analysis of spinal reflexes, Sherrington (1909) reported the preservation of emotional behavior in dogs with chronic spinal transections at lower cervical levels, isolating the sympathetic innervation of the viscera. In some of these animals the parasympathetic vagus nerves were additionally severed. With guarded British reserve, Sherrington commented (1900, p. 402):

> These experimental observations yield no support to the [James-Lange] theories of the production of emotion. On the contrary, I cannot but think that they go some way toward negativing them. A vasomotor theory of the production of emotion seems at any rate rendered quite untenable.

The next physiologist to question the James-Lange theory of emotion was Walter B. Cannon, professor of physiology at Harvard Medical School where, under Cannon's predecessor, Henry P. Bowditch, James had earlier entered his own career. Cannon had graduated from Harvard summa cum laude and in his undergraduate years was so greatly influenced by William James that he seriously considered a career in philosophy. James, however, discouraged him, warning he would find it unrewarding. Specializing in physiology instead, Cannon's observations on the influence of emotional states on the digestive process (1909) led him to a more ranging investigation of bodily changes in different emotions. His findings on "The interrelations of emotions as suggested by recent physiological researches" (1914b) first took issue with the James-Lange theory of emotions, precipitating a recurring cycle of papers opposing the theory, each followed by a defensive rebuttal, extending over the next quarter century.

Successive supporters of the classical theory (and their arguments) included James Angell (1916); Knight Dunlap (1922); Ralph Perry (1926); Edwin Newman, Theodore Perkins, and Raymond Wheeler (1930); Harry Harlow and Ross Stagner (1932); and Karl Lashley (1938). Cannon took on all those contenders through 1931, when he moved to other studies. His place was taken by his protégé, Philip Bard, who joined the fray in 1928 and continued jousting until 1950. In his interrelations paper, Cannon (1914b) had discussed the similarity of visceral effects in different strong emotions and emphasized that indeed, the bodily conditions which had been assumed by some psychologists to distinguish emotions from one another must be sought elsewhere than in the viscera.

In Defense

William James had died in 1910 and could no longer defend himself, but James R. Angell, then professor of psychology at the University of Chicago, jumped to his defense. Angell wrote (1916) that although James and his theory had survived their initial ridicule and Sherrington's striking experiments of 1900, now Cannon's physiological experiments tended also to discredit the Jamesian views and warned that the theory could be forever disproved.

The question which Angell adroitly raised; i.e., whether or not the animals deprived of spinal and visceral inputs – in Sherrington's and Cannon's experiments – actually experienced emotions, was apparently answered, in

1921, by a remarkable case presented by Charles A. Dana, professor of neurology at Cornell Medical College, New York City. Under the title "The anatomic seat of the emotions: a discussion of the James-Lange theory," Dana presented clinical facts demonstrating that somatic skeletal muscles and the sympathetic nervous system made only minor contribution to the arousal of conscious emotional states. He described a woman with complete quadriplegia and loss of cutaneous and deep sensation from the neck down who showed emotions of grief, joy, displeasure, and affection and no change in her personality or character. The only skeletal muscles at her command were the cranial, the upper cervical, and the diaphragm. The parasympathetic system was able to function, but the sympathetic was absolutely eliminated from consciousness. Dana concluded: "It is difficult to understand, on the peripheral theory, why there should have been no change in her emotionality, with the skeletal system practically eliminated and the sympathetic entirely so" (Dana, 1921, p. 636).

Another line of research was opened by Sherrington and Robert S. Woodworth in 1904 with their observation of "pseudaffective reflexes" in decerebrate cats, which initiated the study of central neural mechanisms for emotional behavior in the upper brain stem. Following ablation of the cerebral hemispheres and thalamus, reactions were observed against the background of decerebrate rigidity including some mimetic movements simulating the expression of certain affective states: limb and head and facial movements, pupillary dilatation, rise in blood pressure, sometimes vocalization. In a later study of chronic decerebrate animals in Sherrington's laboratory, H.C. Bazett and Wilder Penfield (1922) found these pseudo affective reflexes more developed after a few days than immediately after operation.

Rebuttals

Subsequent studies in this direction shifted from Sherrington's laboratory at Liverpool to that of Walter B. Cannon at Boston. In 1910, when Alexander Forbes graduated from Harvard Medical School, at Cannon's suggestion he went directly to work under Sherrington. On his return from England, Forbes brought with him a Sherrington guillotine. In his monograph *Bodily Changes in Pain, Hunger, Fear and Rage* (1915), Cannon pointed out that a mechanism of emotional reactions is present below the pons which although released from cerebral control is itself subordinate to whatever influences maintain decerebrate rigidity. He postulated that the thalamic region[3] is normally

[3]In the 1920s, the anatomic distinctions between the thalamus, subthalamus, and hypothalamus were not widely recognized, particularly among physiologists. Cannon appears to use the term "thalamus" generically to mean, as we would say today, "diencephalon." In 1931, Cannon himself stated in a footnote: "the expression 'thalamus' or 'thalamic region' is used to include not only the sensory stations in the diencephalon, but also the motor centers, e.g., the hypothalamus."

under cortical "government" which can be set aside when necessary. The acute preparation allows the "subordinate" activities to appear and prevail over the rigidity that masks them. Cannon proposed creating such a model with the use of the Sherrington guillotine or with a dull stylet introduced through the orbit[4] which would thus eliminate sensation and the need for anesthesia. Of equal importance, prolonged exhibition of the "quasi-emotional" phenomena might produce insights that could be applied to the study of intact animals.

Cannon eventually offered his own theory in a substantive paper (1927) on "The James-Lange theory of emotion: a critical examination and an alternate theory," in which he elaborated, five research findings from physiology which rendered that theory untenable. They were: 1) total separation of the viscera from the central nervous system does not alter emotional behavior; 2) the same visceral changes occur in very different emotional states; 3) the viscera are relatively insensitive structures; 4) visceral responses are too slow to be the source of emotional feeling; 5) artificial induction of the visceral changes typical of strong emotions does not evoke emotion.

Alternatively, since emotional expressions result from the action of subcortical centers, and since thalamic processes are a source of affective experience, Cannon prepared a "theory of emotions based on thalamic processes" (Cannon, 1927, p. 119). In essence, he proposed that in their centripetal course to the cortex, impulses from the receptors excite thalamic neurons, which act in combinations of reaction patterns typical of the several affective states. These neurons do not require innervation from the cortex to be driven into action; on release they discharge precipitously and intensely.

> Within and near the thalamus, the neurones concerned in an emotional expression lie close to the relay in the sensory path from periphery to cortex .... When these neurones discharge in a particular combination, they ... innervate the muscles and viscera .... [T]he *peculiar quality of the emotion is added to simple sensation when the thalamic processes are aroused.* (Cannon, 1927, p. 120).

SIGNS OF INTEREST

During the 1920s, growing interest in the psychology of emotion was marked by a reprinting of William James's and Carl Lange's original papers establishing their theory, edited by Knight Dunlap (1922); by a conference called by the

[4]Cannon's death in 1945 prevented, by only a year, his learning of Walter Freeman's introduction in 1946 of transorbital frontal lobotomy in human patients (the so-called 'ice-pick operation') which, in a number of respects, resembled Cannon's earlier transorbital procedure for decorticating cats.

Laura Spelman Rockefeller Memorial, with a focus on the emotions of children (1925; see Lomax, 1977); and by the appointment in the division of anthropology and psychology of the National Academy of Sciences, of a Committee on the Experimental Study of Emotion, which held three smaller conferences in 1925–1927. At the last conference, it recommended that the division support the following projects (with names of the principal investigators) in this field: a critical survey of the literature of emotion (Margaret Washburn), a thorough investigation of the galvanic response (Carney Landis), a study of the relation of emotion to formation of conditioned reflexes (Karl Lashley), and continued work on oxygen consumption under emotional conditions (Knight Dunlap). Funds were not forthcoming and in 1928 the committee was disbanded.

An International Symposium at Wittenberg College

The most significant event in this era, however, was of independent origin. In October 1927, an auspicious, week-long "International Symposium on Feelings and Emotions" was held at Wittenberg College in Ohio, to celebrate the opening of a new psychology building which included several laboratories for psychological research. James McKeen Cattell, then president of the American Association for the Advancement of Science (AAAS), served as honorary chairman and some 30 outstanding figures from this country and abroad participated. Professor Martin L. Reymert (1883–1953) at Wittenberg, who arranged the program, conducted the meeting with the aid of three undergraduate psychology assistants, one of whom was Donald B. Lindsley. The proceedings, assembled and edited by Reymert, were published by the Clark University Press in 1928 and were reprinted in *Classics of Psychology* (1973).

One of the participants, Margaret Washburn, recalled in her autobiography:

> The Wittenberg Conference ... was a remarkable affair .... [A]ll probably remember how the wonderful energy and efficiency of Dr. Martin Reymert ... made the opening of a psychological and chemistry laboratory at a comparatively small Ohio college a truly international event. It is likely that other speakers beside myself arrived ... wondering whether any one but ourselves would be present to hear us .... In fact, the conference was worth far more than any meeting of the Psychological Association, since many of the leading psychologists of Europe and America sent or presented papers, and the audiences must have averaged five hundred psychologists (Washburn, 1932, p. 355).

In his opening address, Dr. Reymert pointed out:

> [The] history [of our science] clearly shows that the field of feelings and emotions has been puzzling ... from the very beginning of scientific pursuit. Scholars have always been aware of this intangible and intricate 'something' which we may call feeling, as a very essential and perhaps the most dominating aspect of mental life ....

> Many scholars have emphasized the rivalry between heart and head.... [This] struggle accompanies the development of psychological thought from the most primitive forms of philosophy and religion up to the psychological problems of our day....
>
> The fact that science has been so conspicuously slow in trying to dissipate the fog of the emotional states, may have its general explanation primarily ... in the unstable and fleeting characteristics of emotional experiences, and the difficulty of finding expressions for the highly complicated factors of emotional patterns (Reymert, 1928, p. 423).

He expressed a sense of a new trend and raised two goals for the conference: taking stock of psychology's present status and realization of a clearer perspective for its immediate future.

In retrospect, this symposium on feelings and emotions was far more significant in assessing the current status of this field and in pointing directions for the future than were the contemporary meetings on this subject, held at major universities, under the auspices of the Committee on the Experimental Study of Emotion of the prestigious National Research Council.

Cannon's Experiments

At the Wittenberg Symposium, Walter Cannon reiterated his findings that bilateral removal of the peripheral sympathetic ganglionic trunk, from the superior cervical to the pelvic ganglia, a procedure which completely isolates the central nervous system from sympathetic connections with the viscera, leaves the animal capable of displaying all the superficial signs of rage. In addition, he announced a study of the localization of emotional mechanisms in the anterior brain stem, initiated by his young collaborator, Philip Bard, by ablating various amounts of the brain stem in decorticate animals and observing their behavior. A quarter century later, Bard (1950) found that typical sham rage, accompanied by vigorous sympathetic discharge, was displayed after transections that left only the caudal portion of the diencephalon connected with the lower brain stem.[5]

The two approaches pointed to an integrated center for visceral and related discharges in the expression of emotion. In conclusion, Cannon revived the suggestion of Hughlings Jackson, which was supported by Henry Head, that the organization of the nervous system is a neural hierarchy with higher levels suppressing the more primitive reactions. When the "cortical government" is set aside, the subordinate activities are released from inhibition and only slight stimulation may produce extreme effects. In other words, for the James-Lange theory that emotional experiences arise from changes in effector organs is

[5] In a series of earlier studies, Karplus and Kreidl in Vienna had obtained sympathetic responses from electrical stimulation of the base of the hypothalamus.

substituted the idea that they are produced by influences from the region of the thalamus.

When Cannon published a second edition of *Bodily Changes in Pain, Hunger, Fear and Rage* in 1929, he included new evidence of neural "arrangements" in subcortical centers dealing with emotional expression which he elaborated in detail and proclaimed as a source of vivid affective experience within and near the thalamus. He acknowledged the coincidence of bodily changes and emotional experience as naturally misleading, but it was the thalamus that contributed "glow and color to otherwise simply cognitive states."

A year later, the next recurring defense of the James-Lange theory was presented by three psychologists at the University of Kansas – Edwin B. Newman, F. Theodore Perkins, and Raymond H. Wheeler – in an article, "Cannon's theory of emotion: a critique" (1930). After a long literary debate, they summarized eight obscure points of negation and in conclusion, stated, "The valuable work of Head, Cannon and others on the function of the thalamus is all the more, rather than the less, significant if we do not attempt to locate emotion in that particular part of the brain" (p. 326).

Again, Cannon responded (1931) commenting that the misapprehension of the points in the argument made it mandatory to use two diagrams with which the chief features of the two theories could be briefly summarized (Fig. 11.1), diagrams widely reproduced in textbooks to this day. Following this last rebuttal Cannon became increasingly preoccupied with concepts of the emergency function of the sympathetic nervous system and homeostasis, as well as with the field of neuroeffector transmission (see Chapter 5). He retired in 1942 and composed an account of his life and views in the monograph, *The Way of an Investigator,* (1945), adhering to his conviction that emotional behavior is not directly "managed" by the cortex but depends on neuronal patterns in the thalamic region.

Bard to the Barricades

As Cannon had dominated the experimental studies of the emotions during the first 30 years of this century, Philip Bard continued to contribute to this research field and program of his mentor, over the subsequent two decades. At the conclusion of an eloquent tribute to Bard, his outstanding student, Vernon Mountcastle, wrote, "He possessed great charity and respect for the opinions of others and avoided disputations" (Mountcastle, 1977, p. 298). This may have been true of Bard's later years and, certainly, he never displayed the contentious hyperbole of his mentor, Walter Cannon, who in his alimentary days, maintained a continuing feud with "Ajax" Carlson. Next, on moving to the secretion of adrenin, Cannon took on and vanquished two contenders, Stewart and Rogoff. Also, on the loose among the emotions, Cannon concentrated on rage and aggressive behavior, rather than tenderness and love and jousted

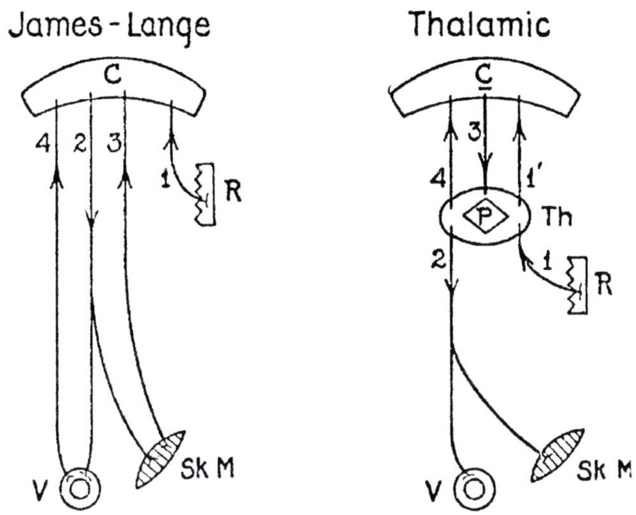

FIG. 1. Diagram of nerve connections in the James-Lange theory.

FIG. 2. Diagram of the connections in the thalamic theory.

R, receptor. *C*, cerebral cortex. *V*, viscus. *Sk M*, skeletal muscle. *Th*, thalamus. *P*, pattern. The connecting lines represent nerve paths, with the direction of impulses indicated in each instance. Cortico-thalamic path 3, Fig. 2, is inhibitory in function.

Fig. 11.1. Walter B. Cannon's attempt to explain the distinction between the James-Lange theory of emotions and his own "thalamic" theory. From Cannon, 1931, p. 282, Figures 1 and 2.

valiantly against William James and Carl Georg Lange, together with the succession of their protagonists.

Admittedly, Philip Bard was a relative light-weight compared with his pugilistic mentor but on moving to Hopkins his attention was called to a paper by Harry Harlow and Ross Stagner on "The theory of feelings" (1932). They wrote deprecatingly of Bard's and Cannon's "erroneous" references to the thalamus as the "seat of emotions" and misuse of the term "emotion," as a display of the psychological naïveté found in physiologists doing psychological experimentation. Bard was condescendingly given some easement, for Harlow and Stagner noted that Bard went only so far as to state that decorticate behavior "resembles" rage, a wise reservation. Although Bard may have "avoided disputations", he was clearly not a man to take such a brash affront without rebuttal. In 1934, he sent a long paper to *Psychological Review* in which he started out softly to correct some misapprehensions. After pointing out that the central mechanisms for sham rage lie within the caudal half of the hypothalamus and the most ventral and caudal parts of the thalamus, Bard found it "somewhat surprising" that Harlow and Stagner used his work to support their idea that the thalamus contains an "excitement center." In his view, they seemed to have overlooked those experiments in which less than the caudal half of the diencephalon remained in their claim that the thalamus is the center for the conscious and reflex expression of the feeling of excitement. On both anatomic and semantic grounds Bard pointed out the fuzzy thinking of Harlow and Stagner in claiming that his experiments actually supported their hypothesis. In Bard's argument the essence of Cannon's theory of emotion lay in a clean-cut differentiation of the neural processes which lead to emotional consciousness and those which result in emotional expression. Bard argued that his thoroughly studied decorticate cats in the chronic condition could "*display* fear, rage, and at least, in the case of the female, sexual excitement" (1934, p. 324).[6]

Lashley's Contribution

More than 20 years after James Angell's paper in 1916, with which this debate began, an article by Karl Lashley in 1938, on "The thalamus and emotions," became the last of these critical attacks by psychologists opposing the contributions of others to this field. Lashley's paper was unique, however, in that it was neither polemical nor devoted to an immaculate preservation of the James-Lange theory. In contrast, Lashley sought an understanding of the neural mechanisms underlying both the subjective experience and the motor expression of emotion, with particular reference to the role of the thalamus which, as was noted earlier, he also identified as "the general region of the diencephalon." Lashley began with Cannon's (1927) definition:

> Afferent impulses initiated by the stimuli capable of arousing emotion are transmitted to the cortex. Within the cortex they are integrated to arouse the appropriate overt behavior. A center in the thalamus is also excited, either directly as the sensory impulses are relayed there, or secondarily by impulses from the cortex. The thalamic center ... discharges somewhat explosively, exciting the effectors in patterns which constitute the 'expression' of the emotion and also discharging to the cortex, where the impulses from this thalamic center add the 'peculiar quality of emotion' to the simple sensation aroused by the direct effects of the exciting stimulus. In addition it is suggested that the flood of impulses from the thalamus constitutes the emotional tension and accounts for the dynamic or motivational character of the emotion. (Lashley, 1938, p. 43).

Lashley then went on to analyze these concepts critically and was led to conclude:

> A review of the evidence fails to reveal participation of the thalamus in [emotion, except that it] contains centers in which some, at least, of the patterns of expressive movement are integrated .... The affective changes resulting from

[6]It is curious, in light of Bard's objection to the term "excitement" in the above issue, to note his own use of it in connection with sexual behavior, which is described as "sexual excitement."

> thalamic lesion are restricted to a small group of somesthetic sensations and cannot be interpreted as a general change in affectivity .... [These] changes ... are primarily in the character of the sensations, in intensity, duration, localization and are therefore not relevant to the problem of affect .... Thus, the only part of the thalamic theory of emotion which has factual support is the localization of motor centers for emotional expression within the hypothalamus (Ibid., p. 60).

THE CIRCUIT OF PAPEZ

The great value of Lashley's penetrating analysis lay in its identification of two major areas of uncertainty in then-current concepts of the neural bases of emotions. The first and more inscrutable concerned the neural basis of the affective aspects of emotional experience, in turn a feature of the more general problem of the basis of all consciousness and subjective experience. The second uncertainty had to do with the forebrain mechanisms and processes normally holding emotional behavior in check, the elimination of which in "decortication" released diencephalic mechanisms and led to the excessive expression of emotional behavior – either as rage or sex. As Lashley (1938, p. 47) commented, "[A]lthough normal inhibition of the thalamic centers for expressive movement is indicated, there is some uncertainty as to the source of the inhibitions."

Research developments in this latter area, just then getting underway, have since uncovered the major functional significance of those basal forebrain structures then identified as the rhinencephalon. In 1937, the year before Lashley's review appeared, James W. Papez (Fig. 11.2) had published a remarkable paper on "A proposed mechanism of emotion" which introduced the concept of what is now called the "limbic system." A few months earlier, Heinrich Klüver and Paul Bucy had first described the abrupt changes in emotional behavior and other functions which followed temporal lobectomy that included the underlying amygdala and hippocampus of the basal forebrain. For an advantage of order rather than a precision of priority, Papez's contribution can be considered first.

James Weneslaus Papez (1883–1958)

As his student and associate, Fred Mettler, pointed out in memoriam, Papez's productivity

> is marked by two distinctly different types of production – meticulously detailed descriptive morphology and generalizations of a surprisingly unfettered and inspirational type .... He himself seems to have regarded his theoretic formulations – especially that on the neural structural substrate of emotional expression – as his more significant work; and this opinion will be confirmed by that

Fig. 11.2. Pearl and James Papez collaborated closely in drawings of anatomic structures, meticulously traced under the microscope. From MacLean, 1978, p. 10, Figure 2.

> wider, predominantly psychiatrically or psychologically-oriented audience for which these compositions filled an existing need (Mettler, 1958, pp. 280–281).

On including Papez among *The Founders of Neurology*, Webb Haymaker (1970b, p. 143) wrote "In his secluded laboratory at Cornell University, in Ithaca ... Papez used to take every hour he could spare from his teaching to look at serial sections of brains .... [T]o what he saw he would add a liberal dash of imagination, which was his forte, and the mechanisms involved in the performance of this or that function would emerge."

From that background in 1937 came "a mechanism of emotions" based on phylogeny evolved through mechanisms concerned in nutrition and reproduction which Papez believed to form the drives necessary for survival. The paper received little notice, and in 1939 Klüver and Bucy reported to the American Neurological Association their preliminary analysis of their findings on "The functions of the temporal lobes in monkeys." Papez's paper was not cited and in the discussion he did not mention it himself. It should be noted, however, that from his knowledge in this field, it was Papez who contributed the most insightful views that were presented at this discussion. "It is my feeling," he commented,

> that these lobectomies reveal ... one, a group of symptoms dependent upon the temporal and parietal association functions, and another a definite group of symptoms which depend upon the hippocampus .... (Papez, 1937, p. 173).

This remarkable suggestion emphasized the important role the hippocampus might play in the integration of psychic functions. Klüver, in concluding the discussion, stated that examination of the brains "revealed that the amygdaloid nucleus and almost the entire hippocampus were removed" together with the temporal lobes.

The other relevant publication that Papez neglected to mention in this discussion was his paper the year before on "Fiber and cellular degeneration following temporal lobectomy in the monkey" (Rundles & Papez, 1938). It explained that the two animals reported in their study had come from the Laboratory of Neurophysiology of Yale University, where they had been part of an extensive investigation of the auditory ability of primates. For this purpose, bilateral removal of the temporal lobes had been accomplished by Professor Dusser de Barenne. The animals remained in good health until killed 6 months and 2 years after operation, respectively; the auditory acuity studies were published, but whether or not any other observations were made on behavior is now lost in obscurity.

Prior Temporal Lobectomies

It seems remarkable to be able to note that a previous instance of the preparation of bilateral temporal lobectomies in monkeys had occurred in this country, at Yale, in 1934, 3 years before that of Klüver and Bucy at the University of Chicago in 1937. Dusser de Barenne died in 1940, before any behavioral studies he or others may have made of the Yale monkeys, beyond the report of their auditory acuity (1935). Rundles and Papez (1938) presented a comprehensive account of the neuroanatomic findings in the brains of these animals, however, and found that degenerative changes were limited to ablations of the temporal neocortex, for the hippocampus, amygdala, and septum were uninjured. Omitting auditory connections, which were substantially spared, commissural fibers of the temporal lobes, in addition to those of the caudal half of the corpus callosum, made up most of the anterior commissure. Neither the pulvinar nor other thalamic nuclei made direct connections with the temporal lobe, the interhemispheric connections of which appeared to be chiefly or totally with other areas of the cortex.

The Classic Paper

In "A proposed mechanism of emotion," Papez (1937) presented some anatomic, clinical, and experimental data dealing with the hypothalamus, the cinguli, the hippocampus, and their interconnections. Taken as a whole this ensemble of structures was proposed as representing theoretically the anatomic basis of the emotions.

> The term 'emotion' as commonly used implies ... a way of acting and a way of feeling. The former is designated as emotional expression; the latter, as emotional

> experience or subjective feeling. The experiments of Bard [1929] have demonstrated that emotional expression depends on the integrative action of the hypothalamus .... For subjective emotional experience, however, the participation of the cortex is essential ....
>
> In order to make clear at the outset the general anatomic picture which I wish to propose as the probable corticothalamic mechanisms of emotion, a diagram has been constructed, showing its main features (Papez, 1937, pp. 726; 727–728).

The balance of Papez's often-termed "circuit of the emotions" was then elaborated with reference to this figure – an elegant half-tone drawing of the medial aspect of the right hemisphere of the human brain, prepared by his wife, Pearl, in 1936 (Fig. 11.3). The basal portion of the temporal lobe was partially dissected and in addition, as the legend states, "In this specimen an unusually large exposed (nude) hippocampus is seen." Having established the anatomic substrate of his circuit, Papez then pointed out:

> The central emotive process of cortical origin may then be conceived as being built up in the hippocampal formation ... transferred to the mamillary body and thence through the anterior thalamic nuclei to the cortex of the gyrus cinguli [which] may be looked on as the receptive region for the experiencing of emotion as the result of impulses coming from the hypothalamic region .... Radiation

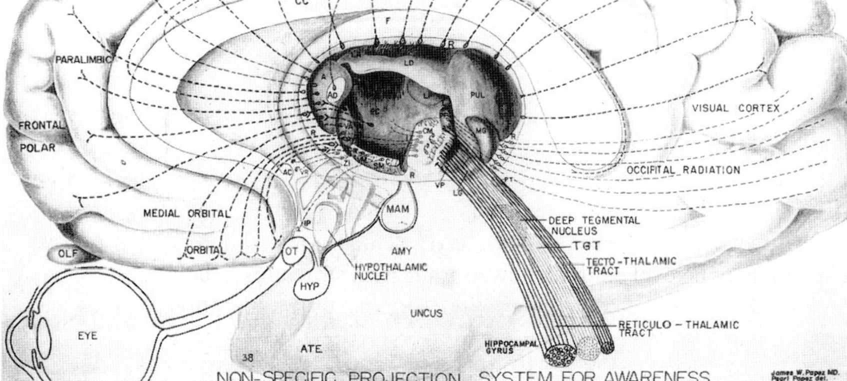

Fig. 11.3. A drawing by Pearl Papez depicting the central reticular path for activating the EEG and consciousness proposed by James Papez. From Papez, 1936, p. 118, Figure 1.

of the emotive process from the gyrus cinguli to other regions in the cerebral cortex would add emotional coloring to psychic processes occurring elsewhere. This circuit would explain how emotion may arise in two ways: as a result of psychic activity and as a consequence of hypothalamic activity (Papez, 1937, p. 728).

Among several features of the components of his circuit, Papez noted that in rabies, a disease characterized by profound emotional perturbance, the lesion tends to reside in the hippocampus. As for the gyrus cinguli, Papez called attention to its older synonym "the gyrus fornicatus" and added his observation that the precuneus shows a greater size difference between genders than any other area of the cortex, being larger in the male, suggesting that representation of the sex organs may be situated there. Later Papez stressed the high development of the mammillary bodies but he failed to note that in contrast to the precuneus, these bodies are typically more highly developed in the female than in the male.

Retirement from Academia

In 1951, at age 68, Papez retired from Cornell, and continued his studies of brain connections, with interpretations of their significance, at the Columbia State Hospital in Ohio. The same year, he contributed a short chapter to a monograph, *The Attitude Theory of Emotion* by Nina Bull, who had naturally consulted with the author of the now widely known theory of the mechanism of emotion. In that chapter he introduced another figure, this time of the medial surface of the left hemisphere of the human brain, drawn by Loyd (sic) S. Papez, his son. Except for being the opposite hemisphere, this is an almost precise replica of the drawing prepared by his wife for the original paper in 1937.

In this connection, another distinguished comparative neurologist, C. Judson Herrick, contributed the introduction to Mrs. Bull's 1951 monograph. "An emotion," Herrick wrote,

is a mechanically made product which may be manifested in two ways – by a change in the pattern of bodily posture and movement or by a change in the pattern of conscious experience. If only the former is demonstrable the reaction is sometimes called a 'sham' emotion. The neurological apparatus employed in these two components of emotion is demonstrably different. The 'sham' emotion may be evoked in the lower brain stem structures ... but the conscious feeling of any but the simplest emotions requires the participation of the higher cortical apparatus also.

The significance of every machine ... is determined by the nature of the product delivered .... Since the human bodily mechanism delivers two quite different kinds of products – observable changes in the physical body and also changes in the quality and content of conscious experience – it follows that

> we must learn as much as possible about these two classes of products before we can hope to explain the ... mechanism that makes them. Investigation of the mechanics of operations of the first class is the province of physiology. The properties of the products of the second class are investigated psychologically. Between these two well developed domains of natural science there is a middle ground of psychoneurology which must be more fully explored before we can hope for an adequate science of psychophysics.
>
> The vital process is an unbroken sequence .... But the stream of consciousness is not an even flow. The awareness emerges only when the vital process assumes a particular kind of pattern in the higher nervous centers. The details of the mechanics of this dynamic pattern are still obscure. That there is a motor factor expressed as some sort of bodily action has been experimentally demonstrated in the mental process of cognition, conation, and emotion. In the latter domain the difference of interpretation of the observed facts as reported by James, Lange, Cannon, Sherrington, and many others are reconciled in Nina Bull's attitude theory .... (Herrick, 1951, pp. xi–xiii).

The Final Assessment

In 1962, Nina Bull assembled a collation of her earlier studies in a book, *The Body and Its Mind* and included a facsimile of a letter from Dr. Papez in 1954 in reply to a request for his judgment on an early version of the Goal Orientation theory. The letter's second paragraph reads,

> "Though bare of neurology, your thesis is a true counterpart and issue of neurological functions; that I would be the first to praise it. You may realize that I have long been a functional neurologist. For only in this perspective have all the mechanisms I have described from long series of researches made any useful sense. Separately they are local gadgets. From neural mechanisms to mental mechanisms seems to be a long and abysmal leap. Yet, I find all the necessary means present. We have to do some rationalization of scientific data; but there is very little that confutes this kind of thinking. The thesis is sound" (Bull, 1962, p. 85).

The letter was signed "James W. and B. Pearl Papez."

In 1958, the year of his death, Papez published a synthesis of his studies in *The Visceral Brain, Its Component Parts and Their Connections*. Its initial paragraph provided a final summation or breviary of his views:

> The visceral brain or rhinencephalon consists of several closely connected mechanisms: olfactory, alerting, visceral, emotional, and autonomic. These parts include olfactory, habenular, visceral, amygdalar, hippocampal, hypothalamic, and pituitary structures. They are closely integrated to coordinate sensory events with bodily [reactions] and visceral needs. They are biologically grounded and are closely bound in action to regulate innate, automatized activities concerned with searching, feeding, sex, fight, flight and the emotion-provoking situations in the body and the environment (Papez, 1958, p. 591).

A recent volume titled *The Hypothalamus* (1974) and edited by Webb Haymaker, Evelyn Anderson, and Walle Nauta was dedicated to James W. Papez, Stephen W. Ranson, and John D. Green, "who contributed to our current knowledge of the hypothalamus." Papez's role has been dealt with here; the other two players, Ranson at his Institute of Neurology at Northwestern University Medical School and Green before and after coming to America, are described in Chapter 5 in relation to their participation in the emergence of neuroendocrinology in America.

Paul MacLean's "Triune Brain"

The next major development in this field came more than a decade after publication of Papez's theory of emotion and as an elaboration of it. A young M.D. from Yale in 1940, Paul D. MacLean, following military service in the Second World War, obtained an early U.S. Public Health Service fellowship providing him with 2 years of study and research with Stanley Cobb at the Massachusetts General Hospital, Harvard Medical School. Toward the end of his postdoctoral experience, MacLean published a paper, "Psychosomatic disease and the 'visceral brain': Recent developments bearing on the Papez theory of emotion" (1949), which enormously extended and enlivened its Papezian forebear. Returning to Yale in the same year, jointly in psychiatry and physiology, MacLean followed this with "Some psychiatric implications of psychological studies on frontotemporal portion of limbic system (visceral brain)" (1952). The initial figure of this second paper reproduced Broca's original sketch (1878) of the medial surface of the otter's brain, depicting "Le grand lobe limbique." In those two publications, MacLean revived the term and established the conceptual identity of the "limbic system" that is current today (Fig. 11.4). In essence, whereas Papez saw through binoculars, MacLean envisioned a wider scenario.

The Argument

MacLean postulated that the phylogenetically old brain (what was then the rhinencephalon) was largely concerned with visceral and emotional functions and had many connections with the hypothalamus for discharging its impressions. Those relationships indicated, he believed, dominance of affective behavior by a relatively crude and primitive limbic system. In his second paper, MacLean presented a synthesis of physiological studies which led him to infer that the limbic portion of the frontotemporal region seemed to concern activities of the animal pertaining to feeding, vocalization, attack, and defense. Pointing to the extensive representation of the "oral senses" in the frontotemporal region, he called attention to the representation of smell in the pyriform area; of taste in the region of the insula; and of the sensory component of the vagus in the rhinal fissure, as well as the amygdala. In support he

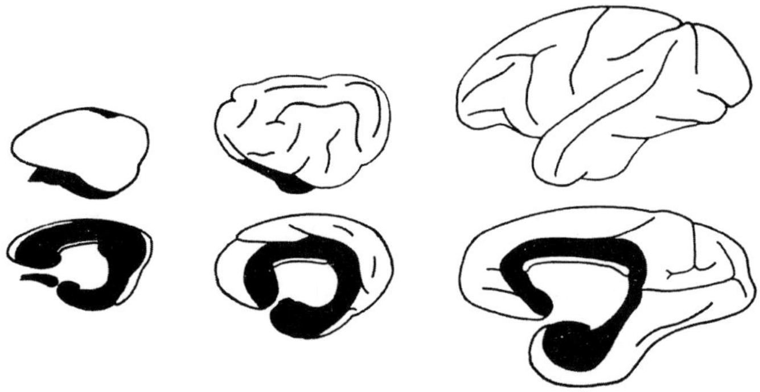

Fig. 11.4. The phylogeny of limbic evolution is illustrated by diagrams (in black) of the limbic systems in rabbit (left), cat (center), and monkey (right). Adapted from MacLean, 1954, p. 106, Figure 6.3.

referred to the epic findings of Klüver and Bucy (1939) that bilateral lobectomy, including the limbic structures, had the effect of transforming wild and ferocious monkeys into animals that are ostensibly tame and docile. Paradoxically, Bard and Mountcastle (1948) had reported that the presumably comparable ablations in the cat resulted in ferocity. In addition to docility, the monkey presented a type of oral behavior, reminiscent of the human infant, as these animals examined everything, noxious or benign, with their mouths; there was also bizarre oral-sexual behavior. The next logical step was to find out what happened when specific brain regions were stimulated, a manipulation made possible with the novel stereotaxic instrument designed by Clarke and Horsley in England.

"Three Brains"

With implanted electrodes in awake cats and squirrel monkeys, MacLean and Delgado (1953) found that electrical or chemical stimulation within the limbic portion of the frontotemporal region evoked "automatism" seemingly related to eating or to situations of defense or attack associated with the animal's search for food and struggle for survival. On occasion there was involvement of the pyriform cortex or hippocampus near the amygdala which appeared to organize many of these stereotyped responses into complex, dynamic behavior patterns. Of particular interest was the finding of organized affective states, including directed attack and defense with appropriate vocalization. Stimulation of the corresponding regions in cats and monkeys elicited similar forms of behavior, and the authors speculated that this division of the limbic system was principally concerned with self-preservation.

In these initial experimental studies, MacLean differentiated his first or amygdala division of the limbic system and led directly to differentiation of his second limbic division, occupied with primal functions required for procreation of the species. In laboratory animals stimulation of the septal region elicited sexual arousal and behavior of an affectionate nature. MacLean held that because of strong interconnections, excitation in one region may spill over into the other, with a resulting combination of oral and genital effects. That relationship was presumably due to their connections with the olfactory sense, occurring far back in evolution and playing a primary role in both feeding and mating, as well as in the fighting that may precede them.

MacLean's third limbic division becomes progressively larger in the primate series and reflects a shift in emphasis from olfactory to visual and other forms of communication. The junctional center of this division was named the mammillary bodies by an early anatomist because its shape was suggestive of two breasts. This now seems provident because it implies that nursing the newborn and the great community of cortical cells connected with the mammae are concerned with maternal behavior. The ascendancy of vision in sociosexual communication in monkeys (both male and female), may be exemplified by electrical stimulation of sites in the third division, resulting in erection of the genital organ, with or without an ejection of urine. Whereas the cat and dog, with a large olfactory apparatus, mark their territories with urine, the male squirrel monkey uses the visceral display of the erected genital as a gesture both of greeting and of dominance.

A wave of renewed interest in evolutionary issues catalyzed by the Darwin Centenary in 1959 was manifested in this country (e.g., Pribram, 1958; Magoun, 1960a). Although MacLean's approach to the limbic system was evolutionary from the start, during the '60s he published a series of papers in which this emphasis was expanded around the concept of "man's three brains." Relatedly, a "logo" was developed (Fig. 11.5) and the heading became elided in the felicitous phrase, "the triune brain."[7] In a focused succession of papers, he marshaled his evidence and that of others, later wonderfully presented in *The Triune Brain in Evolution. Role in Paleocerebral Functions* (1990). A philosophic implication appeared in at least one paper, in which MacLean wrote:

> There are those who argue that one has no right to apply behavioral observations on animals to human affairs, but they may be reminded that man has inherited the basic structure and organization of three, brains, two of which are quite similar to those of animals .... Man's brain of oldest heritage is basically reptilian. It forms the matrix of the brain stem and comprises much of the

[7] The only previous use of the term "triune," though in a somewhat different context, was by George W. Crile who wrote in a monograph, *The Origin and Nature of the Emotions* (1950, p. 127): "Traditional religion, traditional medicine, and traditional psychology have insisted upon the existence in man of a triune nature."

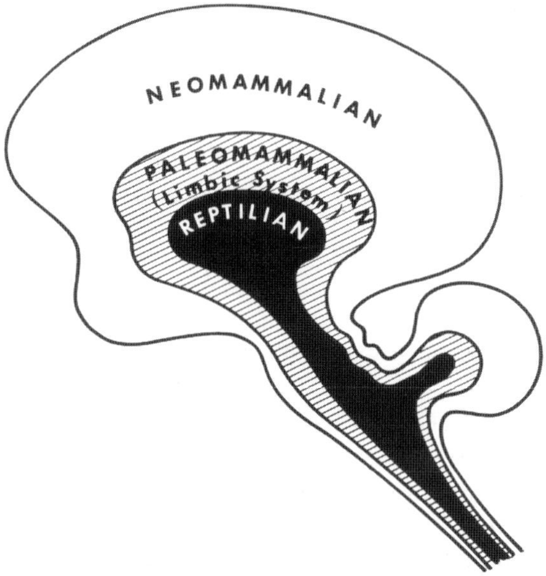

Fig. 11.5. A symbolic representation of Paul D. MacLean's conception of the "triune brain," showing the successive overlay of brain tissue postulated during evolutionary increase in complex behaviors. From MacLean, 1967, p. 377, Figure 2.

> so-called reticular system, midbrain, and basal ganglia. Superimposed on the reptilian brain is a structure inherited from lower mammals. It consists of the two oldest forms of cerebral cortex, together with related nuclei. One might refer to it as the paleomammalian brain. Superimposed on this old mammalian brain, and appearing later in evolution, is a more elaborate neocortex which, together with its related nuclei of the brain stem, reaches the most advanced stages of development in man and can be characterized as the neomammalian brain (MacLean, 1967, p. 377).

Culminating these developments, in 1971 Paul MacLean was appointed chief of a newly established Laboratory of Brain Evolution and Behavior in the National Institute of Mental Health in Bethesda and Poolesville, Maryland.

A FUNCTIONAL TURNING POINT

While the neuroendocrinologists were wrestling with the biochemistry of the sex hormones and their experimental and practical effects on animals, both laboratory and human, the neurologists were discovering that stimulation or ablation of certain brain regions below the cortex and above the brain stem produced marked behavioral changes in experimental subjects. As noted, the prominent behaviors affected included reduction of fear reactions and aggressiveness and increased sexual activity when lesions were made in the rhinencephalon. In his masterly and comprehensive review, "The Hippocampus," John D. Green wrote:

> The real turning point in studies of the rhinencephalon came in 1937 with the observations of Klüver and Bucy who showed that bilateral temporal lobectomy in monkeys gave rise to a bizarre behavioral syndrome. It consists of changes in emotional and discriminative behavior and in sexual and dietary habits (Green, 1964, p. 563).

Klüver and Bucy's Temporal Lobe Syndrome

When the neurologist Percival Bailey joined the University of Chicago in 1928, he helped start an informal neurology club, which met monthly to discuss the research interests of its members. The club included such outstanding local figures as C. Judson Herrick, comparative anatomist; George Bartelmez, embryologist; Anton J. Carlson, physiologist; Arno B. Luckhardt, pharmacologist; Nathaniel Kleitman, an authority on sleep; Karl Lashley and Heinrich Klüver, experimental psychologists; Roy R. Grinker, neuropsychiatrist; Ralph Gerard, a neurophysiologist, and neurosurgeon Paul Bucy. It was to this club in 1937 that the result of bilateral temporal lobectomy in monkeys (the "Klüver-Bucy temporal lobe syndrome") was first reported[8] and they also described that work at the American Physiological Society meeting later the same year (Klüver and Bucy, 1937). The preliminary analysis of their results, reported 2 years later to the American Neurological Association as noted earlier (see p. 290), afforded Papez the opportunity to interpret the findings in light of his prior studies.

A Forthright Confession

In 1940, Bucy again addressed the Chicago neurological club on the "Anatomic changes secondary to temporal lobectomy," verifying that the amygdaloid and hippocampal formations were extensively removed in their operations. At the conclusion of his presentation, Bucy commented,

> It is extraordinarily appropriate that this subject should be presented to the Chicago Neurological Society. Just more than half a century ago [in 1888], Sanger Brown, one of the founders and the first president of this society in 1898, in conjunction with E.A. Schäfer, professor of physiology in University College, London, removed both temporal lobes from a rhesus monkey ... and noted the following operative changes .... 'Prior to the operations [the monkey] was very wild and even fierce, assaulting any person who teased or tried to

[8] Following the presentation, Luckhardt is said to have ecstatically exclaimed, "You have produced an experimental psychosis!"

handle him. Now he voluntarily approaches all persons indifferently, allows himself to be handled, or even to be teased or slapped, without making any attempt at retaliation or endeavoring to escape .... He gives evidence of hearing, seeing, and of the possession of his senses generally, but ... no longer clearly understands the meaning of the sounds, sights, and other impressions that reach him. Every object ... even those with which he was previously most familiar, appears strange and is investigated with curiosity. Everything he endeavors to feel, taste, and smell .... And even after having examined an object in this way with the utmost care and deliberation, he will, on again coming across the same object accidently even a few minutes afterwards, go through exactly the same process, as if he had entirely forgotten his previous experiences' (Bucy & Klüver, 1940, p. 1145).

Bucy acknowledged that there was no doubt that the symptoms observed by Sanger Brown and Schäfer were the same as he and Klüver had observed, but "[u]nfortunately, however ... they did nothing toward elucidating the mechanisms involved ... ." (Ibid.). Bucy also noted that the temporal neocortex (excluding the auditory area) was the only portion of the cerebral cortex that receives only afferent impulses relayed to it from other cortical areas and few, if any, from the thalamus. Furthermore, its efferent connections are primarily with other cortical areas; the temporal cortex is physiologically the most "associational" in function.

Subsequent Analyses

In addition to the early reports of 1937 to 1940, Klüver reviewed and discussed the findings on temporal lobectomy at the Hixon Symposium on *Cerebral Mechanisms and Behavior* (1951),[9] and the Ciba Foundation Symposium on *The Neurological Basis of Behavior* (1958). Culminatingly, at the age of 80, he considered it again in his fascinating Samuel W. Hamilton Lecture on *The Neurobiology of Normal and Abnormal Perception*, at the American Psychological Association in 1963, in which Klüver referred to the temporal lobe syndrome, described about 25 years previously, as so widely known that he mentioned only the outstanding symptoms (Klüver, 1965).

A dramatic confirmation of the syndrome of Klüver and Bucy reproduced in man was reported by Hray Terzian and Giuseppe Dalle Ore (1955). Bilateral temporal lobectomy, including most of the uncus and the anterior part of the hippocampus and amygdaloid nucleus, was undertaken for the relief of psychomotor epilepsy, with paroxysms of aggressive behavior. The postoperative syndrome reproduced all the symptoms reported in monkeys, except the oral tendencies. The patient displayed loss of recognition of people; change in

[9]At the Hixon Symposium, Lashley commented on "some behavioral changes which probably represent the most striking ever produced by a brain operation in animals".

emotional behavior (loss of fear and rage reactions); increased sexual activity, in the form of self-abuse and homosexual tendencies; a voracious appetite; repeated picking up of objects in his field of vision; and serious memory deficiency.

The Release Phenomenon

The next link in the chain of events probing the question of emotion was the outcome of a career-long search. In his autobiographic account, Philip Bard (1973) related the beginning in 1924 of his research in this field as a doctoral graduate student at Harvard Medical School under Walter B. Cannon, who was exploring conditions that induced sympathoadrenal medullary activity. Cannon and Britton (1925) had just completed experiments in which the cerebral cortices were disconnected from the brain stem, producing a state termed sham rage in response to slight provocation. Bard's task was to find the central locus of the mechanism necessary for that behavioral activity. He did this by first decorticating and then transecting the brain stem at different levels and planes and identified the caudal hypothalamus as the responsible region. Bard concluded that the hypothalamus was not an "autonomic center," but was, instead, requisite for complicated patterns of behavior such as emotional expression and defense against heat and cold.

Bard's studies of emotional expression after chronic decortication were initiated by his observation of such a cat at Harvard, prepared by Schaltenbrand and Cobb in 1931. Before leaving for Johns Hopkins in 1934, Bard himself had prepared and studied during long survival periods four cats and three dogs in which the least cerebral ablation had been bilateral removal of all neocortex. The brains of these cats had been serially sectioned and a full histological study carried out by David McK. Rioch in the department of anatomy, Harvard Medical School. In these animals, in addition to ablation of the neocortex, Rioch found extensive removal of the hippocampus, olfactory tubercle, pyriform lobe, and amygdala. Nonetheless, in the report of the findings, Bard (1934) concluded that the state of overactivity or hyperexcitability may be regarded as a "release phenomenon" resulting from removal of an inhibitory control of cortical origin.

The question of the cerebral cortex normally holding in check the hypothalamic rage mechanism was further investigated (Bard & Rioch, 1937), correlating anatomical and physiological findings but the results with long-term chronic animals were inconclusive. The authors suggested that preservation of afferent projections to the diencephalon enhanced the excitability of hypothalamic mechanisms for rage behavior that were already released from cortical inhibition. A year later, Bard (1939, p. 190) appeared to argue oppositely:

> The excessiveness and easy elicitation of the decorticate sham rage seen in acute experiments suggest that the subcortical (hypothalamic) mechanism is in

> a state of overactivity or hyperexcitability. The behavior might therefore be regarded as a release phenomenon denoting removal of a cortical check. The pattern and the threshold of the sham rage were the same after ablation of neocortex without damage to subcortical structure as after decortication plus transection of the brain stem at any level above the caudal hypothalamus. This fact indicates (a) that direct irritation of the hypothalamus is not an important factor in the genesis of sham rage, and (b) that if sham rage is a release phenomenon, the cortex is the source of the inhibitory influence.

Pages later, however, Bard wrote, "On the whole the evidence suggests that the cortex, *or possibly some part of the upper brain stem,* [italics added] does normally exert an inhibitory influence on the hypothalamic mechanism (Bard, 1939, p. 200). A year earlier, in a critical essay (see above) on "The thalamus and emotion," Karl Lashley (1938, p. 46) had first raised the question of the neural origin of this inhibition. "All investigators," he wrote,

> have reported a more ready elicitation of emotional expressive movement from the decorticate than from the normal animal .... Although we may assume that the increased excitability ... is a result of withdrawal of inhibition, a survey of the evidence leaves some doubt as to the source in the normal animal.

Basal Forebrain Structures

It was at this point that Spiegel, Miller, and Oppenheimer presented their findings on "Forebrain and rage reactions" (1940a, p. 127).[10] "Nearly a half a century has passed," they began,

> since Goltz (1892) in his classic account of a decorticate dog described the onset of abnormal irritability, the animal being provoked to a marked reaction of anger and rage even by slight and innocuous stimuli. The question, however, still is unsettled upon the elimination of which parts of the forebrain this phenomenon depends. Since the vegetative component of these fits of rage reactions seems to be due to a release of the posterior hypothalamus (Bard) and the somatic component to a release of the mesencephalic tegmentum and its continuation into the hypothalamus (Hinsey, Ranson, and McNattin), we turned to a systematic study of the effects of [bilateral] ablation of all those parts of the forebrain which send efferent impulses to these subcortical regions (experiments on 58 cats and 12 dogs).

The authors found that only animals with lesions of amygdaloid nuclei or septum showed rage reactions. This major conceptual revision presented by Spiegel and his associates indicated that the source of higher inhibition of hypothalamic mechanisms for emotional behavior was not in the cerebral neocortex, but in nuclear groups of the basal forebrain. Obviously, those findings were not to pass unnoticed by Bard. Eight years later, he and Vernon B.

[10] A full paper was published in *J. Neurophysiol.,* 3: 538–547, 1940b.

Mountcastle presented to a meeting of the Association for Research in Nervous and Mental Disease (1948), the results of their studies in chronic cats. Twenty years after Bard's first paper on this subject, they pointed out that attention to the role played by rhinencephalic structures in central management of emotional behavior had been directed by the work of Klüver and Bucy and of Spiegel, Miller, and Oppenheimer (1940b), and that previously Papez had presented the proposition that hypothalamus, anterior thalamic nuclei, gyrus cinguli, hippocampus, and their interconnections constitute a mechanism for central emotion and participate in emotional expression. Bard and Mountcastle set out to test that "interesting and provocative suggestion" and to extend the earlier acute work with chronic experiments. They studied five cats in which neocortex had been removed but spared rhinencephalic structures and in a long survival period found that the animals exhibited an unexpected placidity to rage-provoking stimuli in both the skeletal muscle and visceral effectors. Extending their work, they demonstrated that cats with bilateral removal of amygdala and pyriform cortex exhibited a progressive lowering of the threshold for rage reactions.

An Interspecies Paradox

In October of 1948, Bard presented these findings at Martin L. Reymert's second Symposium on Feelings and Emotions, held at Mooseheart, Indiana and the University of Chicago. More than 20 years after the first symposium on the topic,[11] Bard summarized his culminating results:

> [I]n cats there originates from or passes through the region of the amygdala an influence which powerfully inhibits the brain-stem mechanisms concerned in the execution of angry behavior.... [A] similar but weaker influence emanates from the limbic cortex; it may be channeled through or near the amygdaloid complex. On the basis of the disclosure by Wheatley (1944) that localized destruction of the ventromedial nuclei of the hypothalamus causes ill temper in cats it may be supposed that these influences extend to that part of the brain stem....
>
> Finally, mention must be made of... an extraordinary discrepancy between certain of the results obtained in cats and... in monkeys. Klüver and Bucy [1939] and Bucy and Klüver [1940] have reported that bilateral extirpation of uncus, amygdala, and hippocampus, together with much of the cortex of the temporal lobes, abolishes or greatly reduces the typical anger and fear reaction of... the intractable *Macaca mulatta*. We have confirmed their results recently in a series of four monkeys, in each of which... we removed on both sides the tip of the temporal lobe, the uncus, the amygdala, and a small segment of the

[11] In the published proceedings Bard remarked, "It may interest the reader... to know that the first public account of my experiments which indicated that 'sham rage' depends on the hypothalamus was given by Prof. Walter B. Cannon in his paper before the First International Symposium on Feelings and Emotions, held [at Wittenberg College] in... 1927."

> hippocampus. At the moment we have no useful suggestion to offer in explanation of these apparently contradictory experimental results, but we are inclined to think that this fascinating problem will be solved in specific anatomical terms (Bard, 1950, pp. 235, 236).

Subsequent Studies of the Limbic Brain

Philip Bard's confidence at midcentury that conflicting and fascinating results would yield to anatomical inquiry was not immediately rewarded; if anything, the puzzle became more confused. Although Herrick had suggested, in 1933, that the rhinencephalon was undoubtedly involved in more complex functions than simple olfactory analysis, it remained for Papez (1937) and Klüver and Bucy (1939) to awaken current behavioral interests in this basal portion of the forebrain. To recapitulate, Papez delineated the anatomic circuit that bears his name and proposed that it might constitute the neural substrate for emotion. Klüver and Bucy's monkeys, with temporal lobectomies that included significant portions of the rhinencephalon, displayed a variety of abnormalities, the most notable of which were docility and hypersexuality. The changes in emotionality, supportive of Papez's proposal, were elaborated and extended by MacLean's (1949) studies of what, following Broca, was termed the limbic system. Further analysis of its role in emotional behavior, replacing the smell-bound thinking of earlier years, influenced a shift from the preoccupation with the neocortex to refocus attention on subcortical elements and stimulated a great deal of research.

The dramatic findings of Klüver and Bucy and subsequent contradictory findings by others attracted experimental efforts to fractionate the syndrome into component elements. As noted, the observations of Bard and Mountcastle (1948) in cats with analogous lesions were conflicting in the first of those changes in that, rather than becoming tame they were excessively aggressive; the opposum, which is sometimes a nasty animal, was calmed down. Brady and Nauta (1953) undertook a systematic analysis of these behavioral changes in tame albino rats, with acute surgical lesions of the septal forebrain region. Two hours after operation, the alert animal had an excessive startle reaction to auditory stimuli and attacked approaching objects immediately, with vigorous biting, including other septal animals. In chronic preparations (1955), such increased emotional reactivity disappeared within 60 days after operation. Soon, Brady's "septal rage syndrome" in the albino rat came to represent for the psychologist what Bard's "sham rage" in the tame laboratory cat had earlier represented for the neurophysiologist.

Changes in Sexual Behavior

With respect to the second prominent change, although Bard (1950) disclaimed seeing sexual changes in operated cats, other investigators were more fortunate. Leon Schreiner and Arthur Kling (1953), at the Walter Reed Army Medical Center, were first to confirm "a state of hypersexuality" following surgical aspiration of the pyriform cortex, amygdaloid nuclei, and adjacent structures in cats. This low-key assessment hardly did justice to the Roman circus that had ensued. When operated cats were placed together vigorous mounting and thrusting pelvic movements on the part of male operated cats and courtship behavior by their female opposites took place. The aberrations in sex behavior by male operated cats extended to a female rhesus monkey, an agouti, and an old hen. Within a week of castration, however, all such activity disappeared; it did not appear in animals castrated before, or at the time of amygdalectomy, but could be restored by injection of testosterone (Schreiner & Kling, 1954), supporting the hypothesis of an indirect, neural influence on endocrine mechanisms. And finally, Green, Clemente, and De Groot (1957) found in cats after electrolytic lesions in the pyriform cortex and amygdala that hypersexuality was never observed in females but in male cats there were examples of abnormal aggressive and sexual behavior. Hypersexuality was observed usually in animals with lesions in the pyriform cortex, whereas lesions restricted to the amygdala, without injury to the pyriform cortex, resulted in no detectable changes in sexual behavior.

It is of interest to suggest here the relevance of these midcentury searches for anatomical explanations of emotions and their expression to the end-century experience of rage and frustration expressed by stabbings and shootings. Humankind's ability to control the "release phenomenon" – hypothesized for suitably operated laboratory animals demonstrating fear, rage, aggression – may depend on the individual's recognition and modification of the "whale" within. The magnitude of that whale was suggested very recently by Jason Epstein (1999, p. 64): "Only when all of us learn to fear our violent nature will we be safe from one another."

Chapter 12
Sex and the Brain

SPOTLIGHT ON HUMAN SEXUALITY[1]

In all biological organisms including the human, two potent forces are paramount for individual well-being and survival, and for species preservation by means of the creation of new organisms in linear replication. These forces are the libidinal/sexual and the aggressive/adaptive, they are inborn and autonomous in living creatures of all complexities of cellular organization, active and regulated at all times by central control systems, and in more complex organisms such as the human, they are subject to some extent to adaptive direction and control. The libidinal/sexual force is associated with the affect of pleasure and affection and serves to establish social and sexual bonding between individuals, thus serving individual well-being and species continuation. The aggressive/adaptive force insures safety and continuity of life for the individual in the face of threats and demands from the environment, leading to protective responses, sometimes violent, and associated with the affects of anger and fear. It is now known, though only in recent decades, that the central locus of activity and regulation of these biological forces lies in the limbic system of the brain.

Studies of these forces, of their propensities in humans and other species, for sexuality and aggressivity, gained momentum in the 20th century. Systematic observation and experimentation for the sake of reducing ignorance and gaining information and understanding of the structural and functional components involved in the phenomena of "sexuality," at biological and behavioral levels, is of quite recent origin as history goes, a matter of less that 200 years. In this overview of key advances in knowledge concerning sexual functions, the first were based on studies of biological organisms other than humans; social tolerance for direct studies of human subjects and modes of procedure for doing so are relatively late developments. Such studies, and public awareness of them have frequently evoked opposition from social conventions and from individual and social anxieties. Explorations in unconventional areas, yielding new

[1] The three introductory paragraphs were contributed by Barbara J. Betz, M.D.

insights, merit illustration in a history of neuroscience. Among key steps marking the 20th century's growth of knowledge concerning sexual phenomena, the North American contributions to the flow have taken two separate channels: the social, observational human stream, and the experimental rapids of ablations and endocrines. This chapter takes each in turn and only hints at the revolutionary social consequences.

Early Attempts to Understand

Before the linkage between the human brain and sex could be discerned from a scientific point of view, the facts surrounding the human sexual response had to be brought into the light where they could be examined. In the United States, a combination of Puritan heritage and Victorian overlay constrained any printed acknowledgment of sex to medical writings in which the pathological aspects were targeted.

Sex As Science

Among those first-published physiological studies was that of Joseph R. Beck in 1872. An Indiana physician, he reported extensively his observations on one female patient (Brecher, 1969). The next noteworthy American study was by Dr. R.L. Dickinson, who published *Topographical Findings in Human Sexual Anatomy* in 1933. Both male and female subjects were observed independently and static measurements made in living subjects. Recognizing their limitations, Dickinson wrote: "The volume on human sex physiology is overdue" (p. 117).

As the 19th century phased into the 20th, experimental psychology in America was initiated by a small number of young psychologists, notable among them John B. Watson. His popularization of the white rat as a suitable subject for experimental studies in animal learning and his introduction of "behaviorism" to American psychology have been discussed (see Chapter 9). A vocal advocate of the application of objective methods useful in laboratory animal experimentation to studies in humans, indications are that Watson took his own advice to a risky extreme, and in 1920 was one of the first Americans to investigate the physiologic aspects of adult human sexual response. The loss of the data was especially cruel, because such investigations seem not to have been made systematically until Alfred Kinsey undertook his studies in the 1930s. In the interim, surveys based on answers to questions became the accepted procedure in sex studies.

John D. Rockefeller, Jr.'s Philanthropy

A major development in sex research in this country followed an investigation in 1910 of white slave traffic in New York City, by a grand jury of which John D. Rockefeller, Jr. was a catalytic member. Rockefeller was convinced that to make a lasting improvement in conditions, a permanent organization should be created that could make continuing warfare against forces of evil. This should be a private organization which would have the freedom from publicity and political bias that a publicly appointed commission could not easily avoid. As an initial step, a Bureau of Social Hygiene was formed in the winter of 1911. Among its members was Katherine Bement Davis, superintendent of the New York State Reformatory for Women.

The Bureau of Social Hygiene

The bureau's first step was to establish a Laboratory of Social Hygiene, adjacent to the reformatory and under Davis's direction. Experts were to study the physical, mental, social, and moral aspects of each person committed to the reformatory. When the diagnosis was completed, the laboratory was to suggest treatment likely to reform the individual or, if this was impossible, permanent custodial care would be recommended.

On entering its labors, the bureau also placed a high priority on making a careful study of the social evils in this country and in Europe. The first task was a thorough and comprehensive survey of the conditions of vice in New York during 1912; the report was the first of four such studies. The bureau had secured Abraham Flexner[2] whose publications on the status of medical schools in this country (1910) and Europe were widely known, to study social evil and the various methods of dealing with it abroad. His report (1914) on prostitution in Europe was easily the most widely read publication in this series. The third assignment was to study European police systems and the fourth report was based on those cities in the United States where special methods of dealing with social evils had been introduced.

Mr. Rockefeller, as chairman of the bureau, believed its worth lay in constructive suggestions of scientific and humane interest in a great world problem. There is no question that he meant what he said for, during the next 30 years, he contributed more than $5.8 million for study, amelioration, and prevention of social conditions, crimes, and diseases that adversely affect society, with special attention to prostitution and its evils.

Research on Sexual Behavior

Ten years later, a completely autonomous development was initiated by an erstwhile graduate student, Earl F. Zinn, who had recently obtained a masters degree in psychology at Clark University. Zinn submitted a proposal[3] to

[2] Abraham Flexner was the brother of Simon Flexner, director of laboratories at the Rockefeller Institute for Medical Research, 1903 to 1935.
[3] Copies of the proposal, subsequent correspondence, and reports of the NRC Committee for the Study of the Problems of Sex are available in the Archives of the National Academy of Sciences U.S.A. to which we are grateful for permission to consult and quote.

Davis to request from Mr. Rockefeller an appropriation to the Bureau of Social Hygiene in the amount of $10,000 needed to work out a plan for research in the field of sex; this was awarded in July of 1921. Zinn also submitted his formulation of a research program in the general field of sex to Clark Wissler, chairman of the Division of Anthropology and Psychology in the National Research Council. The plan was considered by some to be a plot upon their specialty, and Wissler's response stated:

> This was carefully considered by our Division, but without being able to arrive at a definite conclusion since the subject is chaotic and complex. The feeling of our Division is that while psychologists could contribute something to such an investigation, the chief line of approach lies through medical sciences. We are, therefore, passing this request over to that Division of the Council (10 May 1921).

During this time Dr. Davis had become director of the Rockefeller-supported Bureau of Social Hygiene. Through her, Mr. Zinn was able to present his proposal for research in problems of sex directly to Rockefeller, who expressed an interest in them.[4] Zinn was made a member of Dr. Davis's staff to seek the advice of scientists and work out ways to implement the investigative program he had developed. Nothing further came from this approach through the behavioral sciences until a new division chairman was appointed.

Among the more fortunate events of Zinn's project, he presented it to Robert M. Yerkes, then chairman of the Research Information Service of the National Research Council (NRC), the operating branch of the National Academy of Sciences in Washington D.C. Yerkes was fully prepared to recognize the importance of Zinn's project, but, when he presented it to the Division of Anthropology and Psychology of the NRC, he found it as unresponsive as before. The division was not yet sufficiently sure of itself to risk taking on a project that was still very general, concerned with unexplored fields of research on questions about which the public and even scientists were very sensitive. The old taboos on open discussion of sex were at the bottom of the division's reluctance to sponsor the project.

Diplomacy at the Academy

Anxious to get wider support for so bold a program, Yerkes arranged a special conference to discuss its research feasibility with the three respective chairmen of the Divisions of Anthropology and Psychology, Biology and Agriculture, and Medical Sciences. The Bureau of Social Hygiene provided $10,000 for the studies recommended, assuming that the NRC would sponsor

[4]Letter 7 June 1921 Davis to Rockefeller confirming their conversation "this noon" [at luncheon?].

the program. The result was whole-hearted encouragement of the proposal to undertake sex research through the agency of the NRC. In that decision a leading part was taken by Walter B. Cannon, professor of physiology at Harvard, who chaired a subconference to summarize the conclusion of the conference and its resolutions. He wrote of the importance of factual knowledge of "[t]he impulses and activities associated with sex behavior and reproduction" unavailable because of

> the enshrouding of sex relations in a fog of mystery, reticence and shame .... The committee is convinced, however, that with the use of methods employed in physiology, psychology, anthropology and related sciences, problems of sex behavior can be subjected to scientific examination .... For these various reasons the committee recommends that the National Research Council be advised to organize and foster an investigation into the problems of sex.[5]

An Influential Committee

Fortified by the support of this advisory conference, the National Research Council organized a Committee for Research in Problems of Sex, within the Division of Medical Sciences. Its initial five members, in addition to Davis, Cannon, and Yerkes, were Frank R. Lillie, professor of zoology at the University of Chicago and Tomas W. Salmon, professor of psychiatry at the College of Physicians and Surgeons, Columbia University. At the committee's meeting with the chairman of the division of medical sciences, Yerkes was elected head and Zinn was appointed its executive secretary. Stated concisely, the main purpose of the committee was: "to conduct, stimulate, foster, systematize, and coordinate research on sex problems to the end that conclusions now held may be evaluated and our scientific knowledge in this field increased as rapidly as possible" (Aberle & Corner, 1953).

The fields of study and research in this uncharted area were divisible, in the thinking of the committee, into the anatomic, physiologic, anthropologic, psychologic, and sociologic aspects of sex. In addition, Lillie outlined problems in biology he had been investigating since 1915 and Karl Lashley, then in the department of psychology at the University of Minnesota, drew up a plan for research in neurobiology and psychobiology – the only time that reference to an area of neuroscience was made in this program.

An Early Survey

In 1925, the ubiquitous Zinn outlined another study of sex problems, this time in marital relations. Its objective was to determine the factors, particularly

[5] From the Report of the Committee on Desirability and Feasibility of Scientific Research of Problems of Sex dated 21 October 1921. NAS-NRC Archives, p. 1.

sexual, that make for marital maladjustment, and to trace their genesis and the conditions of their development. The assumption was that adult heterosexual adjustments center in the marital relationship and failures in them could be most readily observed there. Through the application of methods of the clinical psychiatrist, at least a partial view of ontogenetic psychosexual development could be obtained and plans formulated for their study. This approach led the committee to believe that the proposed investigation would provide an orientation for the development of a systematic human program. The nature of the problem indicated that a well-trained, experienced psychiatrist with strong research interests was needed to conduct this research. Among others, Dr. G.V. Hamilton of Santa Barbara, California was invited to meet with the committee, was selected to direct the program, and a suite of offices in New York City was leased.

Sex in Print

In the Spring of 1928, Hamilton submitted a comprehensive report on his findings to the Bureau of Social Hygiene and the advisory committee. It consisted of a factual presentation of the results of his examination of 100 married men and 100 married women, together with many descriptive correlations suggestive of explanatory possibilities. His manuscript for a book, *A Research in Marriage* (1929), evoked laudatory comments by the committee members.

By the decade of the 1930s, two popular books had helped banish the reticence to publicly discuss sexual behavior. The translation into English of T.H. van de Velde's *Ideal Marriage* (1930) with its explicit descriptions of a variety of coital positions, expedited the coming of the sexual revolution.[6] Perhaps more significantly, the welcome accorded the publication in 1931 of Dickinson and Bryant's factual discussion of contraception indicated the end of the invisibility of human sexual behavior. Although at long last it had become possible to publish openly and in descriptive terms the behavioral interaction between men and women, scientific, physiologic data were scarce.

NEUROPHYSIOLOGY OF SEX

The Kinsey Interviews

Seventeen years after the NRC committee's support of Hamilton's earlier study, it approved an initial grant for the now famous studies of human

[6]Copies of the book disappeared routinely from the libraries of women's colleges and were reported to reappear in the hands of students at nearby men's colleges.

Fig. 12.1. The achievement of Alfred C. Kinsey was to move from the early fumbling studies of human sexuality to the collection of standardized data from various groups of subjects. Photographed in 1953.

sexual behavior by entomologist Alfred Charles Kinsey (1894–1965; Fig. 12.1) and his staff at Indiana University.

The Persona

To identify an overriding feature of Kinsey's personality, he was a collector: as a boy, he collected stamps. After obtaining his B.S. in psychology at Bowdoin College, and as a doctoral student in entomology at Harvard he began collecting gall wasps. On joining the University of Indiana as an assistant professor, he extended this activity widely over the United States and Mexico. Ultimately he gave the 4 million specimens of wasps he had assembled to the American Museum of Natural History – the largest collection it ever received. In a subordinate direction during this period, Kinsey acquired two-and-a-half acres surrounding his home and, implementing the "bloom" in Bloomington, he devoted some two acres to flowers, particularly day-lillies and irises, the ranging varieties of which he promptly began to acquire. In time, he possessed nearly all of them and his garden became a show-place to which people drove for hundreds of miles to enjoy during its blooming season.

Beginnings

In 1938, Indiana University took a bold step and began a "marriage course" and, prophetically, Kinsey was chosen its coordinator. With only limited familiarity with this broad field, he decided to learn as much as could be obtained from the histories of his students. Applying the pattern developed with his wasps, a year after he began he had 350 histories and was looking forward to his first 1000. Together with this, he also undertook 280 counseling sessions. Two years after its introduction, Kinsey was pressured into resigning from the course, but continued his conferences with students. The search for more histories commenced to absorb him, however, as did the realization that he was working in an important, but unexplored field of investigation. The scientist, collector, and teacher then came together in Kinsey – as an investigator in human sexual behavior.

The work preceding his own that impressed Kinsey more than any other was G.V. Hamilton's study. Hamilton had used personal interviews rather than questionnaires, but the questions were written on cards to prevent interviewer bias. Hamilton had studied only 100 married males and 100 married females, half of them pairs of spouses. The group was mostly upper-class and all were from New York City. But Hamilton's research had been forthright, direct, objective, and unafraid – all qualities that Kinsey admired.

In contrast to Hamilton's methodology, Kinsey's contacts in Indianapolis, Chicago, and St. Louis began to spread, like the branches of a tree, to 700 histories recorded by 1940. By then Kinsey recognized that his project could not grow unless he obtained financial aid. He applied to the NRC's Committee for Research in Problems of Sex, and a 1-year grant of $1,600 was made to aid in collecting and analyzing more histories. When he applied for its renewal a project site visit was made by Robert Yerkes, Lowell Reed, and George W. Corner, then on the committee, and Kinsey insisted on interviewing each of them on their own sex life. They reported so favorably on Kinsey's methods, as well as his records and staff, that the second grant quadrupled the first. The visitors thought that Kinsey was up to something remarkable and beginning in 1947, $40,000 was a constant annual figure until the Rockefeller Foundation cut off all support in 1954. The devotion of the committee to this project was measured by the fact that, from 1947, Kinsey had been receiving about half of its entire annual resources.

Incorporation of the Institute

In 1947, Kinsey's sex research program was incorporated as The Institute of Human Behavior, based at, but autonomous from the University of Indiana. Kinsey transferred the library and its collections to the institute, together with all other sex-related activities. A year later, Kinsey's first magnum opus, *Sexual Behavior in the Human Male* (1948) was published by Saunders, which ordered an initial printing of 10,000 copies. This may have been the most spectacular publishing error of the decade, if not the century as public curiosity built up rapidly and in the first 2 months, the phenomenal sales topped 200,000 copies. The second blockbuster, *Sexual Behavior in the Human Female*, was published in 1953. Kinsey wrote both books himself, using all the editorial help available. The royalties from sales were returned directly to the institute's budget for continuing its research.

Following a first wave of minor assaults on the *Human Female* study, Kinsey and his institute's research suffered a devastating blow. A reactionary Republican congressional committee investigated the financial backers of the institute and the Rockefeller Foundation's new president, Dean Rusk, thought that obviously a "get Kinsey" move had been launched. In an effort to extricate the foundation from the embarrassment of having financed Kinsey's research, in February 1954, Rusk terminated all further support either to Kinsey or his institute. Although the dimension of royalties from the volume on women was not certain, Kinsey believed that the institute could support itself for a time on royalty income, but not long enough to complete the next volume, on sex offenders, much less his planned investigations of physiologic and neurologic problems involved in sexual responses. From that time Kinsey's health declined from a cardiac disorder, which was aggravated by stress over the budgetary deficits in his program and he died in 1956.

Subsequent Events

Kinsey's Institute for Sex Research continued at Indiana University under the direction of his assistant, Paul H. Gebbard, M.D. Belatedly for Kinsey, its financial rescue came with the establishment of the National Institute of Mental Health that, in a sense, replicated the Rockefeller Foundation, but in greater dimensions. Wardell B. Pomeroy continued at the institute as director of field research until 1963 when he started a private practice in New York City and a decade later published the classic biography of Kinsey and the Institute for Sex Research (1972), from which most of this account has been drawn.

In 1963, an Institute for the Advanced Study of Human Sexuality was founded in San Francisco, and 15 years later was approved by the State of California, as the world's first and only Graduate School of Sexology, to grant four degrees in human sexuality. The school's dean, Pomeroy, had been primary co-author of the late Alfred C. Kinsey's two classic studies on sexuality in America.

Under Paul Gebbard as director, the Indiana institute was steered on a low-profile course for more than two decades. Then, in 1982, June Reinisch, a young psychologist with impeccable credentials, including 7 years of teaching and research at Rutger's University, was chosen as director. The institute was renamed "The Kinsey Institute for Research in Sex, Gender, and Reproduction," to reflect Reinisch's broader interests as a scientist.

Achievements of Masters and Johnson

The next significant publication in the field of sexuality was *The Human Sexual Response* (1966) by William H. Masters and Virginia E. Johnson, which sold more than 250,000 copies in the United States and was translated into nine foreign languages. It had two advantages over Kinsey's earlier studies: first, it included data relating to concurrent sexual responses in the male and female – a combination that still continues to dominate human sexual activity; and second, it was not assembled from a plethora of verbal interviews, fascinating as those might be. Rather, in scrupulous detail, it described the physiologic responses of the human body to erotic stimulation during both masturbation and coitus. The study made it possible to follow the entire human sexual cycle, from the first stirrings of erotic desire, through orgasm and to ultimate subsidence. This study was conducted as objectively as 19th-century physiologists had followed the digestive cycle from mastication to defecation. Moreover, it was not based on oral recollections, but on direct laboratory observation of more than 10,000 male or female orgasms. Whereas scores of eminent physiologists in many countries contributed through the decades to the gradual exploration of the digestive cycle, Masters and Johnson accomplished their initial work on the sexual cycle, alone, in 12 years of concentrated effort.

Destiny by Training

William Howell Masters (1915–2001; Fig. 12.2) was born in Cleveland, Ohio, the son of parents in comfortable circumstances. He attended a preparatory school in New Jersey and secured his Bachelor of Science degree at Hamilton College in 1938. Masters then enrolled in the University of Rochester School of Medicine and Dentistry, not to prepare for the practice of medicine, but toward research in the biological sciences. In his first year, he worked in the laboratory of one of the country's foremost anatomists and an authority in the biology of sex, George W. Corner, the first major event in the chain that was to lead to the Masters–Johnson reports. Corner recollected: "I remember Bill Masters as a very serious and intelligent young man of more independent character than many" (quoted in Brecher, 1969, p. 285). Corner assigned Masters to a problem in sex research: the estrous cycle in the female rabbit and the ways in which it differs from or resembles the menstrual cycle in the human female. That laboratory experience and Corner's general influence led Bill Masters to narrow his goal from biology in general to sex research in particular. Although Corner approved this decision, he said Masters should wait

Fig. 12.2. The team of W.H. Masters and V.E. Johnson refined human sexual studies by direct neurophysiologic measurements of the complete arousal, climax, and subsidence cycle.

until he could secure the sponsorship of a major medical school or university with ample support for his project.

In 1942, the co-investigators were married, and a year later Bill received his M.D. degree and did a residency in obstetrics and gynecology at Barnes Hospital and Maternity Hospital, Washington University School of Medicine, St. Louis. Successively rising to associate professor at the medical school, in addition to his changing clinical activities, during the 6 years from 1948 to 1954, Masters published 25 contributions to the medical literature, 14 of which were on a single research project – hormone replacement therapy for aging and aged women. This work was important for its later application in clinical practice.

Initial Studies

In 1954 an investigation of the anatomy and physiology of the human sexual response was initiated within the framework of the department where a closely coordinated clinical research program in problems of human sexual inadequacy was instituted in 1959. Since January 1964, these programs have been continued under the auspices of the Reproduction Biology Research Foundation. During the decade of the '80s, the anatomy of human responses to sexual stimuli was established, and such psychologic variables as intensity and duration of individual reaction patterns were observed and recorded.

Interrogation in depth of medical, social, and psychosexual backgrounds of both laboratory-study and clinical populations has been a concomitant of the basic science and clinical investigative programs since their inception.

Although the Kinsey work has become a landmark of sociological investigation, it was not designed to interpret physiologic or psychologic response to sexual stimulation. Those fundamentals of behavior could not be established until two questions were answered: What physical human reactions develop in response to effective sexual stimulations? Why do men and women behave as they do when responding to it? If human sexual inadequacy were to be treated successfully, the medical and behavioral professions must provide answers to these more basic questions.

The original techniques of defining and describing the gross physical changes which occur during human male and female sexual response cycles were primarily those of direct observation and physical measurement. Because the integrity of human observation for specific detail varies significantly, regardless of the observer's training and considered objectivity, reliability of reporting was verified by many of the accepted techniques of physiological measurement and the frequent use of color cinematographic recording in all phases of the sexual response cycle. Masters and Johnson came to the conclusion that

> the primary physiologic reaction of either male or female to effective sexual stimulation is superficial and/or deep vasocongestion, and that the secondary reaction is one of increased myotonia ... hyperventilation, tachycardia, muscle spasm, pelvic-musculature contraction, etc., are of secondary import.... The parallels in response emphasize the physiologic similarities in male and female ... rather than the differences (Masters & Johnson, 1966, pp. 284–285).

Those findings awaited analysis of their basic physiologic processes for explanation.

EXPERIMENTAL ATTACK

Parallel with the accumulation of data on human sexual responses, discoveries and interest in the endocrine glands were accelerating and the subdiscipline of neuroendocrinology was taking root (see Chapter 5). Several decades of experimental laboratory work were to elapse before the story of sex and the brain returned to the human track.

Some Contemporary Investigators

Between 1922 and 1935, Calvin P. Stone, (1892–1954), professor of psychology at Stanford University, published a dozen papers on the sexual drive and copulation ability of the male albino rat. In his chapter in *Sex and Internal Secretions* (1939), Stone collated studies that contributed to more recent interests in the neural substrates of sex. He showed in young male rats that as much as 50 percent or more of the cerebral cortex may be removed before puberty without preventing the subsequent appearance of typical patterns of copulatory behavior. In the female rat, Stone also found normal copulation, gestation, and maternal behavior were still preserved after ablating one-quarter of the total cortex. Failures in maternal behavior were increasingly prevalent, however, when more that 30 percent to 50 percent of cortex was destroyed; when lesions were more extensive, no adequate maternal behavior was displayed.

Stone's findings were corroborated by Karl S. Lashley, then at the University of Chicago, who wrote (1933, p. 11):

> I have observed that nearly complete decerebration in the male rat abolishes the reactions to the female in heat. Lack of recognition of the adequate stimulus seems to be the primary factor in the loss .... [Moreover,] more complicated activities involved in rearing the young (nest building, collecting and cleaning the young, nursing, etc.) are interfered with by cerebral lesions and that the behavior is progressively simplified with increasing size of lesion.

Five years later and from Harvard, he elaborated with the cautionary statement that although mating behavior had not been observed in rats lacking more than half of the neocortex, it was difficult to keep such animals vigorous which might explain their lowered vitality (Lashley, 1938).

Another Laboratory

Next in this series was Frank Ambrose Beach (1911–1988; Fig. 12.3). While earning a master's degree in psychology at Kansas State Teachers College he read a number of Lashley's papers describing the effects of cortical lesions of various sizes on maze learning in rats. The idea of such brain surgery excited Beach and he eventually became able to make reasonably clean lesions in the neocortex without an undue mortality rate. He then attended the University of Chicago for doctoral study with Lashley who the next year accepted a research professorship at Harvard. Beach completed his dissertation on "The effects of cortical lesions on the maternal behavior of the rat", and was then invited to spend a year with Lashley at Harvard. In 1937, Lashley recommended Beach for assistant curator of experimental biology at the American Museum of Natural History of New York City, where he established a Department of Animal Behavior at the museum and in 1942 was appointed full curator and chairman of the department, while still continuing his research (Beach, 1937) on the effect of cortical lesions on the copulatory behavior of rats. A study of 52 animals operated in the pattern of Lashley,

Fig. 12.3. The contribution of Frank A. Beach to neuroscience included career-long research on reproductive behavior patterns in rats, which revealed the complexity of those patterns, and he brought animal behavior into the American Museum of Natural History. Photograph from McGill, Dewsbury, and Sachs, 1978, frontispiece.

indicated that an intact cortex was important in the inherited train of activities of a mother rat raising her pups. A mother with 50 percent of cortex extirpated bilaterally was still able to carry her pups around the cage, but could not assemble the scattered litter in a single corner and then hover over the young in a nursing posture. Such results did not confirm those theories that held instinctive behavior to be mediated exclusively by subcortical mechanisms.

Continuing in this pattern, Beach's subsequent study (1942) of the copulation behavior of rats revealed that all males with lesions of less than 20 percent of their cortex continued to copulate normally. As the lesions were extended, however, there was a corresponding and proportional decrease in the incidence of postoperative copulations; removal of more than 60 percent of the male neocortex abolished all sexual congress.

By 1942, Beach had concluded that injury to the neocortex in the male rat reduced the ease with which sexual arousal occurred, but did not interfere with the motor pattern of copulation. The probability of postoperative coition was found to be inversely related to the amount of neopallium destroyed. When lesions did not exceed 40 percent of the total cortex, postoperative mating was often spontaneous and, in other cases, could sometimes be elicited after administration of testosterone propionate and the pattern and vigor of the response were normal. As for female mating behavior, Beach (1944) reported that after complete decorticating, in general the female rat exhibited characteristic copulatory responses, delivered viable young, and occasionally removed the fetal membranes but seldom did they bring the young together or build nests for them.

In 1946, Beach accepted a professorship at Yale, then after 14 years he moved to the University of California, Berkeley. While still at Yale, he was appointed to the NRC Committee for Research in Problems of Sex and in 1957 became its chairman. As noted, for many years the committee had been the only continuing source of support for sex research and its grants so served an important function. By 1960, federal support for research on sexual behavior in this country amounted to $957,000, compared with $57,000 from the committee. When the committee reluctantly decided to terminate its program it agreed to conclude in glory, with a two-stage conference on "Sex and Behavior" as proposed and designed by its chairman. Both meetings were held a year apart on the Berkeley campus of the University of California. Their presentations and discussions were published in *Sex and Behavior* (1965) with Frank A. Beach as editor. The book was "dedicated to the memory of Robert Mearns Yerkes in recognition and appreciation of his twenty-five years of service and leadership as Chairman of the National Research Council Committee for Research in Problems of Sex."

Sex at Ranson's Institute

Putting aside the ill-defined and variously located sex centers postulated above – in the mammillary bodies, the caudal hypothalamus, the upper midbrain, and even the reticular substance of the mesencephalon, its major and essential integrative neural substance is now known to lie in the preoptic and anterior hypothalamus, together with its connection with the anterior pituitary gland through the infundibular stalk. The research that brought all this to light, for a change was not at Harvard or Bard's laboratory at Hopkins, but at Ranson's Institute of Neurology at Northwestern University Medical School in Chicago. In 1938, in a paper by C. Fisher, H.W. Magoun, and S.W. Ranson, it was noted that in 85 cats in which diabetes insipidus had been produced by lesions in the anterior hypothalamus,

> We have never observed any of them to come into heat so that pregnancy has never occurred .... Although gross atrophy of the ovaries does not appear to develop in diabetic cats, it is possible that there is some abnormality of ovarian function (p. 6).

Following up these studies of disturbances in sexual behavior, J.M. Brookhart, F.L. Dey, and S.W. Ranson (1940, 1941) turned to guinea pigs. In 17 females, ventromedial lesions were made in the hypothalamus between the chiasma and the pituitary stalk, leaving the latter uninjured. None of those animals showed estrous behavior in response to manual stimulation of the vulva or the lumbosacral region of the back, nor would they accept a male; all gave avoiding responses to such stimulation in the manner of a normal anestrous female. In subsequent trials, eight of these animals were injected with double the normal dosage of estrogen and progesterones but failed to show estrus, and four others were injected with quadruple the usual dosage but also failed to come into heat. In other animals, similar lesions at the tuberal level, interrupting connections with the pituitary, but leaving the latter uninjured, were followed by continuous anestrus, with extreme atrophy of ovaries and uteri. Those animals also failed to display estrous behavior, nor could it be induced by injection of ovarian hormones. When placed for long periods with males, none of these females became pregnant. In contrast, females with more caudal hypothalamic lesions ran regular cycles, displayed normal estrous activity, mated, and bore normal litters. Three male guinea pigs with similar lesions in the rostral hypothalamus (Brookhart & Dey, 1941) failed to display courtship or mating behavior when placed with estrual females, suggesting that the hypothalamic substrates for sexual behavior were similarly distributed in the two sexes – in the midventral portion of the anterior hypothalamus instead of the mammillary bodies or the mesencephalic tegmentum. The more recent discovery of releasing factors elaborated by hypothalamic neurons and conveyed to the anterior pituitary by portal vessels in the infundibular stalk has at last accounted for the integration of neural and endocrine functions in sexual behavior (see Chapter 5).

Triune Concepts

The involvement of levels higher than the hypothalamus in sexual responses became the target of studies by neurologist Paul D. MacLean when he inaugurated a program of brain, evolution, and behavior at the National Institutes of Health in Bethesda in 1957. His triune brain concept (see p. 295) of a profusely interconnected arrangement of phylogenetically evolved neural substrates of behavior required circuits for complex patterns such as the sexual response in advanced vertebrates. Utilizing the small squirrel monkey, MacLean and his associates (summarized in MacLean, 1973) demonstrated the role of limbic system components, especially the cingulate gyrus, in representation of the monkey's genital display response to presentation of an image of a "rival." An additional goal was to look for connections between the limbic system and vision, mankind's most developed sense. "[T]he question was this: how does the brain transform the cold light with which we see into the warm light with which we feel?" (Ibid., p. 42).

Fig. 12.4. The persistence of George W. Corner and associates in elucidation of the suppression of ovulation by a hormone produced in certain cells in the ovary led to the successful development of an oral contraceptive agent. Photograph from Ramsey, 1994, p. 56.

The Progesterone Story

George Washington Corner (1889–1981; Fig. 12.4) had been called in 1924 to the University of Rochester, New York, which had organized a new school of medicine, as it's founding professor of anatomy. "I was now at the time of life", he later wrote, "at which, in the biological and medical sciences at least, the capacity and opportunity for research are highest. My program of investigation of the reproductive system now matured and became fruitful" (Corner, 1958, p. 46). In 1927, Corner began studies with rabbits of the function of the corpus luteum of the ovary. He showed that it produced a hormone which caused changes in the lining of the uterus that facilitated the survival of young embryos. With the considerable skill of Willard M. Allen, a medical student with a brilliant record in chemistry, they extracted and gradually purified the hormone, now called progesterone. On injection into spayed rabbits, it brought about natural progestational changes including inhibition of ovulation (Corner & Allen, 1929).

Confirmation

Through the '30s and '40s, some additional noteworthy work called attention to the effectiveness of progesterone in inhibiting ovulation. At the University of Pennsylvania, A.W. Makepeace, G.L. Weinstein and H.H. Friedman (1937) found that the inhibition was not exercised directly on the ovarian follicles but more centrally in the ovulation mechanism. From Harvard, E.B. Astwood and H.L. Fevold (1939) reported that progesterone did not inhibit ovulation in hypophysectomized immature rats. At the University of Wisconsin, R.H. Dutt and L.E. Casida (1948) believed that the effect of progesterone was directly on the pituitary gland in inhibiting the production or the release of the luteinizing hormone. And finally, at the Worcester Foundation for Experimental Biology, Shrewsbury, Massachusetts, Gregory Pincus and M.C. Chang (1948) affirmed the potency of progesterone as an inhibitor of ovulation in the rabbit: a single injection of 30 milligrams of progesterone may prevent ovulation in that species for as long as 24 days. In 1945, when George Corner, who had first identified and demonstrated the effects of progesterone (1928), reviewed the prospects for its clinical use, he suggested that it might control excessive menstrual bleeding, as well as painful menstruation. Strangely, since he knew that progesterone also suppressed ovulation, Corner did not mention the possible use of this in contraception. Neither did any of the nine scientists at Pennsylvania, Harvard, or Wisconsin. Most strangely, neither did the key scientist and his associate in this field at the Worcester Foundation, who in the early '50s would make the foundation a world center for their achievements in preparing a practical oral steroid contraceptive.

And Then THE PILL

In 1927, Hudson Hoagland and Gregory Pincus obtained their Ph.Ds together at Harvard. Hoagland then spent 2 years as an instructor in physiology and biology in W.J. Crozier's department of general physiology and next accepted a position as professor and head of the department of biology at Clark University in Worcester, Massachusetts where a year later he established a laboratory of neurophysiology. After 6 years, Pincus's appointment at Harvard was not renewed and Hoagland arranged for his position as visiting professor at Clark.

A Foundation Established

By 1944, their research at Clark had overgrown the capacity of the converted barn that had served as their laboratory. With Hoagland and Pincus as codirectors, an autonomous Foundation for Experimental Biology was established and an attractive residence in a suburb of Worcester was converted into laboratories. By 1970, the foundation's annual budget had grown to $5 million and some 300 persons were employed.

> Dr. Pincus and Dr. M.C. Chang were both outstanding in the field of reproduction physiology.... During the late forties and early fifties we had become increasingly aware of the unprecedented upsurge in population growth rates in the underdeveloped countries, and Pincus and Chang started work on what they hoped might be a practical oral steroid contraceptive (Hoagland, 1975, p. 155).

Fig. 12.5. The influence of Katherine Dexter McCormick on the progress of neuroscience was two-fold: she provided support and approval of the first American anthropoid colony (at the McCormick estate in Santa Barbara) and gave generous and sustained underwriting of contraception research (at the Worcester Foundation for Experimental Biology, Massachusetts).

Initially, the support of this research seemed impossible to finance. In earlier years, Hoagland had become acquainted with Mrs. Stanley McCormick, when they were both involved with research in the substrates of schizophrenia.

> [O]ne day she came to my office" Hoagland wrote, "and I learned of her deep concern about the 'population explosion.' She had heard of Pincus's work ... and hoped that he might wish to seek an oral contraceptive. [As these associations developed], Mrs. McCormick was the primary source of funds that led to the discovery ... of 'the Pill' (Ibid., p. 155).

A Loyal Benefactress

Katherine Dexter McCormick (1875–1967; Fig. 12.5) was the widow of Stanley McCormick, a son of Cyrus, founder of the International Harvester Company. They had married in 1904 and 2 years later Stanley was diagnosed as schizophrenic.[7] When Mrs. McCormick visited the foundation in 1953, she promised a gift of $10,000 on the spot, and during her lifetime gave

[7] As described in Chapter 10, Stanley was cared for in a baronial mansion named Riven Rock, in Montecito, near Santa Barbara in Southern California. One of the full-time psychiatrists in Stanley's case was G.V. Hamilton, who established a monkey colony on the estate where Robert M. Yerkes studied learning ability of an orang-utan. Later Hamilton carried out the study that was published as *A Research in Marriage* in 1929.

$150,000 to $180,000 annually and bequeathed an additional $1 million in her will.

After Pincus and Chang found a group of 19-norsteroids that would block ovulation in rats and rabbits without untoward side effects, Pincus asked Dr. John Rock at Harvard Medical School to test the preparation on ovulation in a small group of volunteer women. After the apparently favorable outcome,[8] Drs. Pincus and Celso-Ramon Garcia, with the cooperation of the Planned Parenthood Organization of Puerto Rico, studied some 15,000 women volunteers in Puerto Rico and Haiti in tests of the effectiveness and safety of the pill, with the well known outcome: its widespread acceptance has hastened cultural and social changes wherever it became available.

Revising a Definition

By end-century, the experimental and clinical knowledge of human sexual function – a confluence of behavioral and neural sciences – had created a new attitude about sex and gender. Eleanor Maccoby at Stanford University, who "could be called the mother of contemporary gender theory in developmental psychology" (K. Hall, 1999, p. 1681), wrote that gender is socially derived from a "two-cultures" mode of upbringing – boys expected to be boys and girls as girls – yet is based on biological predispositions derived from hormonal effects with timing and quantity modulation components. Uneven rates of development of language, for example, in spite of equal blood concentrations of androgens between birth and puberty, "hints at a neural substrate that enables girls to progress faster (or at least differently) with respect to language" (Maccoby, 1998, p. 107). Such sex-dimorphic behavior was explained by the "organization theory" that hypothesized a sensitization function of the nervous system during prenatal development that finds expression later when the influence of androgens is minor.

Maccoby's thesis was predicated on early work in reproductive biology such as that of C.A. Pfeiffer (1936) at the University of Kansas, who showed that sexual differentiation occurs neonatally in the rat, not at maturity as had been assumed. The contributions to sorting out the complexities made by Herbert Evans (see Chapter 5), Frank Beach, and the group in Ranson's laboratory, among others and studies of human conditions have led to hypotheses about the underlying factors. In connection with observations on girls born with adenogenital syndrome, whose masculinized behavior persisted in spite of surgical correction, tests on monkeys (Phoenix, Goy, Gerall, & Young, 1959) presented the possibility that a prenatal exposure to excessive

[8] Four decades later, epidemiologic evidence had appeared showing that suppression of ovulation was related to ovarian cancer and John Rock was unfairly criticized for not having conjectured this possible side effect when the pill was tested (Gladwell, 2000).

androgens had predisposed the individual to later behaviors. Those positive findings were confirmed (Money & Ehrhardt, 1972) and organization theory became the explanation for the behavioral effects of hormones and their influence on the uncommitted embryonic brain. The theory gained wide acceptance especially in America, based less on neural and psychological research than on proliferation of clinics, media promotion, and peer approval to which it gave rise (Wijngaard, 1997). Relatedly, with increasingly rigorous separation of gender from sex, scientists in reproductive biology and developmental psychology have called for a redefinition: "Gender is much more than two" (K. Hall, 1999, p. 1682) and gendered behavior is dynamic and not a "two-cell grid" of hetero- or homosexuality (Laqueur, 1999, p. 547).

Chapter 13
"The Root of Consciousness"

It is tempting to avoid "the hoary mind-and-matter problem . . . ." (C.J. Herrick, 1955, p. 60), but the question of how the mind relates to the brain and its expression in consciousness is of such ancient lineage and continuing complexity that it would be inappropriate not to attempt an updating in 20th-century neuroscience. An especial urgency is added by the fact that so many distinguished neuroscientists have reflected on this intimidating challenge.

HISTORICAL PRELUDE

Foreign Samplings

The antiquity of the mind-brain question is easily demonstrated by brief quotations from the earliest records. The biblical Book of Job recorded two sonorous queries: "Where is the place of wisdom? And where is the place of understanding?" From the Hippocratic writers in the 5th century B.C. only a single discussion of the function of the brain and nature of consciousness has survived:

> Some people say that the heart is the organ with which we think and that it feels pain and anxiety. But it is not so. Men ought to know that from the brain and from the brain only arise our pleasures, joys, laughter and tears. Through it, in particular, we think, see, hear and distinguish the ugly from the beautiful, the bad from the good, the pleasant from the unpleasant . . . . To consciousness the brain is messenger (quoted in Penfield, 1975, pp. 7–8).

Evading the Issue

As The Enlightenment took shape, for some writers the issue of brain and mind was too complex to be tackled. The 17th-century English philosopher, John Locke, in the introduction to *Essays on Human Understanding* (1700) wrote: "I shall not at present meddle with the physical consideration of the mind, or trouble myself to examine wherein its essence consists ...." (Book I, Chapter I). The problem did not go away, however, with such a brushing-off.

An Anatomist's View

During the early decades of the 19th century, the most important contribution to the anatomy of the brain after Gall were made by the German, Karl Frederick Burdach (1776–1847). Basing his conclusions on 1000 anatomical-clinical observations, he left very precise descriptions from data obtained earlier to which he added several new descriptions, in particular those dealing with the basal ganglia and the internal capsule (Hécaen, 1977). In his views on thalamic functions, Burdach wrote, "It is the root of consciousness which is further developed in the corona radiata and ultimately perfected in the covering organ" (quoted by A.C. Meyer, 1970, p. 557). This was the only detailed, anatomical consideration of the seat of consciousness in the aggregate of neuroscientific studies throughout the 19th century.[1]

Many of Burdach's descriptions and terms have stood the test of time and some of his theories of brain function are of interest today. His concept of the interaction between thalamus, corona radiata, and cortex to some degree anticipated Jacksonian ideas, as well as those of Meynert and Monakow at the end of the 19th century. What he wrote about the visceral function of the hippocampus and the mammillary body, and of the instinctual components of frontal lobe function are still contemporary problems. The general ideas of Burdach and of other "Naturphilosophen" strike us now (as they did Neuburger in 1897) as remarkable because so much of what first appeared to be speculation was later confirmed, and remains alive (A.C. Meyer, 1971).

The Perennial Dilemma

By the opening of the 20th century, the anatomical knowledge of the cerebral cortex and the mechanisms of the higher brain stem had turned a suspicion or mere notion into a vital question: How **IS** the brain related to the mind? The answer was consistently elusive, however. Sir Charles Sherrington succinctly identified the problem of mind as being "difficult to bring into a class of physical things." And elsewhere he concluded discouragingly: "We have to regard the relation of mind to brain as still not merely unsolved, but still devoid of a basis for its very beginning" (quoted by Penfield, 1975, p. 3).

[1] A discussion of modern views of consciousness not limited to American neuroscientists is in Marshall and Magoun, 1998, pp. 201 passim.

AMERICAN PERSPECTIVES

William James and Jacques Loeb

Stream of Thought

By the end of the 19th century, the great American psychologist, William James, looked on the connection between mind and brain as "the ultimate of ultimate problems" and noted the renown and profit that would accrue to its solution (James, 1899, p. 468). Somewhat disdainfully he described the problem as a "breath moving outwards, between the glottis and nostrils ... the essence out of which the philosophers have constructed ... consciousness" (James, 1904, p. 491). But in *Principles*, he made it clear that his psychologist's idea of consciousness was based on a fact: thinking of some sort goes on and thus the seamless "stream of thought" was synonymous with "stream of consciousness." References to James, his classic textbook, and their influence are scattered throughout this section.

Consciousness as Associative Memory

Among those scientists brash enough to address consciousness was the foreign-born general physiologist, Jacques Loeb whose experiments on parthenogenesis conferred a notorious eminence in the United States (see Chapter 7). Before emigrating, he had become involved in the mind-brain question. During 5 years as assistant to Goltz, Loeb developed a materialistic conception of higher nervous activity based on the idea of consciousness as a metaphysical term for phenomena determined by mechanisms of associative memory that arose only in the cerebral hemispheres of the vertebrate brain and increased in functional significance as the neocortex became elaborated in the animal series. Thus, after experimental decortication, an impairment of behavior in surviving animals was imperceptible in the shark, but became increasingly more pronounced in the frog, pigeon, and dog.

In his monograph on *Comparative Physiology of the Brain and Comparative Psychology* (1900), Loeb described his experiment in this period, undertaken in rebuttal to "those who believe in a psychic localisation in the cerebral hemispheres" (p. 265). He noted that after a unilateral lesion in the dog's cortical foreleg center, Fritsch and Hitzig had observed that no attention was paid by the dog to its abnormally placed opposite paw and concluded that: "The animals evidently had only an imperfect *consciousness* of the condition of this limb." Goltz had shown that removal of the entire left hemisphere, in a dog taught to dig its food out of a heap of pebbles with its right paw, did not abolish this capability. Opponents then proposed that after the operation the foreleg center in the remaining hemisphere had taken over the psychic functions for both legs.

I made an experiment," Loeb wrote "to which this objection was not possible. A dog was taught to walk on its hind-legs, when it wanted to be fed. Then the hind-leg centres were removed in both hemispheres. In spite of this loss the dog was still able to walk on its hind-legs. When I offered it food or whenever it expected to be fed it rose voluntarily on its hind-feet. *The conscious actions or associations for the use of the hind-legs had not suffered*, but there was decidedly a muscular disturbance .... I showed this dog at the naturalists' meeting in Berlin in 1886. The day after the demonstration I showed the brain of the animal that had been killed in the meantime. The hind-leg centres had been removed completely (Loeb, 1900, p. 268).

Analogously, Loeb objected to Munk's interpretation of disturbances of vision following lesions in the dog's visual cortex as "psychic blindness," and concluded that the process of association allowed the hemispheres to act as a whole, not as a mosaic of independent parts. Loeb's first published papers, in 1884 and 1886, reporting this work, evoked a prompt denunciation by his opponents, but William James at Harvard wrote a kind letter endorsing his work, initiating a long friendship between them.

Toward the end of his monograph on the brain, Loeb introduced the mechanisms of consciousness and the phenomena of associative memory as the great discoveries to be made in the field of brain physiology and psychology. It was apparent to him that histology, ablation, or measurement of reaction times would not suffice for the solution of this problem, but that physical chemistry and protoplasmic physiology must be combined. Near the end of his career, Loeb attempted to bring the new tools of vacuum tube amplification to bear on the questions of this initial preoccupation with the physiology of the brain.[2]

Evaluating the Subconscious

The seriousness with which psychologists of the early 20th century held the problem of consciousness may be judged from at least two events. In 1904, when Edward Titchener addressed the International Congress of Arts and Sciences in St. Louis on "The Problems of Experimental Psychology," his list of ten items culminated with "The Problems of Consciousness." And in 1924, a terse questionnaire was dispatched by the chairman of the NRC-NAS Division of Anthropology and Psychology, A.E. Jenks[3] to 30 division members requesting opinions on the subconscious as creator or carrier of ideas; the responses were immediate.

[2] Magoun's manuscript for this chapter does not extend beyond this point and the following addendum is inserted for historical completeness.
[3] This month-long exchange of letters from the archives of the National Academy of Sciences presents a fascinating sample of the thinking of some of the era's most influential experimental psychologists in the United States.

A Wide Spectrum of Opinion

Adolf Meyer (7 February 1924) was noncommittal and felt the concept "has been overworked." John Watson (8 February 1924), writing on stationery of the J. Walter Thompson Company in New York, confessed to having recently lectured on the topic, in which his "angle" was: "To the behaviorist, of course, there is no 'conscious' and naturally there can be no 'subconscious.' " He then wrote at length about Freud's "ridiculous, mystical, unscientific" idea of repression of consciousness. Both Edwin Boring and G. Stanley Hall (8 and 9 February 1924, respectively) were quite sure the subconscious creates, amplifies, and preserves.

The most analytical reply, perhaps not surprisingly, was from Karl Lashley (9 February 1924); also his was the longest. He brushed aside the concept as having no accepted evidence that the subconscious performs any function.

> The best evidence available indicates that the preservation of existing thought or action need involve nothing more than a static condition of neural tissue such that particular conditions of excitability or resistance persist and determine the reaction when adequate stimuli recur.

Then came the unkindest cut: "The subconscious presents for some tender minded psychologists a means of escape from the reality of difficult problems of cerebral physiology and as such can be dealt with adequately only in the category of myth and fantasy." Another skeptic was R.S. Woodworth (10 February 1924) who stressed the physiologic aspect of the preservation of thought or action. Knight Dunlap, still at Hopkins, contrasted the definition of the subconscious by psychologists with that of the Freudians and opined that initiation, development, and preservation of all thought and action were functions of the nervous system. Robert Yerkes (16 February 1924), too, claiming no special competence on the subject, because he had not thought about it very much, wrote that he would have to take a neurological point of view. Margaret Washburn's rather quaint reply (18 February 1924), self-typed in keys badly in need of cleaning, revealed the perseverance of introspection and her training with Thorndike: she focused on the ability to dissociate and its dependence on "physiological peculiarity."

Perhaps the most significant opinion was that of Shepherd Ivory Franz (17 February 1924), although he deemed it "not fashionable at the present moment." He had to answer in the negative because the question seemed to set consciousness apart from other mental states. Franz then suggested a revised form of the question that disregarded consciousness as an entity. The same stance was taken by an unnamed anthropologist.

An Anthropologic View

At the end of the month, February, Jenks dispatched another letter to respondents to the first query, with a rephrased question that he trusted made the problem clearer. The new version came from "a Government anthropologist who is also a member of this Division," and asked "Can a new cultural element, or an element having cultural potentialities, be introduced among men without reaching the level of consciousness?" The replies second time around were even lengthier and attested that Jenks had set in action a good deal of conscious thinking without a forthcoming consensus, as is the situation eight decades later.

The anthropologist in question was John R. Swanton at the Bureau of American Ethnology of the Smithsonian Institution. After reviewing the replies he wrote (27 March 1924) his partner in inquiry, Jenks, that they illuminated the "necessity of a saving grace of common sense." The answer from Yerkes (5 March 1924) particularly appealed to him as being to the point:

> The facts which I have at hand do not justify a dogmatic answer .... The way I should answer it would depend on the definition of consciousness. I have no doubt whatever that cultural elements become effective without reaching the introspective level but I seriously question whether any act without achieving the level of 'awareness.'

Yerkes seemed to view consciousness in two ways; in any case, the NAS division at its meeting the following month was to debate what could be done to stimulate research on fundamental problems and consciousness would surely qualify. Whether this temporary brush with the problem, although it stimulated some otherwise neglected thinking by psychologists, had any influence on the direction of research is doubtful. A more fruitful inquiry was to come from the active mind of one of the century's most productive investigators in biomedical fields.

A NEUROSURGEON'S OBLIGATION

Among the diverse groups of scientists and physicians concerned with the enormous task of exploring the "nerve-cell jungle" of the brain, only neurosurgeons have the unusual opportunity and privilege of observing directly the living human brain and the behavioral responses to its stimulation, all in the course of bringing therapeutic relief to their patients. This attitude characterized the great Wilder Penfield, whose dedication to the needs of his patients did not preclude recognition of an opportunity to advance the knowledge of brain function. In contrast to most investigators, the mind-brain question was a career-long interest and Penfield studied it anatomically, physiologically, pathologically, and surgically.

Wilder Penfield's Many Contributions

Neurosurgeon Wilder Graves Penfield (1891–1976; Fig. 13.1), an American-Canadian deeply involved in this special field of brain science (see Chapter 7),

Fig. 13.1. Even before he became founding director of the Montreal Neurological Institute, neurosurgeon Wilder Penfield was pondering the seat of consciousness. With open-brain stimulation of conscious patients preceding excision of epileptic-generating tissue, he concluded that a subcortical center was possible. Photograph from Denny-Brown, 1975, p. 264.

provided further insight into "the physiology of the mind." In *The Mystery of the Mind* (1975) he cited Hughlings Jackson's early 19th-century pronouncement: " '[W]e are differently conscious from one moment to another,' " and continued with his own insights:

> [Mind] is a function, presumably, of synaptic activity, now here and now there. It seems to me more probable that its representation is in the cortex than in the diencephalon, having regard to the relative numbers of neurones available. That considerable areas of cortex may be injured, or removed, without loss of consciousness may depend upon a large amount of equipotentiality in what you call the uncommitted cortex.
>
> The synaptic activity associated with consciousness is continuously present except during sleep. The explanation for this appears to be that the reticular formation in the brain stem in some way facilitates, or 'drives' the higher centers, and that in sleep the activity of the reticular formation is inhibited. Here the relationship of consciousness to the brain-stem seems well established (Penfield, 1975, pp. 96–97).

Penfield's late-life summary of the attributes of consciousness, emphasizing the associative (or noncommitted) cortex, was the reflection of a neuroscientist for whom the mind-brain question had been a career-long interest, in contrast to others for whom the problem received only transient attention (Marshall & Magoun, 1998).

Auspicious Training

After several years in neurosurgery at Presbyterian Hospital, New York, and just before moving to Montreal in 1928, Penfield spent 6 months with Professor Otfried Foerster in Breslau, Germany, collaborating in the operation and study of a dozen cases in which brain scars were removed for the relief of epileptic seizures.

Penfield was not unappreciative of this momentous development in Foerster's clinic.

> [D]uring such operations, which are carried out under local anesthesia," he wrote, "the opportunity presents itself to study the living brain of man as never before in history. I never even dreamed of this in Oxford. Here is a human neurophysiology waiting to be studied with the help of conscious patients.... Here is the chance to go beyond the achievements of Pavlov and Sherrington and to apply their pioneer thinking to the study of conscious man. A wonderful scientific harvest is waiting in this new field (Penfield, 1967, p. 249).[4]

Early in his career, in a Harvey Lecture at the New York Academy of Medicine in 1936, Penfield pointed out that "Long-continued unconsciousness appears clinically in patients who have a lesion... somewhere above but not far removed from the midbrain ... ." (p. 66).[5] In conclusion he went a step further:

> [T]here is much evidence of a higher level of integration within the central nervous system that is to be found in the cerebral cortex.... I would suggest that this region lies, not in the new brain, but in the old, and that it lies below the cerebral cortex and above the midbrain.... All parts of the brain may well be involved in normal conscious processes but the indispensable substratum of consciousness lies outside of the cerebral cortex, probably in the diencephalon (p. 68).

The "Centrencephalic System"

By the 1950s, Penfield had refined those ideas and wrote:

> More recently, after anxious consultations with Herbert Jasper and also with Stanley Cobb, I proposed the term 'centrencephalic system,' an expression that was intended... as a protest against the supposition that... cortical association

[4] See also Chapter 4.

[5] A slightly edited version, more clinical in tone, is in *Arch. Neurol. Psychiat.*, 40: 417–442 (1938) and a very polished form of the quotation was used by Penfield in *Brain*, 81: 231–234 (1958).

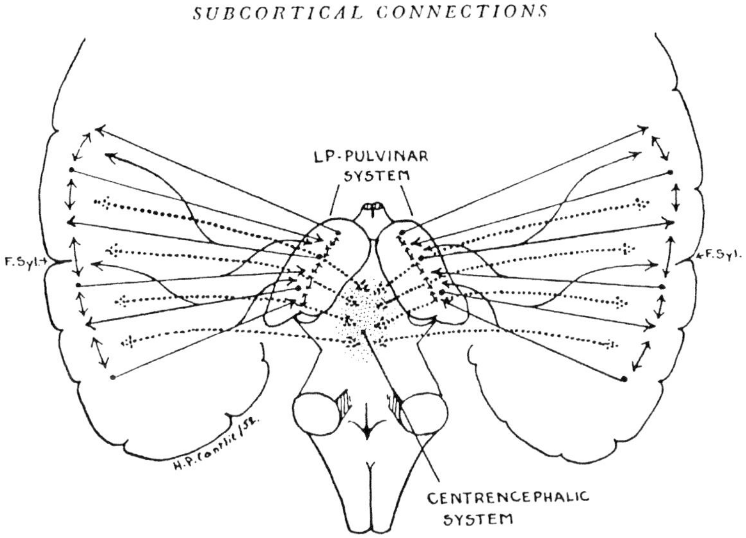

Fig. 13.2. Diagram of Wilder Penfield's conception of "centrencephalic system" of consciousness situated below the neocortex. From Penfield and Roberts, 1959, p. 217, Figure X-14.

> or corticocortical interplay was sufficient to explain the integrated behavior of a conscious man (1958, p. 740).

Concerning the subcortical site of the highest brain mechanism (Fig. 13.2), Penfield explored the nature of its basic processes.

> This mechanism, as it goes out of action in sleep and resumes actions on waking, may switch off the mind and switch it on ... but to expect the highest brain mechanism, however complicated, to carry out what the mind does, and thus perform all the functions of the mind, is quite absurd .... Fresh explorers must discover how it is that movement of potential becomes awareness, and how purpose is translated into a patterned neuronal message. Neurophysiologists will need the help of chemists and physicists in all this, no doubt .... Electricity was first revealed to science while it was being conducted along the nerves of living organisms. Physicists might well consider our questions seriously today, if only out of gratitude!

Penfield's anxious consultation with his colleague, the Harvard and Boston City Hospital neurologist, Stanley Cobb, concerned the latter's evocation of William James, in whose writings Dr. Penfield found considerable inspiration (Feindel, 1975) – a half-century earlier James had described consciousness as "a stream of thought." Cobb (1952, p. 176) wrote:

> It is the integration itself, the relationship of one functioning part to another, which is mind and which causes the phenomenon of consciousness. There can

be no center. There is no one seat of consciousness. It is the streaming of impulses in a complex series of circuits that makes mind feasible.

Relation to Psychiatry

Wilder Penfield's electrical explorations of the conscious human brain, provoking memories of past events and emotions, had implications far beyond the neurophysiologic and neuroanatomic domains of cerebral function and localization. For those psychiatrists interested in the "psychic dynamics" of analysis, Penfield's findings contained the possibility of testing the usefulness of affective recall in treatment of their patients. The suggestion was made at a meeting in 1951 of the American Psychoanalytic Association by Lawrence A. Kubie, M.D. (1896–1973) in his presentation on concepts of brain organization relating to psychoanalysis (1953). In casting about for an explanation of the mechanism of consciousness, Kubie envisioned Penfield's centrencephalic concept as a rostral extension of Magoun's ascending reticular system in the lower brain stem.[6] In Kubie's words: "The ascending reticular substance in the brain stem must from now on be included in our thinking about the energetics of the central nervous system" (Kubie, 1953, p. 34). And Penfield boldly predicted that "Time will doubtlessly show the functional importance of this system during centrencephalic integration" (Penfield, 1975, p. 18). Both neurosurgical and neurophysiological evidence continued to accumulate to reinforce Kubie's pronouncement and link conscious, preconscious, and subconscious (the "visceral brain" of MacLean) states with structures below the neocortex, in contrast to the weak support for cerebral cortex involvement (Marshall and Magoun, 1998).

Anatomy of Consciousness

Identification of the subcortical substrate of consciousness was accomplished by studies of long-term coma corroborated by post mortem confirmation of brain disease or trauma (French, 1952; Barrett, Merritt, & Wolf, 1967). The results of one such study were shown in a remarkable exhibit (Fig. 13.3) at the 1948 meeting of the American Medical Association. Prepared by two neurologist-psychiatrists, George N. Thompson and Johannes M. Nielsen, it announced the "Area Essential to Consciousness: Cerebral Localization of Consciousness as Established by Neuropathologic Studies" and consisted of enlarged photographs of human coronal and sagittal sections. The authors localized a system essential to "crude consciousness" at the junction of

[6] Two years earlier, Magoun and associates (Moruzzi & Magoun, 1949; Lindsley, Bowden, & Magoun, 1949) had published a formulation of the diffuse ascending reticular system in the lower brain stem stimulation of which aroused a sleeping cat or monkey and desynchronized its slow-wave, high-amplitude electroencephalogram (see Chapter 5).

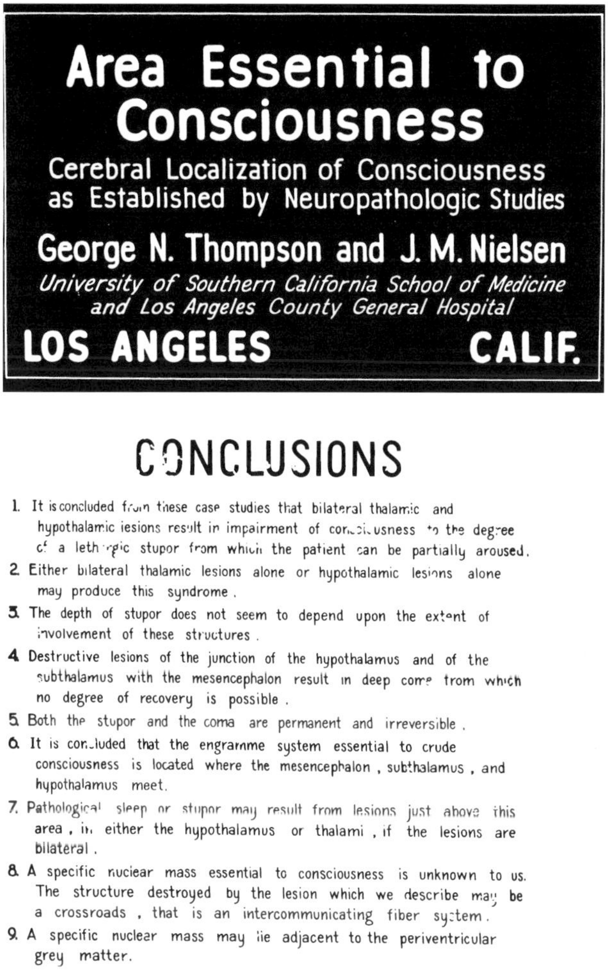

Fig. 13.3. The first and last panels of a series exhibited in midcentury by Drs. George N. Thompson and Johannes N. Nielsen, identifying (see text) a brain region with consciousness. Courtesy of George N. Thompson, M.D.

mesencephalon, subthalamus, and hypothalamus, an expansive region that leaves a good deal of latitude for more precise studies.

Midcentury Outlook

The confused state of definitions and theories concerning consciousness that prevailed at midcentury was a sign of the lack of progress in finding testable explanations for the phenomenon. It is instructive at this point to look at a symposium on "Brain and Mind" organized for the annual meeting of the American Neurological Association in 1951. The topics of the papers (and their presenters) were: cortical evolution (Von Bonin), ascending reticular system (Magoun), corticofugal projections to brain stem (Jasper, Ajmone-Marsan, & Stoll), locus of mind (Cobb), and memory (Penfield). Even more interesting than the neurological views was the inclusion with their publication of a masterly dissection of the state of knowledge of consciousness by the eminent neurologist-historian, Francis Schiller. His "Special Article" was "an attempt to bring some order into the various problems" encountered by a consideration of consciousness and even to find a common ground for "philosopher to meet physiologist" (Schiller, 1952, p. 199). He took a common-sense look at definitions and causal theories and produced an open-ended statement of possibilities that was almost prescient. "This matrix," he wrote about the base from which patterns of integration were formed, "is still multipotential and capable of change and growth; it also has the capacity for forming new, and as yet unstable, patterns" (Ibid., p. 224). The analogy of the role of stem cells in specialized cell formation that is unfolding in current research is inescapable.

ASKING THE RIGHT QUESTIONS

Neurophysiologic Aspects

A more daunting aspect of consciousness than its anatomic substrate is the neurophysiology of the brain's psychic dynamics mentioned earlier. Pioneering in this effort, neurophysiologist Benjamin Libet and his associates (1964) at the University of California, San Francisco measured the time before a conscious patient could sense a threshold electrical stimulus applied to the sensory cortex. As Libet later wrote: "The most striking finding was the requirement of a surprisingly long train duration (about 500 msec) of repetitive pulses at liminal intensity" (1996, p. 105). The surprise initiated a dedicated series of manipulative studies of the experimental parameters and led 30 years later to the proposal of a testable field theory of mind-brain interaction (Libet, 1994).

The Bicameral Mind

Although popular and scientific interests in consciousness were not yet in ascendancy during the 1970s, at least one academic philosopher's thinking was directed toward explication of how the process might have arisen. From Princeton University, Julian Jaynes published *The Origin of Consciousness in the Breakdown of the Bicameral Mind* (1976). His bold and novel concept, in brief, was that very early hominids possessed a two-chamber mentality with (usually) the right serving as a theocratic authority whose directives were mediated via the anterior commissure to the left hemisphere. Precivilization signaling was within small groups, was confined to intragroup business, and became language when calls were made intentionally (to convey information) and varied in intensity. The postulated change from six millennia of directed action to subjective experience in the second millennium B.C. characterized the beginning of civilization and Jaynes has marshaled an array of artifactual and literary evidence in support of his thesis. Deeming vision "the very ground and fabric of consciousness" (Jaynes, 1976, p. 269), he sought "to reveal the plausibility that man and his early civilizations had a profoundly different mentality from our own, that in fact men and women were not conscious as are we . . . ." (p. 201).

The response to Jaynes's speculative concept was largely negative. He was accused by a colleague of confusing the idea of consciousness with the phenomenon itself (Block, 1981); others had to be persuaded to consider the idea seriously (Dennett, 1998). A genuine concession was that consciousness could not be brushed aside as the behaviorists had attempted, but "We are going to have to talk about the ephemeral, swift, curious, metaphorical features of consciousness" (Dennett, 1998, p. 123), and that Jaynes had served brain science well by asking the right questions but had the answers wrong. The present-day literature has allowed Jaynes's ideas to be neglected.

A Split Brain: Two Conscious Selves?

A vast and unexplored landscape opened with the courageous undertaking of a team representing psychology, surgery, and neurology that joined forces in the attempt to alleviate intractable epilepsy grand mal in the patient "W.J." The behavioral effects of commissurotomies[7] have been well documented (see Chapter 15) but less explored are the philosophic aspects of that gross manipulation of the human brain.

Sperry's Analysis

Like Penfield, Roger W. Sperry held an early interest in the relationship of brain and mind (personal communication to L.H.M.). With his previous experience (in association with Myers, 1956) on split-brain cats and monkeys, Sperry was prepared to lead the team in a program investigating the mental capabilities of commissurotomy patients in studies of the "other side of the brain" (Bogen, 1969a, b; Bogen & Bogen, 1969). With the revelations stemming from Bogen's patient W.J. (Bogen & Vogel, 1962) and subsequent cases, Sperry found that he had had to revise his earlier views of the mind-brain relationship.

[7] Commissurotomy (transection of the corpus callosum and anterior commissure) is not to be confused with hemispherectomy; they have some similar as well as distinctly different effects.

In searching for the clues to what might be responsible for consciousness, Sperry laid his bets on specialized circuits in the brain and overall constant neural activity "of the alert waking type. Take away the specific circuit, or the background, or the orderly activity of either one, and the conscious effect is gone" (Sperry, 1966, p. 307). Although the post-operation psychological testing found the two hemispheres extremely unequal in control of behavior, he was convinced of their equal endowment with consciousness; nonetheless, as Sperry has said, "a unified brain is more than its two halves."

The Contrarian

Although the thorough psychological and neurological testing of commissurotomized patients virtually proved their having duplex consciousness, the unitary view of John Eccles (1966) was that the nondominant hemisphere lacked conscious experience, mutely carrying out its specialized complex activity as does a computer. Eccles's main argument was based on the dominant (left) hemisphere's command of language with which it could express itself, and only through which the subordinate hemisphere could be expressed. His self-confidence in organizing and editing the proceedings of the workshop of the Pontifica Academia in 1965 assured the prominence of the topic if not his point of view.

Current Thoughts

The exacerbation of interest and speculation about mind-brain mechanisms at end-century produced both popular and scientific writings that ranged from "thorough and profound" to "intellectual pathology" (Searle, 1995). Writers of all persuasions – philosophers, journalists, neuroscientists – continue to probe and tease this amorphous question and many express hope or despair about its solvability. The fascinating and advantageous diversity of thinkers is manifest in this brief survey of neuroscience in the American 20th century. Some of the disciplines mentioned (and a few of the figures who represented them) were philosophy (William James), history (Francis Schiller), neurophysiology (Benjamin Libet), neurosurgery (Wilder Penfield), neurology (Stanley Cobb), and psychiatry (David Rioch, Lawrence Kubie). The burst of speculation and contradiction created a moraine deeper, wider, and denser than that left, for example, by aphasia at the end of the previous century – a conglomerate of terminology, definitions, clinical interpretations, and conceptualizations. Two of the most prominent current writers (they are both Nobel laureates) not afraid to hazard a possible solution to the mind-brain question have revealed themselves to be deeply appreciative of the enormous complexities.

A Working Hypothesis

Even though Gerald Edelman in *Bright Air, Brilliant Fire* (1992), did not explain why the brain is sentient or aware, his attempt was authoritative and scholarly. With his goal "to dispel the notion that the mind can be understood in the absence of biology" (p. 211), he presented a working hypothesis: Sheets of points on the brain correspond to the layout of distant receptors; those momentary maps are closely related to each other and occur all over the brain; at birth the sheets are made up of neuronal groups which during development are strengthened by a selective mechanism or they die; the groups exhibit "reactive reentry"[8] or parallel pathways active in both directions. In short – perceptive maps, neuronal groups, active reentry. In Edelman's view, the above qualifications constitute a primary (core) consciousness characteristic of subhuman animals. The human animal, in addition to the primary core has a higher-order consciousness that entails symbolic language and self-awareness.

Somewhat resembling Edelman's idea of "perceptual categorization," to explain how multiple inputs to the nervous system might be handled, the proposal of Francis Crick in *The Astonishing Hypothesis* (1994) was based on how stimulus inputs to different parts of the brain might create a unified experience. The mechanism of "binding," introduced by neuroscientists as a functional concept mediating the synchronized activity of neuronal groups, was extended to the mind's conscious experience. Overall, both Edelman and Crick are a part of the strong move to put consciousness back into nature, a move that commenced with the rise of physiological psychology in late 19th century and has been subject to cycles of relentless chipping away. Whether or not the problem is solvable inserts another factor into the equation.

Final Words

An appropriate closure to the place of consciousness in any history of neuroscience comes from philosophy, where the first speculations on the nature of consciousness and mind took shape. The chronic absence of contacts between philosophers and neuroscientists was noted by John R. Searle in his opening remarks as chairman of the Ciba Foundation's symposium, *Brain and Mind* (1979). An eminent British-American philosopher, Searle's views on the significance and timeliness of the philosophy of mind to his discipline had great import. In his words, "We happen to have been forced into a situation where the philosophy of mind has become central at a time when ... neuroscience stands on the brink of some very important discoveries" (Searle, 1979, p. 3). Those discoveries have materialized in diamonds and two decades later Searle again assessed the progress made in solving the brain-mind problem. The intervening new knowledge reinforced his conviction that "To understand the mind and consciousness we are going to have to understand in detail

[8] Not to be confused with "feedback."

how the brain works" (Searle, 1995, p. 56). He wrote that the discovery of the body image in brain was "one of the most exciting in the history of the field" (p. 60) as it constitutes the self and needs memory for continuity. As for resolution of the question, Edelman, Crick, and their associates seemed to be on the right track to explain how neurobiological processes cause conscious experience.

Searle stopped there, yet his later conclusions (Searle, 2000) were up-beat and it is not beyond imagining that neuroscience, with its predisposition to see life processes from at least two vantage points, may eventually find a way to "What is Life?"

Part 3
Communicative Sciences

Chapter 14
Infrahuman Communication

Among the "higher" brain activities – defined as functions that are complex in integration and rich in structural foundation – the communicative sciences are paramount. They encompass sending and receiving signals by several modes – auditory, visual, tactile, and olfactory. Verbal language is claimed as the unique achievement of humankind and communication through speech, its formation, articulation, and comprehension, merits first place in the communicative sciences. The keynote was sounded by the Darwinian crusader, Thomas Huxley:

> Our reverence for the nobility of manhood will not be lessened by the knowledge, that Man is, in substance and in structure, one with the brutes; for, he alone possesses the marvelous endowment of intelligible and rational speech…. (Huxley, 1863, p. 132).

The 19th century's historical interest in problems of speech stimulated the field of neurology and propelled it toward the multidisciplinary neuroscience, taking second place only to the then-dominant neurophysiology. For a brief period in that century the localization of speech to a specific brain region was a major question in the broader problem of localization of function.

Among other influences contributing to the development of the communicative sciences was the preoccupation of the early experimental psychologists with vision and audition – the psychophysicists Helmholtz and Fechner come to mind – fields which provided a 19th-century example of the "new" interdisciplinary biopsychology. Subsequent involvements with psycholinguistics and theories of language evolution and acquisition provided a broad base for study of speech pathology and therapy, audiology, ophthalmology, and special education. As with the neural and behavioral sciences, there has been and continues to be an abundance of developmental and adult problems and their gradual elucidation has created a wealth of historical material.

Yet another characteristic of the communicative sciences and information transfer in several modes is the interplay of human activities with those of other animals. The urge to find out just how animals communicate and to breach the human-infrahuman barrier has compelled advances in knowledge of communication behavior in diverse species. That conspicuous element of the research in the field has been a source of wide media attention as the public is drawn to the spectacle of communication with nonhuman species.

Semiotics

The bewildering diversity and massive confusion of information exchange in the biological context was given some organization and meaning by Thomas A. Sebeok (1920–2001), distinguished linguist and eloquent student of language at the University of Indiana, Bloomington. From Greek roots, "semiotics" was launched in 1963 and defined as " 'the scientific study of signaling behavior in and across animal species' " (Sebeok, 1972, p. 61). Subsumed by the umbrella term were anthrosemiotics and zoosemiotics, or human and nonhuman transmission of signals and inevitably their crossover in human-to-animal interaction (Sebeok & Umiker-Sebeok, 1980). In a tightly edited volume, Sebeok (1977) brought together descriptions of the many, sometimes bizarre, biological modes of communication and advocated the study of zoosemiotics for full understanding of human communication. In a positive review of that massive book, the sociobiologist, Edward O. Wilson (1978), predicted future research on animal communication would be as fruitful as in the 1960s and '70s and would be along six parallel lines: specific environmental influences, meaning context, sites of neural screening, evolutionary optimum design, decomposition of complexity, and message classification.

In Sebeok's (1977, p. 1056) words, "the subject matter of semiotics is ... messages – any messages whatsoever" and the messages consist of signs which constitute a code or set of rules "whereby messages are converted from one representation to another" (Cherry, 1966, p. 305; quoted in Ibid., p. 1055). Until the origin of language is solved, Sebeok predicted, the human-animal dichotomy would persist (see also Lieberman, 1975), but well into the 1980s, zoosemiotics had not provided a comparative perspective for language. The implications of "inner" and "outer" semiotics were fascinating a quarter-century ago and today provide both a focus and a source of research.

DOLPHIN BRAINS AND COMMUNICATIVE ABILITY

Evolution of the Cetacea – dolphins, porpoises, and whales – from land animals and the concomitant enormous expansion of their neocortex in the last 20 million years sets the scene for comparisons within the human context. They have been celebrated in mythology and in an extensive anecdotal literature (Kellogg, 1928), and their brain "has fascinated and baffled anatomists of almost three centuries, as has their behavior the naturalists of two millennia" (T. Edinger, 1955, p. 37). Scientific work on cetaceans, however, was sporadic and the resultant accumulation of knowledge became incremental.

Anomalies of Cetacean Brain Anatomy

Whales were among the first groups of mammals that "abandoned the normal, terrestrial way of life" (T. Edinger, p. 37). In the process, major feats of adaptation occurred in response to severe environmental pressures and created the new field of paleoneurology.

Cerebral Cortex

In the dolphin brain the frontal lobe is undeveloped, placing the motor cortex at the extreme rostral end of the cerebrum, where it presents a broad and flat surface. There are differentiated regions for movements of the eyes and flippers demonstrated by electrical stimulation under local anesthesia (Lilly, 1958), but no evidence of hemispheric specialization or dominance. Compared with primates, the major proportion of the cetacean cerebral hemisphere consists of an enormously enlarged parietal lobe, the absence of olfactory nerves, but a well developed rhinencephalon and limbic system (see Fig. 14.1); the somewhat primitive appearance of the neurons was confirmed (Jacobs, Morgane, & McFarland, 1971).

Auditory System

The most notable morphological difference from other mammals, in addition to shear size, is the greatly enlarged auditory system, consisting of the inferior colliculus, medial geniculate body, and the temporal lobe. Those structures were described systematically by neurologist O.R. Langworthy (1931, 1932) and further elaborated by anatomist L. Kruger (1959). That arrangement is consistent with the development of sound-production and analysis specialized for echolocation – the perception of objects by their echoes, or sonar. Realization of the cetaceans' great capability in use of this ultrasonic communication system was possible only after captive specimens were successfully trained and free-swimming schools methodically observed.

The priority for suggesting the capability for echolocation in cetaceans goes to Arthur F. McBride, curator at Marine Studios in Florida and a life-time observer of porpoise behavior. In an unpublished note written 2 July 1947 he described the ability of *Tursiops truncatus* to detect temporary avenues of escape from fine seines even in murky water at night. In his words, "[M]ight we not suspect that the above described behavior is associated with some highly specialized mechanism enabling the porpoise to learn a great deal about his environment through sound?" [1] Five years later, that possibility was formally recognized in a study on bottlenose dolphins from each side of the Florida coast and Bimini, British West Indies (Kellogg, Kohler, & Morris, 1953). With frequency analysis of amplified sounds from a hydrophone, the scientists distinguished high-frequency whistles and clicks from the background noise and suggested

[1] McBride's observation was brought to publication in a letter by W.E. Schevill (1956, p. 154).

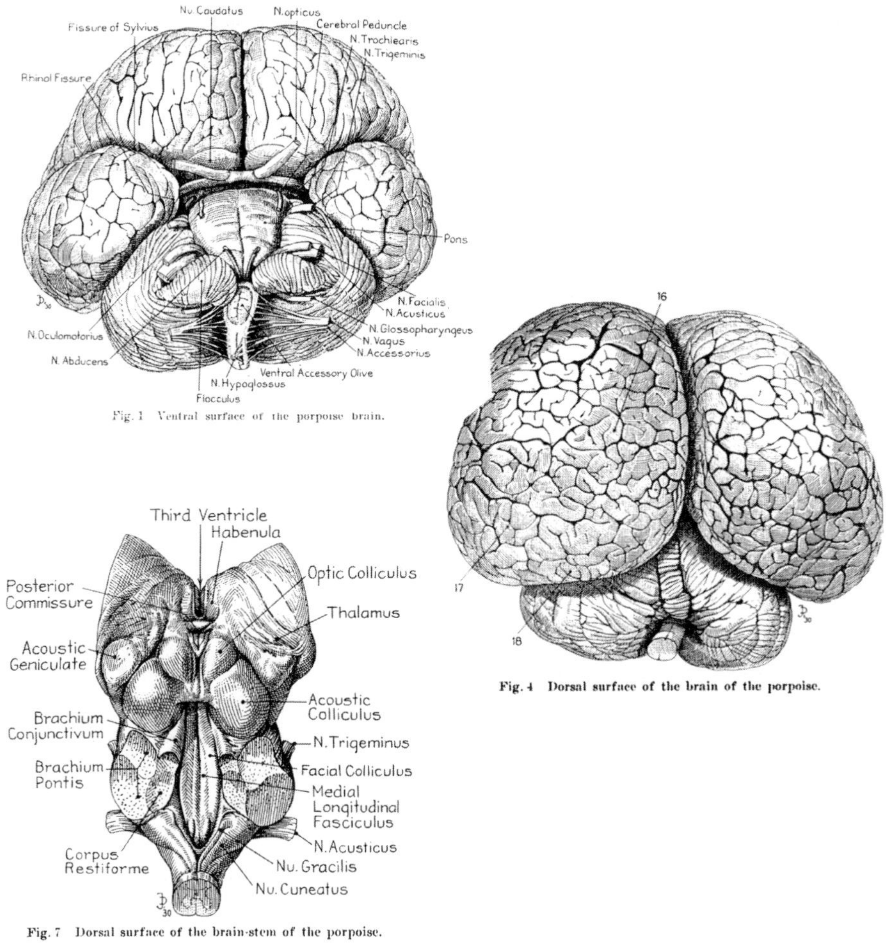

Fig. 14.1. Drawings of the ventral and dorsal surfaces of the dolphin (*Tursiops truncatus*) brain and of the ventral brain stem showing the highly developed auditory apparatus. From Langworthy, 1932, pp. 442, 448, 454; Figures 1, 4, 7.

tentatively that "the sounds emitted by these remarkable animals can conceivably be used for echoranging or echolocation in water" (Ibid., p. 239). Apparently not as familiar with dolphin behavior as was A.F. McBride, they concluded that the final proof of use of a dolphin "sonar system" would be to test it after vision had been eliminated.

Behavioral Studies

False Starts

During the first half of the last century, at least two teams of American scientists attempted to study the physiology of captive dolphins and failed because of ignorance of the adaptive mechanisms developed in the evolution of an air-breathing mammal living in the sea. Cetaceans accomplish the vital function of respiration by a pattern of brief surface breaches during which clearing the blowhole of water and inhalation of air occurs at intervals ranging normally from 1 to about 4 minutes (Kellogg, 1961). The initial failures were attributed to the general anesthetic agents' action on the cortex, where the dolphin's voluntary respiratory center probably is situated (rather than in the brain stem as in most mammals breathing automatically), thus interrupting the sensory control exercised by Cetacea.

In 1928, six adventurous investigators traveled to a fishing station at Cape Hatteras, North Carolina where 10 young male dolphins were available. As related by a participant, their plans for physiological observations were thwarted by "technical difficulties" (the animals died when anesthetized) but the brains were removed and fixed in formaldehyde for shipment to Baltimore. There the Johns Hopkins neurologist-psychiatrist, Orthello Richardson Langworthy (1897–1996), made extensive studies on Weigert-stained sections "prepared under the direction of Dr. Adolf Meyer," who had already worked up a porpoise brain (Langworthy, 1932, p. 441). In his first report, Langworthy (1931) noted the foreshortening of the skull to accommodate the blowhole, which represented the migration and merging of the nostrils, and the absence of the olfactory nerves and bulbs. Under the light microscope, he found the cortical layers poorly differentiated and the cell arrangement primitive. The detailed report (Langworthy, 1932) confirmed the unusually large cochlear portion of the auditory nerve and suggested the possibility that in the porpoise it had taken over equilibrium functions. Probably all optic fibers decussate and reach the thalamus on the opposite side, as the eyes migrated from their forward positions to the sides, precluding binocular vision. Langworthy concluded that in spite of size and "marked infolding of the cerebral cortex" (Langworthy, 1932, p. 476), it was not yet possible to suggest a correspondingly large intelligence.

The second team attempt to study the anesthetized dolphin was made in 1955 and was briefly described: Eight scientists from five laboratories[2] had high hopes of success at Marine Studios, St. Augustine, Florida, but as in 1928, "Our neuroanatomical specimens were magnificent and our neurophysiological findings were zero" (Lilly, 1958, p. 500). A member of the team, neuroanatomist L. Kruger, then at Hopkins, carried out an extensive histological investigation of the cerebral structures for sensory perception. He confirmed the earlier reports of an absent olfactory system and enlarged auditory structures, and noted in addition the relatively large size of specific

[2]The group consisted of Lawrence Kruger, Vernon Mountcastle, Jerzy Rose, Joseph Hind, Clinton Woolsey, Leonard Malis, Karl Pribram, and John Lilly.

thalamic nuclei reflecting the specialization of cetacean sensory systems (Kruger, 1959).

Successful anesthesia of Cetacea was obtained eventually with a mixture of 50 percent nitrous oxide in oxygen delivered by endotracheal intubation in a preset timing pattern of inflation and deflation. In addition to the usual observations, the animal's condition was monitored by rectal thermometer, electrocardiogram, and tongue color (Nagel, Morgane, & McFarland, 1964).

Work of John C. Lilly

To one participant in the 1955 expedition, the lure of behavioral studies with dolphins was irresistible. John Lilly (1958) circumvented the general anesthesia problem and found that under local anesthesia metal guides could be easily placed in the skull through which electrodes were inserted. With electrical stimulation he found in two dolphins evidence of positive and negative reward systems: It was easy to train them to self-stimulate and terminate rewarding and unpleasant stimuli respectively. The presence in Cetacea of subcortical reward and punishment systems near the head of the caudate nucleus was conducive to their facile operant conditioning (Lilly, 1961) and opened the possibility for a series of studies. These were carried out at the Communication Research Institute, St. Thomas, Virgin Islands, for which Lilly had secured funding from the Office of Scientific Research and Development (OSRD) through the National Science Foundation, the Office of Naval Research, and the Air Force Office of Scientific Research, sources indicative of the national interest in cetacean research.

Through the work at St. Thomas, it was found that bottlenose dolphins produce and can hear sounds to at least 200 kilocycles, but usually communicate with one another at frequencies of 8 to 10 kilocycles. The institute produced the first evidence that clicks and whistles are generated separately as well as simultaneously and that the animals have two independently controllable sonic emitters (Lilly & Miller, 1961a). Amplified and recorded on magnetic tape, "[t]hese sounds are classified as vocalizations used for communication", specifically with other dolphins (Lilly & Miller, 1961b, p. 1693). In addition, the animals were found to be capable of engaging in vocal transactions with human trainers, even to mimicking the same number and duration of bursts as the trainer (Lilly, 1965).

Cetacean Intelligence

The obvious intelligence of Cetacea invited comparisons, and among the first to attempt such studies, the Canadian psychologist, Donald Hebb and his collaborator, A.F. McBride (1948), were able to report only incidental observations. After a lengthy discussion of the difficulties inherent in intelligence measurements for comparative purposes, they concluded that none was possible. John Lilly, in contrast, attempted to equate brain size with capability for language development. In his paper presented at the International Symposium on the Physiological Basis of Mental Activity in Mexico City, October, 1961, he first proposed the possibility of such a relationship and offered evidence suggesting "that in mammals, at least, a threshold brain size is necessary for the development of language, i.e., language as the normal adult human being knows it" (Lilly, 1962, p. 424). Lilly was to follow this speculative line of thought with Communication Between Man and Dolphin: the Possibility of Talking with Other Species (1978), including those from outerspace.

Hawaiian Spinner Dolphins

By far the most comprehensive American work on cetacean communication was by Kenneth Norris and his associates. They chose the Hawaiian spinner dolphin as their subject[3] and the study of free-swimming, wild behavior as their ultimate goal, which they finally achieved with documented observations of dolphin schools in their native habitat.

Kenneth Stafford Norris (1926–2000)

Considered a primary founder of marine mammalian science, Norris (Fig. 14.2) was born in California and educated in zoology at the University of California at Los Angeles. He abandoned his early studies of circadian rhythms in desert reptiles and moved south, taking his doctorate at the Scripps Institute of Oceanography at La Jolla, the well-spring of his career-long interest in the large denizens of the sea. After a series of appointments along the coast of California, Norris developed a base at Kealake'akua Bay, Hawaiian Islands, on a quiet, sandy body of water frequented by spinner dolphins. There he and his team developed techniques and approaches to observe freely swimming schools with emphasis on their means of communication and aerial behavior. Summarized and illustrated in detail in Norris, Würsig, Wells, and Würsig (1994; Fig. 14.3), the authors concluded that "spinning behavior serves to produce a matrix of acoustic markers that define school shape … ." (p. 121). The markers included spins, bubble clouds, and several types of leaps and slaps, all conjectured as a means of defining the school's swimming envelope and for socializing.

From sound emissions recorded with hydrophones placed strategically in the dolphins' diurnal pattern of activity, mimicry of a characteristic signature-whistle of an individual animal (proposed as a form of recognition by Tyack, 1986),

[3]Individual spinners leap vertically above the water surface and can execute as many as four rotations before falling back (Norris, Würsig, Wells, & Würsig, 1994).

Fig. 14.2. A relative rarity – a native-born Californian – Kenneth S. Norris was instrumental in establishing marine mammalian biology as a science. He was photographed ca. 1980. From Norris, Würsig Wells, and Würsig, 1994, p. 7, Figure 2; with kind permission.

Fig. 14.3. By close observation of free-ranging spinner dolphins, as shown in a spin, above, as well as in schools, Kenneth Norris and his colleagues discerned many details of cetacean communicative behavior. From Norris, Würsig, Wells, and Würsig, 1994, p. 10; with permission.

was seen as "the keystone" of a sensory integration system (Norris, Würsig, Wells, & Würsig, 1994, p. 349) for communication of information essential to individual and species survival. From submersible observation chambers, small planes, and a steep seaside cliff they observed behaviors they believed to be a manifestation of "obligate cooperation" among these "quintessential group animals."

Historically, the long trail of research on Cetacea has served to reaffirm the importance of gestalt studies. Norris and his associates made the point that only in the natural-group context is it possible to discern the true behavior of societal animals such as apes, monkeys, and elephants. In their words: "Such societies seem to be knitted together by communication systems that carry a rich freight of information" (1994, p. 347). The extent and content of that information carried by biological sonar is for the most part not yet known, but what has been discovered is evidence of a high degree of intelligence in those mammals.

ECHOLOCATION

In the third of a series of NATO-sponsored conferences on animal sonar systems, held in Denmark in 1986, the tantalizing question of the evolutionary origins of the echolocation developed by flying mammals played a recurring note. One suggestion put forth was that

> leaps after aerial insects were gradually perfected under the constraints of natural physical laws, such as conservation of angular momentum, until mid-leap maneuverability crossed the border from acrobatic to aerobatic .... Vision seems a more likely candidate to provide the appropriate resolution and range, just as it does with many microbats today (Pettigrew, 1988, p. 649).

Examination of how the sonar and visual systems interact was suggested as lending clues to sonar insect-capture.

In a boost to progress of the field, the communication capabilities requisite for the phylogenetic emergence of echolocation were delineated at the same conference by Evans and Norris (1988, p. 665).[4] The echolocating animal must have

(1) near-field hearing for feedback control of emissions,
(2) far-field hearing for incoming signal analysis,
(3) sharp onset signals and high-speed processing ability,

[4]These authors inadvertently invoked the magical number 7 as originally expounded by G.A. Miller in 1956 (G.A. Miller, 1994).

(4) directionality of sound emissions,
(5) high acoustic sensitivity with overload protection,
(6) a stable time-base for uniform analysis, and finally
(7) the outgoing sound must not neurally suppress the echo.

Animals feeding in open space and in darkness were most dependent on those acquisitions. Their eventual attainment of a point of convergence in evolutionary time was considered as instrumental in allowing (or even driving) the adaptation of some land mammals to aquatic life (Norris, Würsig, Wells, & Würsig, 1994) and a separate but parallel development was taken by the dark-flying bat. If that be the case, then bat sonar evolved to detect targets in light-free caves, as three other flying vertebrates – swiftlets, oil-birds, and megabats – have developed independently under similar conditions.

Bats and Sonar

First Detection

The use of biological sonar by bats was published by G.W. Pierce and D.R. Griffin in 1938, although the animals' prowess at avoiding objects in flight had been reported earlier in England (Hartridge, 1920). By the 1950s, Griffin's work accounted for most of what was known about bats. That body of knowledge and the large literature on echolocation seemed to indicate that "the bat's sonar may be a comprehensive mode of perception fully up to the task of substituting for vision as a means of experiencing important aspects of the near and moderately-near environment" (Simmons, 1973, p. 157).

Donald Redfield Griffin

As a graduate student at Harvard with Lashley in 1934, Donald Griffin was advised to consult George Washington. Pierce (1872–1956), an electrical engineer who was recording insect sounds with a sonagraph newly developed by Bell Laboratories. The resulting collaboration produced the first evidence (Pierce & Griffin, 1938) that bats emit ultrahigh-frequency sound. With a fellow graduate student, Robert Galambos, a study was made of the ability of *Myotis lucifugus* to avoid stationary objects in the dark (Griffin & Galambos, 1941). Measurements from their movies of bats (Fig. 14.4) in flight[5] revealed the animals' amazing abilities in detail and as with bird song (see below), the use of the new spectrograph made possible the fine analysis of their sound emissions. Griffin, a self-styled experimental naturalist in his autobiographic account, had been banding "as many bats as I could catch" (1985, p. 124). In

[5]Now in the Robert Galambos Collection, Neuroscience History Archives, Brain Research Institute, University of California, Los Angeles.

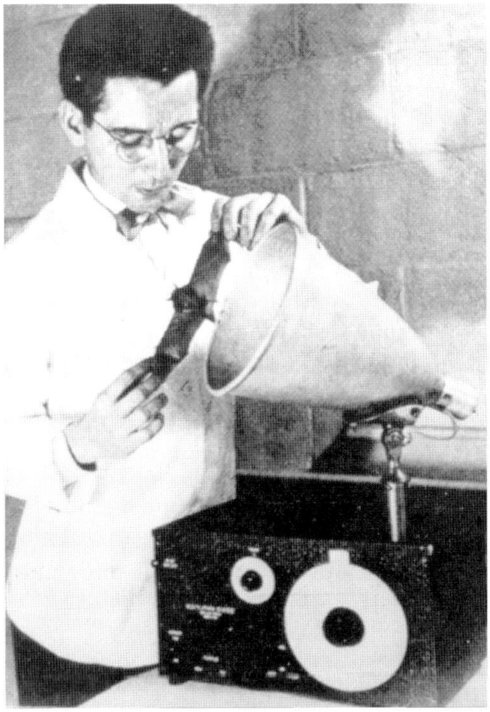

Fig. 14.4. A cinema frame filmed ca. 1940 at Harvard University of Robert Galambos recording sounds emitted by a captive *Myotis lucifugus*. Reproduced with kind permission.

pursuit of his interest in bird navigation, he enlisted the enthusiastic collaboration of Alexander Forbes, neurophysiologist at Harvard's department of physiology and avocational small-plane pilot and explorer. They tracked ground-released gulls so successfully that Griffin later purchased a small plane and learned to pilot it himself. Discouragingly, however, the data collected on the homing flights of the birds did not resolve the contemporary argument over whether homing was accomplished by random scattering or by systematic exploring.

Griffin's 7-year tenure at Cornell University and experiments on stationary-obstacle avoidance were interrupted by the Second World War and participation in improving speech reception in noisy environments. While at Cornell, his ideas inculcated by early exposure to Jacques Loeb's viewpoint of behavior as being governed by innate tropisms were shaken by Karl von Frisch's demonstrations of the honeybee's use of symbolic communication of distance, direction, and desirability of food through "waggle-dances." Griffin furthered Von Frisch's discoveries by persuading Cornell University Press to publish the German's "messy" manuscript of his lectures given while on tour of the United States.

Shortly thereafter, Griffin's scientific thinking received an additional jolt. By chance carrying out an experiment with free-flying bats catching insects at night over a pond, he discovered that they were using echolocation to catch their targets, an observation not possible with captive bats avoiding stationary obstacles (Griffin, 1953). Returning to Harvard, he continued his comparative studies of the characteristics of echolocation utilized by the many different groups of bats. In ever more refined detail, the work spread to other laboratories, including that of J.A. Simmons (1973) at the Auditory Research Laboratory, Princeton University, who compared four species of echolocating bats. He found from simultaneous target-range discrimination measurements an acuity of 1 to 3 centimeters that was independent of absolute range. His subjects could fine-tune their signals to produce long-duration and short-duration CF/FM sonar and perceive by time-delay the distance to target. The data suggested a neural mechanism for a "matched-filter, ideal sonar receiver which functionally cross-correlates a replica of the outgoing signal with the returning echo to ... determine its arrival time (Ibid., p. 157). Those studies were forerunners to sorting out differences in orientation sounds used by different groups of bats (Grinnell & Brown, 1978).

Later Directions

Griffin did not neglect his earlier interest in bird navigation and was never convinced that the earth's magnetic field is involved (1985). His move to Rockefeller University in 1965 to establish a research program in ethology and neurobiologic mechanisms underlying animal behavior signaled a shift of his bent away from the physical characteristics of echolocation and toward the sources of behavior. That program, envisioned by Fairfield Osborn of the New York Zoological Society and Detlev Bronk at Rockefeller, was the genesis of later studies by others, for example, of bird song. Opportunities for ethological studies greatly expanded in 1970 with a field station established at Millbrook, New York. Griffin's thinking culminated in his belief in the possibility that animals possess consciousness and have subjective feelings, summarized in *Animal Thinking* (1984), in which he elaborated and encouraged a new subdiscipline of cognitive ethology.

Continuing the thread of inquiry and talent, when graduate student Alan Grinnell discussed his plans about possible experimental approaches to the problem of animal communication with Robert Galambos, he was urged to talk with Donald Griffin. The outcome was 15 years of productive investigation of bat behavior and sensory modalities that helped bring animal communication into the mainstream of neuroscience research. Grinnell focused his attention on the physical and anatomic aspects of bat sonar communication (e.g., Grinnell & Brown, 1978) and later shifted his focus to the physiologic mechanisms of the neuromuscular junction.

BEAUTY OF BIRD SONG AND SINGING

The mythic reputation of whales and dolphins brought them to American scientific attention during the 1920s, as described above, an interest that crystallized before midcentury when the cetaceans' proficiency in the use of sonar became clear. The century also saw another modality of animal communication take center front on the broad stage of neuroscience, the singing of birds. The major share of the sustained interest in bird song and singing has been British, in keeping with the source of our most eloquent paeans to those tiny songsters.[6]

Song birds are among the few animal groups that learn to vocalize and in that sense are analogous to humans learning to talk. The vocalizing prowess of song birds not only became a field of serious scientific study, it also led to discoveries in learning and developmental behavior far beyond those revealed by work with dolphins. Some contributions to understanding animal behavior that have emerged from research on bird song were listed by one of its most distinguished investigators, Masakazu Konishi (1985) at Caltech: species-specificity in animal signals, an innate predisposition to learn, sensory templates for motor development, maps of neural substrates of behavior, and elevation of the instinct-learning (nature-nurture) semantic question to a higher level. Research on bird song and singing has and will continue to provide opportunities to investigate learning, memory, developmental plasticity, sensorimotor coordination, and sexual dimorphism. A trend toward fuller integration of perspectives in animal communication studies was evidenced by W.J. Smith's (1991) close analysis of the processes of signaling. An ethologist based at the University of Pennsylvania, Smith identified at least five general uses for signaling, including territorial probing and defense, tracking and coordinating conspecifics, "tonic" effects of repetition, interchanges between parents and offspring, and social interactions. He clearly distinguished two separate components of avian signaling: song, made up of primary units, and singing, characterized by formal sequential order, and he deemed them manifestations of "a fundamental signaling device of wide applicability. Singing performances incorporate not just songs, but many kinds of component signals. We must recognize this ... if we are to uncover ... the many ways they contribute to communication" (W.J. Smith, 1991, p. 250).

[6] Shelley's "The Lark"; Byron's "The Prisoner of Chillon." In a rough estimate of relative research interest in the United Kingdom and the United States, we compared the number of references cited by Catchpole and Slater (1995) in comparable serials, *Nature* and *Proceedings of the Royal Society of Science* versus *Science* and *Proceedings of the National Academy of Sciences, USA*; there were 37 percent more in the British publications than in the American.

American Contributions

The Sound Spectrograph

Before significant progress could be made in the study of auditory transmission as a channel of communication by animals, a device was needed to capture and record the very high-frequency sound waves they characteristically emit. About midcentury such an instrument, the frequency spectrum analyzer, was developed by Bell Laboratories in New Jersey, the research arm of the Bell Telephone Company. Its pioneer use is credited (Catchpole & Slater, 1995) to Canadian zoologist James Bruce Falls (1963) and refinement and testing of the playback technique to W.A. Searcy and P. Marler (1981). After an extensive series of field studies of song recognition, in which playback methods were widely used (in contrast to their sparse application to research in mammalian sounds), Falls later wrote (1992) a critique of the method, its limitations and possibilities. In general usage, for example in the hands of W.H. Thorpe in England, the sonagraph revolutionized the study of sound and made bird song an efficient medium for investigating processes of learning. The result was that "[T]he present literature on learning is dominated by studies on song birds" (Catchpole & Slater, 1995, p. 32). And again (p. 45), "The development of bird song has probably been studied in more detail than any other aspect of animal behavior."

Species Recognition and Territorial Defense

One of the first elements of bird song to be investigated using playback of modified magnetic tapes was to tease out the important factors in conspecies recognition. Peter Marler (1970) suggested that the most reliable features for recognition were the more invariant elements of a song (those that vary little and are relatively constant between individuals). In indigo buntings, the time intervals between song units were found (Emlen, 1972) to be more important clues to conspecies recognition than was the sequence of units, thus constituting evidence of a temporal element in avian learning as seen in their temporal patterns.

Red-winged blackbirds were frequent subjects of studies of natural bird song because the males combine visual and vocal displays in the maintenance of small, defined territories in easily observed habitats (often among reeds growing at the edge of ponds or ditches). An experimental testing of the relative importance of song and epaulet coloration in defense of nesting territory suggested a three-tier system of communication: an "advertising song" aimed at potential but distant trespassers, a visual warning to closer conspecifics aided by spread of the brilliant epaulet, and finally the signal for impending chase and attack (Peek, 1972). Other field studies with this species suggested that an advantage in territorial defense was conferred on birds with larger song repertoires, perhaps because it indicated higher status or evolved as a reduction in monotony (Yasukawa, 1981).

Structural Arrangements
From sonagrams, C.H. Greenwalt (1968) discerned that the membranes on each side of the syrinx, positioned at the immediate entrance to the bronchi, could act together or separately, and so must have independent neural control; from his data, he proposed the "two-voice" theory of bird song. After the observations were confirmed in England, Fernando Nottebohm (1971, 1972) demonstrated the presence of separate sound sources by unilateral transection of the hypoglossal nerves or removal of the cochlea in chaffinches. A conspicuous feature of sensorimotor control of bird song was the left-side lateralization at the level of the motor neuron with ipsilateral predominance of projections to the many syringeal muscles. Nottebohm formulated a theory of learning (Fig. 14.5) that postulated innervation by the left hypoglossal nerve ahead of the right hypoglossal, a primacy that conferred greater vocal control during ontogeny and was further developed by vocal learning. The neurology of the forebrain nuclei involved in the efferent pathways was mapped in the canary (Nottebohm, Stokes, & Leonard, 1976). Whether or not a left-right bias in motor control exists, as in the human brain, is unclear (discussed by Arnold, 1982).

Twenty years later, single- and multiple-unit analyses of avian vocalizations were being carried out by Daniel Margoliash and collaborators at the

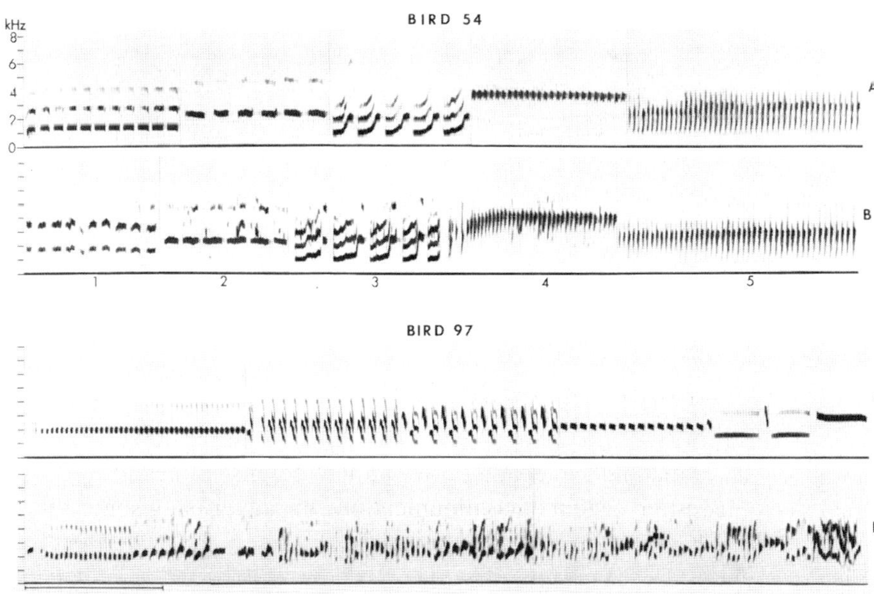

Fig. 14.5. Segments of song recorded spectrographically before (A) and after (B) lesion of the right hyperstriatum ventrale, pars caudale (HVc) in canary 54 (upper traces) and the left HVc in canary 97 (lower traces). The striking song deficit after the left lesion suggested left hemispheric dominance. From Nottebohm, Stokes, and Leonard, 1976, p. 274, Figure 12.

University of Chicago. A temporal hierarchy in neural control of the syringeal muscles was demonstrated in adult, male zebra finches fitted with delicate implanted electrodes (Yu and Margoliash, 1996). From responding neurons in the HVc and RA nuclei (see Fig. 14.6), correlations of spike train histograms, sound spectrograms, and oscilloscope traces of neuronal activity prompted the conclusion that "These data imply a hierarchical organization for the forebrain control of bird song production … ." as Tinbergen had anticipated in 1950 (Ibid., p. 1874).

Memory, Learning, and Feedback

As noted, the enormous literature on this topic is rife with research on bird song and singing (see Arnold, 1982; Konishi, 1985; Nottebohm, 1991). Prodromal to the refocusing of interest from innate behaviors to the developmental questions was a critique of Lorenzo's and Tinbergen's theory of instinctive behavior by an early and prominent animal behaviorist, D.S. Lehrman (1953). He attributed the rapid spread of their traditional concept of innate patterns in part to its warm reception by U.S. ornithologists, and deplored the resulting distraction from the more fundamental problems of developmental psychology. Gradually, interest in stereotypic, inflexible behavior slipped into proper perspective and amazing behaviors were uncovered.

A few American highlights convey the depth and scope of those investigations and the intense interest in their psychologic revelations. Among Nottebohm's contributions was the suggestion that song learning was due to the interaction of two "critical periods," one for learning an auditory template and the other for motor learning (1969). From studies on birds raised experimentally in auditory isolation the conclusion was that "[w]hen birds develop song without any song tutor, they use an innate template or an internal reference for the control of voice by auditory feedback" (Konishi, 1985, p. 134). But when a tutor was present with which the isolate could socially interact with visual contact, it was found that one song was transmitted culturally across three generations of birds (R.B. Payne, 1981). Konishi produced a flow chart (Fig. 14.6) of a neural model of song learning showing motor and auditory neurons mixed in the same singing center nucleus. From his own work and that of Katz and Gurney (1981) with multiunit recordings, the auditory neurons were shown to be inhibited during motor neuron firing. Relatedly, in a reassessment of vocal learning in birds, Nottebohm (1991) suggested that the ability to modify vocal sounds through auditory feedback might be an evolutionary device to protect the cochlear cells from damage by the loud, low-frequency sounds of self-vocalizations. In his words, the "roots of vocal learning, the most momentous breakthrough in animal communication, might be found in a modest brainstem [sic] reflex" (p. 210).

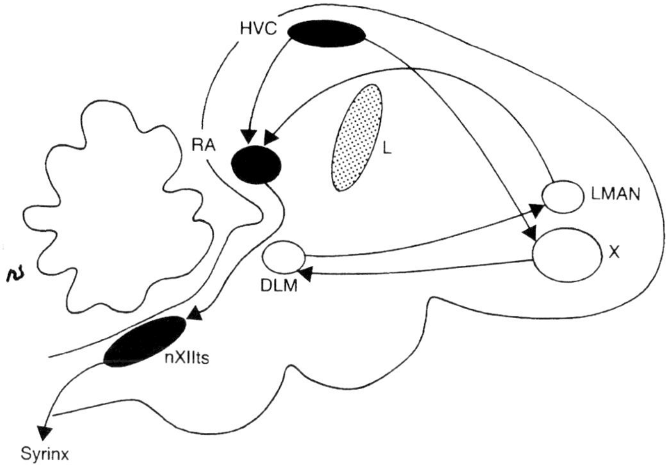

Fig. 14.6. Diagram of the main neuronal pathway for bird-song control [HVC – higher vocal center, RA – robust nucleus of the archistratum, L (stippled) – main auditory field, nXIIts – tracheosyringeal portion of the hypoglossal nucleus, DLM – dorsolateral nucleus in the thalamus, X and LMAN – groups of forebrain nuclei]. From Doupe and Konishi, 1991, p. 29, Figure 2.4.

As the methodology became more complex, so did the import of the discoveries. For example, the influence of the behavioral state was demonstrated: The RA neurons were shown to spontaneously produce bursts of spikes (without vocalization) and to respond to sounds during sleep (Dave, Yu, & Margoliash, 1998).

The Neuroendocrinology of Song

Preceding the midcentury ascendency of neuroendocrinology as a subdiscipline of neuroscience (see Chapter 5), and contributing importantly to that rise, the gonadal effects on behavior in birds already had been shown. The new direction, moving from studies of the constituent elements of bird song to its hormonal control, was an unplanned research strategy that in hindsight seems only natural.

Androgens and Vocalization

An American clue to the relationship between the gonads and communicative output was provided by C.R. Carpenter (1932), a prominent pioneer in studies of infrahuman primates (see Chapter 10). He reported that crowing of roosters diminished after castration, an observation well known in the husbandry of the succulent capon. A few years later, there appeared a short paper from Rutgers University announcing that four of five female canaries had burst into male-type song after injections of testosterone propionate (Leonard, 1939). Decades later it was shown that the HVc nucleus of female canaries increased in size after administration of testosterone (Nottebohm, 1980), a change accompanied by neurogenesis (Goldman & Nottebohm, 1983). The effect of androgen was confirmed by a student of Nottebohm's, A.P. Arnold (1975), who extended the studies to include courtship and aggressive behavior. Arnold's work during the next quarter century helped clarify several aspects of the relationship between gender and androgen accumulation in brain regions of the zebra finch (Arnold, 1980). The distribution of androgen at different levels of the neural projections to syringeal muscles suggested that the hormone could modify more than one function in that specialized brain circuit. The unexpected and unique qualities of bird song "... [have] made the song system an important model for studying neural control of species-specific and learned behaviors in general" (Arnold, 1982, p. 90).

Exploitation of the triad of bird song, developmental psychology, and gonadal physiology provided fascinating examples of the interplay of behaviors between the sexes under experimental conditions and were presumably illustrative of free-ranging conditions. Under the aegis of the Rockefeller University program mentioned earlier, female canaries exposed to playback of tapes of elaborate repertoires of male song built nests faster and laid more eggs than did canaries hearing playback of more restricted repertoires. In a broad sense, this was held to demonstrate an attentiveness to the attributes of male song, and must have influenced the evolution of singing behavior (Kroodsma, 1976).

And Finally Genes

Inevitably, research on the motor control of bird song and the neural pathways involved in its perception moved into the molecular era of the gene, a general area pioneered by Seymour Benzer, molecular biologist at Caltech in Pasadena. Although his work has been largely with Drosophila, it was among the first to prove a direct relationship between genes and behavior (Benzer, 1973; Miller & Benzer, 1983). An early finding in birds, again from the Rockefeller University Laboratory of Animal Behavior, was the experimental induction of gene expression in the forebrains of zebra finches and canaries (Mello, Vicario, & Clayton, 1992). A rapid increase in a transcriptional regulator of an immediate early gene occurred after exposure to tape-recorded conspecific songs in a primary brain region (field L) known to be effectively activated by that stimulus. Analysis of the cells involved suggested the level of induction to be proportional to the number of cells recruited and a role for "genomic responses" in song pattern recognition and auditory assessments. Repeated playing of the same song brought about a rapid decrease in the gene's response (Mello, Nottebohm, & Clayton, 1995).

In an end-century summary of genetically triggered sexual differentiation of brain and behavior, Arnold (1996) traced the theoretical background of gonadal hormone control of structure, which he and Nottebohm (1976) had been the first to report in the vertebrate brain. He cited the results of treatments with several steroids in avian species that did not support the traditional hormonal theory of sexual dimorphism (Arnold, Wade, Grisham, Jacobs, & Campagnoni, 1996) and argued for the plausibility of its direct genetic control through gene product(s). Proof of such a mechanism would require multiple lines of evidence independent of hormonal influence, experiments which are underway in several laboratories.

COMMUNICATION BY INFRAHUMAN PRIMATES

The human urge to communicate with other species of animals amounted to an obsession in some observers of animal behavior, who set out to learn the meaning of the sounds produced and to mimic them as closely as possible. During the late 19th century an American naturalist/explorer consigned his energies to that endeavor.

Richard Lynch Garner (1848–1920)

His Life

According to his obituary in the New York Times of January 24, 1920, Garner's life was unusually adventuresome. Born in Abington, Virginia, he ran away from home to join the Confederate Army while underage and was captured and held in three successive Northern prisons. After returning home and completing his education, he founded the high school at Williamstown, Kentucky.

From his youth, Garner had dreamed of visiting the jungles of Africa to learn about the life of the apes and in preparation he visited the zoological gardens of the eastern and midwestern United States and patiently observed, recorded on a phonograph cylinder, and learned to imitate the "words" that monkeys and a few apes used to communicate with each other. Although his "investigations [were] chiefly in the direction of learning one tongue" (Garner, 1892, p. 131), "[i]n quest of the great secret of speech," he took many "detours" for comparative studies. With the latest phonographic equipment then available, he recorded in addition the sounds made in captivity by many other species of animals including tigers, lions, leopards, as well as cats, dogs, birds, humans, and musical instruments. By meticulous analysis of those recordings, Garner searched for (there were no precedents to guide him) the components of animate and inanimate sounds, looking for any common characteristics; he varied the speed of the revolving cylinders, ran them backwards, and experimented with playback techniques.

From those experiments Garner became convinced that "all mammals possess the faculty of speech in a degree commensurate with their experience and needs ... and seem to be as capable of producing and controlling sounds as the brain is of thinking ... ." (Garner, 1892, p. 173). This combination of vocalization and brain power reflected the contemporary concept of speech as a measure of intelligence. To Garner speech was "only one mode of expressing thought, and ... the sounds must be voluntary, have fixed values, and be intended to suggest to another mind a certain idea, or group of ideas ... ." (Ibid., p. 169).[7] He learned to reproduce calls of captive monkeys and demonstrated to his own satisfaction that they distinguished colors, numbers to at least 3, and cause and effect (Ibid., pp. 32; 204). The speech of simians occupied such an important place in Garner's thinking that he chided Darwin for having "so nearly omitted the question of speech from a work of such ample scope, such minute detail, and such infinite care as characterizes the 'Descent of Man' ... ." (p. 154).

Garner made several expeditions to tropical Africa, the last of 4 years duration under the auspices of the Smithsonian Institution and from which he had returned only 6 months prior to his unexpected hospitalization and demise. With the idea of living as closely as possible to the animals he wished to study, he had a steel mesh cage constructed in demountable sections which he occupied (Fig. 14.7) for weeks at a time in habitats frequented by his target subjects "in the freedom of their native jungle," a favorable experimental condition "not hitherto ... enjoyed by any other student of Nature" (Garner, 1896, p. vii).

Research on Vocalization

Garner's main interest lay in the vocalization of nonhuman primates. In the course of recording phonographically and learning to imitate simian communication, he apparently became sufficiently proficient in mimicking some calls that he could elicit responses from wild individuals. By moving his cage into new territory remote from his earlier locale, Garner established the fact that previously recorded calls held significant meaning for a separate genus. He was reported to claim that apes have some 20 words, including those meaning Where are you, Here I am, and Wait for me. See also Chapter 9.

Other early findings may be credited to this pioneer in the communicative sciences. He suggested (1892) that a succession of recordings of infants' vocalizations from birth to 3 years would reveal how speech develops in the human. And he voiced his intention to "try to secure photographs of the mouths of the great apes while they are in the act of talking ... ." (Ibid., p. 223) with the objective of furnishing the eminent Alexander Melville Bell (father of

[7] About 15 years later, Robert Yerkes cautiously attributed the capacity for ideation to animals, in this case to two pigs (Yerkes & Coburn, 1915).

Fig. 14.7. A demountable steel cage for protection enabled Richard L. Garner and his native companion to study animal behaviors and vocalization in African forests. From Garner, 1896, facing p. 22.

the inventor of the telephone) some "new and novel subjects" for his study of the coordinated movements necessary to produce speech. In the judgment of a distinguished younger student of apes, Garner, although untrained, made many solid observations and "the writer humbly confesses that the more he learns about the great apes ... the more facts and valuable suggestions he discovers in Garner's writings" (Yerkes, 1925, p. 169).

Apes in Captivity

A Question of Language

Robert Yerkes, America's most persistent and eloquent champion of anthropoid apes as ideal research subjects (see Chapter 10), wrote a slim volume in which he briefly described previous "temporarily fruitful" research stations for psycho-biological study of primates. He also recounted his acquisition of a famous pair of chimpanzees, felicitously named "Chim" and "Panzee" and most of the text was given over to musical notations of the apes' vocalizations set down by his skilled collaborator, Blanche Learned. That novel but laborious method of recording ape sounds yielded the conclusion that

> Although the young chimpanzee uses significant sounds in considerable number and variety, it does not ... speak. Consequently there is no chimpanzee language, although there certainly is a useful substitute which might readily be developed ... if the animals could be induced to imitate sounds persistently (Yerkes & Learned, 1925, p. 60).

The efforts expended in trying to coax Chim or Panzee to utter human-like word-sounds were further elaborated in the benchmark tome, *The Great Apes*, by Ada and Robert Yerkes. They wrote that the problem seemed to be the chimpanzees' lack of a tendency to mimic auditory stimuli, in contrast to their ready reaction to visual stimuli. "If a language can be developed by them it is more likely to be constituted of visual or kinaesthetic than of auditory elements .... Speechlessness notwithstanding, intercommunication is highly complex and useful in the chimpanzee" (Yerkes & Yerkes, 1929, p. 309). Herein lay the spark, fed by the work of Romanes and Köhler from abroad, and Garner's continued studies, that ignited the bonfire of teaching American Sign Language (ASL) to clever chimps. This became the method of choice after many unsuccessful experiences such as described by Kathy and Keith Hayes (1951). After 2 years of trying to teach "Vicki," a chimpanzee, to communicate in English, her repertoire was limited to "mama," "papa," "cup," and "up," all explosively delivered as Vicki strained to manipulate her lips and tongue.

Teaching Symbolic Language to Chimpanzees

Although a nonstop group effort was required plus high degrees of motivation and patience, to some psychologists and anthropologists it seemed reasonable to look to our closest relatives for a precursor of some element of the ability to use language (Terrace, 1979). "Washoe" in the hands of Allen and Beatrice Gardner (1969) at the University of Nevada, "clearly earned her place in history as the first chimpanzee to communicate with the words of a natural human language" (Terrace, 1979, p. 10). During 3.5 years, she learned 132 words expressed in ASL (and understood the meaning of three times as many), but the data did not clarify whether she used words grammatically or at random. Nonetheless, the Gardners' experience influenced concept and method of all subsequent studies (Wallman, 1992). Washoe was further trained by R.S. Fouts (1972), then returned to her owner.

Do You Speak Yerkish?

Studies of anthropoid communication were thrust into the digital age by an experiment with "Lana," a chimpanzee born at the Yerkes Primate Center in Atlanta, Georgia. Using an artificial language called "Yerkish" in honor of its source, D.M. Rumbaugh (1974, 1977) and colleagues found that Lana could learn to manipulate the abstract symbols on a computer keyboard to obtain services such as food and drink from invisible trainers. The advantages of the computer, including no articulatory burden, an immediate permanent record of the transaction, and absence of possible human clues, significantly aided the reorientation of research interest to the nature of referential communication with single symbols (Wallman, 1992).

A different method was used in training another young chimpanzee. "Sarah" was taught with a synthetic language developed by David Premak (1971) and his group at the University of California, Santa Barbara. She learned to associate magnetized plastic chips of different size, color, shape, or texture with words representing objects, actions, or concepts and achieved a remarkable display of ability to "read" sequences of chips and demonstrated simple categorization of objects. Again, no firm conclusions from the data were forthcoming and the experiments with Sarah gradually moved from language to cognition (Wallman, 1992), as the field shifted from psycholinguistics to comparative cognitive science in the 1980s.

Premack, who embraces the school of ethologists that believes in the utility of teaching a form of human language to apes so they can reveal their thought processes, affirms the possibility of a deeper understanding of the animals' capacity beyond language acquisition (Beer, 1986). But the training must be optimal to accomplish such a goal: failure is ambiguous, "[i]t could indicate an incapacity of the species, or merely that the training was improper. Only when we know what constitutes proper training can we be certain who failed – teacher or 'pupil'" (Premack, 1971, p. 821). He was critical of the inflated claims of previous reports that chimpanzees used rules in their manipulation of symbolic representations of words and believed that there is more thought beyond language learning in human children than the apes displayed. Premack's experiments, most recently set forth in 1986, led him to declare the deficiencies of earlier work and to point new directions for future studies. The criticisms were ratcheted up by R.J. Sanders, a psychologist who worked with the chimpanzee named Nim Chimsky which "failed to use sign order as a syntactic device" (Sanders, 1985, p. 197), but was very good at imitating the trainers as a way to get rewards.

The Nim Project

The winds of criticism blew to a storm with publication in 1979 of Herbert S. Terrace's Nim, a detailed account of data collected on the personal development of a captive chimpanzee similar to those amassed by psycholinguists on human language development. With the assistance of a large battery of teachers and analyzers, Nim's training sessions were videotaped, analyzed, and categorized daily. The data formed a fascinating and reasoned natural history

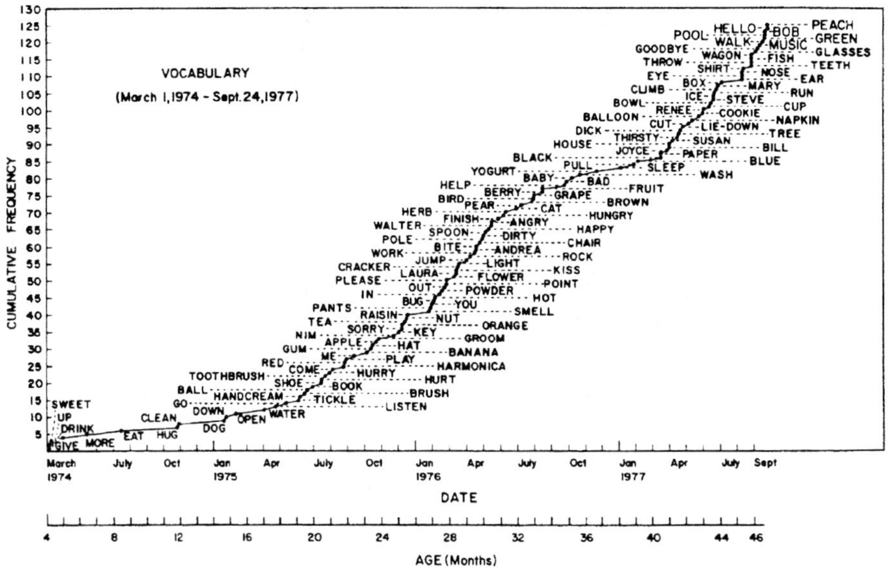

Fig. 14.8. The capacity of a young chimpanzee to acquire American Sign Language (ASL) was demonstrated by "Nim" in the vocabulary sequence shown above. From Terrace, 1979, p. 138.

of a complex project in estimating animal intelligence – administrative details, observations, analysis, theory, and conclusions. One example from many (Fig. 14.8) showed the sequence of acquiring new signs, with spurts and plateaux on a steadily rising slope, as in any learning curve. Nim's eagerness to learn is palpable from his expression (Fig. 14.9).

Comparisons of the Nim data – distributional, statistical, and semantic analyses – with those of a child learning to talk revealed similarities (and some surprises) but again, no clearcut revelations. Terrace wrote (1979, p. 221):

> [U]ntil it is possible to defeat all plausible explanations short of the intellectual capacity to arrange words according to a grammatical rule, it would be premature to conclude that a chimpanzee's combinations show the same structure evident in the sentences of a child.

He stated that the "uncontaminated" experiment has yet to be carried out, then divulged his feeling that it would be worth trying: "Communication with another species at the level of human language would be as exhilarating as receiving a message from outer space" (p. 226).

Gorilla gorilla Weighs In

A primate nicknamed "Koko" is the recipient of the longest, uninterrupted, and most dedicated effort to communicate on human terms so far attempted.

Fig. 14.9. "Nim's" eagerness to learn is unmistakable. From Terrace, 1979, p. 89.

Born in the San Francisco Zoo in 1971, Koko was adopted by an eager graduate student in anthropology at Stanford University, and after surmounting recurring financial and administrative hurdles, has been the subject of the only ape-language experiment with a "constant, uninterrupted record of progress" (Patterson & Linden, 1981, p. 209). The record of continuous instruction from one teacher now extends for almost 30 years (see www.koko.org) and Koko is credited with having learned the meaning and used appropriately about 250 ASL signs as well as understanding additional verbal communications.[8]

The acceptance of Koko's achievements as a demonstration of language capability, comparable in any way to that of humans was mixed: "Project Koko began during turbulent times in the behavioral sciences, and it was only because [of] previous pioneering work with chimps that I had any chance of being taken seriously" (Patterson & Linden, 1981, p. 15). Noting that "[l]anguage is a fugue in which the linguistic elements are but one voice" (Ibid., p. 207), Patterson wrote that the school of linguists that posits syntax as the crucial voice displays a certain circularity in its arguments. To nonlinguists

[8] Koko is talked to simultaneously with signing.

the evidence showed that apes use sign language "spontaneously, appropriately, and creatively" (Ibid., p. 210). At the least, Koko has contributed (and continues to do so) to the demystifying of language. The controversy that flared in the 1970s over interpretation of apes' use of symbols in communication seemed occasionally to stoop to hair-splitting, as in the exchange of letters in Science between Terrace and associates with Patterson (1981), who vigorously defended her interpretations of Koko's achievements.

Species Differences

In spite of sweeping criticism of methods and interpretations from respected figures in the field, the controversy did not die (Gibbons, 1991) but received new fuel from reports of experiments with a different species, the pygmy chimpanzee. The lessons gained from studies with Lana mentioned above were extended at the Yerkes Center with an animal perhaps even more closely related to modern humans than the species usually studied (Savage-Rumbaugh, Sevcik, Rumbaugh, & Rubert, 1985). A pygmy chimpanzee, "Kanzi" spent his first 2.5 years with his mother, while she was being trained in the use of Yerkish. Although he gestured and vocalized freely and seemed not interested in the lexigraph, when permanently separated from his mother, Kanzi displayed remarkable spontaneous linguistic ability with the lexigram keyboard, without instruction. In the absence of training and food rewards, Kanzi gradually learned symbol meanings as well as comprehension of spoken English without the conditioning of having been taught. The team working with Lana and Kanzi and their respective kin emphasized the differences between two species of chimpanzees (*Pan troglodytes* and *Pan paniscus*) and considered the explanation of the latter's linguistic competence a unique and unprecedented challenge for both philosophers and psychologists.

The more readily available *Pan troglodytes*, however, has held its own valuable niche as provider of amazing discoveries. Ever more complicated training and testing of the 20 or so chimpanzees versed in the use of symbolic representation of language continues to yield data documenting the animals' capacity to learn. To cite only one example, unprompted recall and reporting of hidden objects were achieved reliably by an 11-year-old chimpanzee experienced in the use of lexigrams (C.R. Menzel, 1999).

Among Free-Ranging Animals

While the language-learning controversy was raging, other investigators more adapted to outdoor living plunged into the jungle or savannah in search of natural communicative activities characteristic of wild-living creatures. In the manner of their prototype, John Garner, but without a cage, they recorded for later analysis the spontaneous sounds and calls of animals within a group and utilized the new playback technique in experimental manipulations. Among those intrepid outdoorsmen were R.M. Seyfarth, D.L. Cheney, and Peter Marler (1980). Observing vervets, a small monkey in Kenya, Africa, their objective was to determine whether or not nonhuman species in natural conditions could "use symbols to represent different objects without support from contextual cues" (Ibid., p. 801), as had been demonstrated in captive apes. From responses to played-back alarm calls to the presence of eagle, python, or leopard, the researchers concluded that the calls were not used for generalized arousal as earlier interpreted by Struhsaker (1967), but had specific semantic meaning.

The Gombe Chimpanzees

Arguably the most motivated of those true pioneers was a self-proclaimed naive graduate student at Stanford University, Jane Goodall. Using her savings from waiting tables for transportation and with the support of Louis Leakey, she arrived in Tanzania in 1960 to begin what extended to a 25-year study of chimpanzees in Gombe. Leakey reasoned that chimpanzee behavior in natural settings might reflect the behavior of early man (Goodall, 1986). "[The] sophisticated social knowledge [in a community] is the matrix in which communication is embedded" (p. 116). Among the daily stream of data recorded manually by pairs of trained assistants were notes on communicative activities – visual, olfactory, tactile, and auditory.

The communicative role of vision and smell seems obvious, and that of touch, as manifest by grooming, "is a crucial component of messages intended to reassure or appease distressed or tense individuals" (Goodall, 1986, p. 124). The auditory mode is more direct in its effects. "Chimpanzee vocalizations are closely bound to emotions. The production of a sound in the absence of the appropriate emotional state seems to be an almost impossible task for a chimpanzee" (Ibid., p. 125). The experiences of the Gardners and the Yerkeses described earlier bore out Goodall's observation. She published a preliminary list relating chimpanzee calls identified by ear to their presumed emotions and found 15; to that number Peter Marler, using a sound spectrograph for analysis, added 17 more. The sequence of calls carried observable meaning and they were often interspersed with nonvocal sound signals such as drumming, stamping, skin scratching, or swishing a tree branch.

Goodall's magnificent contribution to understanding behavior of humankind's closest relative includes the formation of dialects (pattern variation at the group level) and identification of a capacity to combine information separated by time and space. She emphasized that "It is the different ways in which signals are combined that, along with the behavioral context in which they are perceived, enables receivers to interpret correctly the meaning of a given communication sequence" (Goodall, 1986, p. 139). Her perseverance in a longitudinal study of a community of chimpanzees was the means of discovery of territorial warfare among the subhumans. Goodall postulated that chimpanzee behavior shows precursors to pseudospeciation[9] and viewed their lack of language as the only buffer between them and warring humans.

In his rugged study of the mountain gorilla, G.B. Schaller (1963) taped and made sound spectrograms of their vocalizations, developing summary tables of the signals. In addition, he was able to distinguish two displays: chest beating and a strutting walk.

Other Species

Several series of well planned experiments on an animal not usually the subject of experimental work, the African grey parrot, have found their way into the scientific literature and are the basis of continuing inquiry. "Alex" was purchased and trained in the Chicago area by anthropologist Irene M. Pepperberg. She and her students used as far as possible a natural learning environment (not laboratory operant conditioning) and succeeded in engendering referential, vocal learning in the parrot. Two experiments yielded results comparable to those being reported for mammals, using a model/rival approach and the same/different concept. Alex's reliability in counting to six and sorting objects into categories, compared with those in children and chimpanzees, she felt demonstrated an evolutionary convergence of specific mental abilities. In open-minded judgment, Pepperberg declared:

> My goal is ... to present my data to encourage a reevaluation of animal capacities. For far too long, animals in general, and birds in particular, have been denigrated and treated as merely creatures of instinct rather than as sentient beings .... [A]ll we need to examine these capacities are some enlightened research tools (Pepperberg, 1999, pp. 327; 328).

Her suggestion had had a substantial basis from neurological quarters 15 years earlier. Norman Geschwind and Antonio Damasio collaborated in a summary of what was known about the neural basis of language and their concluding remarks enthusiastically anticipated further studies:

> We believe ... that animal models will lead to experiments that will cast significant light on the evolution of language and other human functions, as well as new approaches to the study of disordered dominance, either developmental or acquired (Damasio & Geschwind, 1984, p. 145).

Another species in which communicative ability has attracted study is the elephant. The use of low-frequency sound by these large mammals was documented by investigators associated with the University of Virginia in captive

[9] Transmission of individually acquired behavior from generation to generation.

individuals (K.B. Payne, Langbauer, & Thomas, 1986) and later in the field (Poole, K.B. Payne, & Langbauer, 1988). The latter studies were extended and subsequently described in an exhaustively anecdotal but fascinating publication (K.B. Payne, 1998) and contained a suggestion but no evidence that in digging for water in a floodplain elephants might use sonar echoed from their footfalls (p. 251). Cooperative work (by personnel from the universities of Virginia and Mississippi and Etosha National Park, Namibia) related optimum atmospheric conditions (early evening, low humidity, calm air) on the savannah habitat to the estimated extreme range of 10 km for sounds with frequencies of 14 to 35 Hz at sound pressures to 103 dB (Garstang, Larom, Raspet, & Lindeque, 1995). From the multiple parameters of their findings, those authors suggested that "the use of low frequencies as a means of communication by terrestrial animals should be treated in a broad biophysical context" (Ibid., p. 950). Again, the accelerating trend toward studies embedded in an ecologic matrix of many subdisciplines became discernible.

New Directions

Late-Century Evaluations

Looking at 20th-century American research in communicative sciences from the vantage of hindsight, one sees a melange of endeavors on the part of many disciplines to comprehend the capabilities of divergent species in the broad arena of language. The elusive proof that animal subjects understood meaning and had syntactic skills was challenged vigorously; not only could ape signing be demonstrated as imitative behavior but the theoretical foundation for animal language was nonexistent (Sebeok, 1977). In a careful screening of the available data, the unresolved issues of proof of mental continuity between beast and man were summarized as (1) cueing, (2) changes in criteria, (3) observers' expectancy effects, and (4) the difficulty of replication (Ristau & Robbins, 1982) That examination went further in seeing occasional clues to the unavoidable question of animal consciousness: "It is not clear that we can progress very far by avoiding the issue of consciousness, too." The ape language projects and other research "nibble at the edges of the problem … ." (p. 216).

Additionally, the constrained laboratory approaches favored by experimental psychologists were eroded by the animal behaviorists' interest in the natural repertoires of signals and their functions among free-ranging populations (Tuttle, 1986). At century's start, "The Yerkeses were constrained by the prevailing Scale of Nature, leading from the lower creatures through the apes to us, the highest form of earthling." Of further historical interest would be what attracted the attention of behavioral researchers who today "attempt to understand the adaptive complexes of each species in its own right" (Tuttle,

1986, pp. 304; 311). The general effect of the negative thinking about apes' use of symbolic language was to discourage new projects in the old mold and hasten an innovative approach to the problem based on many perspectives.

Comparative Developmental Evolutionary Psychology

By the fourth quarter of the century, abandonment of the stimulus-response hegemony and its replacement by cognitive science had forced a changed emphasis in research programs from language acquisition studies in apes to the broader consideration of animal intelligence, including human. In a clarification of the issue it was noted that:

> The disagreement is not so much about what the apes do as it is about the relationship between what apes do and human children do .... Ultimately, because they are the experts in language, linguists and psycholinguists must have the last word in the matter of ape language abilities (S.T. Parker, 1990, p. 37).

But it was more than the linguists' domain that was important. By bringing together representatives of many subdisciplines with interests in brain and behavior – comparative, developmental, and evolutional biology – the scope of psychology was broadened and the communicative sciences' domain redefined. The power of this new mix, as it related to information processing by infrahuman animals, was evidenced by publication of a great variety of research work in monographs and articles. One such volume originated by invitation from Departments of Biological Sciences at the Universities of Arizona and Nebraska, where the domains of behavioral ecology and comparative psychology have been and continue to be particularly dynamic. The editors' goal of cross-fertilization of new ideas was expressed thus:

> When disciplines interact and begin to share facts, ideas, methodologies and hypotheses, new interpretations and theories are formulated. These fresh, new viewpoints may be so utilitarian that they are immediately incorporated into existing research programs (Balda, Pepperberg, & Kamil, 1998, p. viii).

That approach embodied the essence of neuroscience and was as properly applicable to the study of human speech as it was to interspecies and intraspecies communication. Historically, it strengthened the emergence of a new frontier in neuroscience – cognitive development (S.T. Parker & McKinney, 1999).

Chapter 15
Speech and Human Language

The full picture of the human communicative sciences encompasses both the initiator and the receiver of the signal. The former has two tasks – to compose a meaningful message in proper symbolic representation and semantic organization, and to control the throat and facial muscles that articulate it. In turn, the recipient must perceive (hear) the message and comprehend it. Such a sketchy two-way matrix does not even hint at the many perturbations and embellishments that may besiege or grace, as the case may be, this major higher nervous activity.

APHASIAS

Recent Background

In a repetition of the familiar pattern, during the early modern period interest in speech was generated by medically trained men who had patients unable to talk – who were aphasic and often had other dysfunctions. There was also the more general attraction of the biological bases underlying human behavior which gained ascendancy after the phenomenon had been freed from the theorizing of philosophers and theologians. The argument over whether language was God's gift to mankind or was a product of man's reason was resolved in 1770 by a German philosopher-historian, Johann Gottfried Herder (1744–1803), who proposed that both language and reason were natural, i.e., innate, entities (O. Marx, 1984; Fromkin & Rodman, 1978).

Initial Continental and English Contributions

During the 19th century, among medical circles speech and its dysfunctions held a sustained interest, especially in Central Europe. Franz Josef Gall (see Chapter 1) countered the philosophers' generalizations with a firm idea of the "internal faculties" situated in specific organs of the brain. He, the Doctors Dax, Bouillaud, and Aubertin all supported the involvement of the left inferior frontal cortex in speech. Decades later, with the case of "Tan" and other patients demonstrating motor aphasia correlated with disease of the left inferior frontal gyrus, Paul Broca (1861) established that region as the seat of articulate speech. Carl Wernicke (1874) added sensory aphasia following injury to the left posterior first and second temporal gyri, adjacent to the auditory receiving area. A period of diagram-makers ensued in the attempt to explain disconnection syndromes after injury to the supramarginal parietal gyrus connecting the sensory temporal area with the motor frontal cortex, a paradigm of the reflex arc. Sigmund Freud (1891) proposed a number of speculative concepts in what Henry Head (1920) termed "the chaotic period" in this field marked by a proliferation of "simplistic and diagrammatic accounts." Jules Dejerine (1893) clarified sensory aphasia which followed injury to parastriate occipital cortex, adjacent to the visual area, with impaired reading of a written language (alexia). He also identified disconnection syndromes after injury to the parietal angular gyrus connecting the sensory occipital area with the motor frontal cortex. Pierre Marie iconoclastically agreed only to the importance of the left Sylvian cortex as involved generally in the mental comprehension of language (Cole & Cole, 1971).

The contribution made by Hughlings Jackson, "the Nestor of English neurologists" (Collins, 1898, p. 135), was to propose an understanding of aphasia based on the physiology and anatomy ("external") and the psychology ("internal") of normal speech. Those early holistic concepts had been defended in a memorable debate with Broca at a meeting of the British Association for the Advancement of Science at Norwich in 1868. Jackson's concept, enlightened later by Henry Head (1920) and the work of others, thus combined mental and motor activities, or internal (intellectual) and external (expressive) processes, which could be validated by evidence from close examination of speech disorders, autopsies, and observations of language learning. Those conceptualizations reinforced the idea of speech as a natural, rather than a social science (O. Marx, 1984) and at the turn into the 20th century, aphasia and language studies enjoyed a widespread attention from Continental, British, and American investigators.

The American Scene

During the late 19th century, physiology of the nervous system in the United States was in a backwater and lacked productive research activity. In contrast, and rooted in their experiences during the Civil War, the neurologists maintained an interest in the system's neuroanatomy and "[a] succeeding professional development of great consequence was the emergence of neurology as a clinical specialty" (Frank, Marshall, & Magoun, 1996, p. 555–556). The American Neurological Association (ANA) was organized in 1875 with annual meetings, and "a handful of practitioners in the larger cities… began to specialize in organic and functional diseases of the nervous system. Among this group of men who practiced psychiatry to make a living and neurology because they were devoted to it", three may be highlighted: Moses Starr, Charles Mills, and Adolf Meyer.

Moses Allen Starr (1854–1932)
Starr (Fig. 15.1) had plans for a career in classical culture when he graduated from Princeton and embarked for Germany to study Greek and Roman history. In Berlin, however, several visits to Helmholtz's laboratory revived a latent interest in natural science. He returned to his native New York, graduated from the College of Physicians and Surgeons (P&S), Columbia University, did a residency at Bellevue Hospital, then returned to Europe to work at Heidelberg, Vienna, and Paris. On return to New York he set up a laboratory in his home and in 1884 published an essay on the sensory tracts of the central nervous system, elucidating some of the then-current questions of myelination.

Starr's regard as an American pioneer in the field of cerebral localization stemmed from his participation in a symposium on that subject with the famous English neurologist David Ferrier and neurosurgeon Victor Horsley, who were delegates to the 1888 Congress of American Physicians and Surgeons in Washington, D.C. He became professor of nervous diseases

Fig. 15.1. Moses A. Starr supplemented his neurologic practice and teaching with anatomical research on sensory tracts and cerebral localization. Photograph from Denny-Brown, 1975, p. 112.

Fig. 15.2. One of the earliest (i.e., oldest) of the American contributors to professional neurology, Charles K. Mills combined human brain stimulation studies of speech with neurological diagnosis.

at P&S (1888–1918), president of the ANA (1897), and held memberships and offices in international, national, and local organizations. His lasting influence was as an educator: "Following in the footsteps of Sequin [at P&S] ... he represented the teaching vanguard of early academic neurology" (Pearce Bailey, 1975, p. 113).

Charles Karsner Mills (1845–1931)

The second neurologist whose career helped shape one of the early trails to a unified American neuroscience was considered the "dean" of neurologists of his era (McHenry, 1969, p. 81) and filled the position well. Mills (Fig. 15.2) was twice elected president of the ANA (1886 and 1924), was one of the first physicians to devote full time to neurology, and in 1877 was appointed neurologist to the Philadelphia Hospital, "the first position of its kind in Philadelphia and probably the earliest in the country" (Weissenberg, 1931, p. 172). The specialized collection of neurological material at that hospital was perhaps unique and served the teaching needs of the large cluster of medical institutions in that locale.[1]

[1]The role of Surgeon-General Hammond in establishing a specialized Army Hospital at Turner's Lane and its use for study by S. Weir Mitchell are described in Chapter 1.

Growing up near Philadelphia, Mills entered the Union Army during high school in what was called the "war class" because so many of its members saw military duty. After graduation in 1864 and teaching for a few years, he gained the M.D. (1869) and Ph.D. (1871) at the University of Pennsylvania. Mills's long career of teaching, practice, and publication commenced and ended at his Alma Mater. He was especially gratified to be appointed professor and head of neurology and commenced assembling members of a renowned "school of neurology."[2]

Mills's many publications included a widely regarded textbook (1898), which was unique for that time in that it included a section on "Chemistry of the Nervous System" and made reference to the work of Thudicum, surpassing Hammond's text as "the foremost in the Nineteenth century" (McHenry, 1969, p. 331). Mills was primarily interested in cerebral localization and he and collaborators used faradic current to stimulate the human cortex. He contributed several studies of the cerebral "zone of speech;" among them and in association with a younger colleague, William Spiller, was "The symptomology of lesions of the lenticular zone with some discussion of the pathology of aphasia." This classic paper on localization of a center for word deafness on the superior temporal areas of both hemispheres came to a conclusion representative of many aphasia studies of the time: "[W]e find it somewhat difficult to form positive conclusions … ." (Mills & Spiller, 1907, p. 649).

Adolf Meyer (1866–1950)

A third noteworthy figure in aphasia studies of turn-of-the-century United States was a Swiss-American psychiatrist who attained great prominence in his profession. Meyer (Fig. 15.3) was born near Zurich and in due time (1890) received the M.D. from its university, having worked with August Forel on the reptilian brain. After postdoctoral visits to neurological laboratories in five European countries, the ambitious young physician sailed for the United States, settling for an honorary (nonsalaried) position at the new University of Chicago. There he apparently had a key role in the momentous introduction of the Norway albino rat to psychological studies (see Chapter 9). In a short time he moved on to other locales and positions that offered greater opportunities for work with neurological and mental patients.

Meyer's impress on the development of psychiatry was enormous. His great opportunity knocked when he was invited in 1910 by Johns Hopkins to become founder of the Phipps Clinic and departmental chair and the time arose for innovations in patient diagnosis and education of a new breed of psychiatrists. Meyer's insistence on the close relation of mental disorder to

[2] It was also an elitist group as attested by Samuel Ingham (see p. 390).

Fig. 15.3. The broad influence of Adolf Meyer on the development of psychiatry and its methods was assured by his long tenure at Johns Hopkins Medical School, 1910 to 1941.

the patient's life history and his teaching of a dynamic psychiatry paved the way for an acceptance of psychoanalysis (Rioch, 1975).[3] His influence spread through an "attitude of orderly enquiry ... that has permeated all the major developments in the field" (Ibid., p. 158).[4] The orderliness stemmed from a conservative Swiss background and the dynamics from uncommon energy and motivation.

Adolf Meyer's interest in aphasia commenced in 1895 when he was in charge of a group of mental patients at the Kankakee State Hospital of the Illinois State Hygiene Department. Through several moves that provided additional study material, he continued intensive programs of research and

[3] Among Meyer's pedagogic innovations was brain modeling, a student exercise later incorporated by David Rioch in a famous brain anatomy course at Harvard Medical School. The end-product was a larger-than-life, three-dimensional model of tracts and nuclei rendered in color with Plasticine, the synthetic, malleable material widely used in creative elementary school programs.

[4] That statement was evidenced by a recent myopic analysis of the history of American neuroscience (Cowan, Harter, & Kandel, 2000) that identified the diverse group of investigators assembled by neuropsychiatrist David Rioch, a protégé of Meyer's, at the Walter Reed Army Institute of Research in Washington, D.C. as one of three pillars of American neo-neuroscience.

writing, and "[h]is contributions to our knowledge of aphasia and his critical, analytical review of that field, culminating in the Harvey Lecture of 1910, are classics" (Rioch, 1975, p. 156). The Harvey Society presentation was a medical treatise, fully illustrated (48 figures) as he worked his way through historical cases in the literature and from his own practice. He cited both positive and negative support for his views and an innovative hypothesis of segmental organization of the brain. In 1898, Meyer had published essays criticizing methods and deductions of contemporary neurology; these were analyzed in detail by the neuroanatomist, Jerzy Rose, in partial celebration of Meyer's 70th birthday. "Instead of the time-honored purely topographically division [of the central nervous system] he suggested combining the neurons into systems if they can be shown to be interrelated in their activity" (Rose, 1951, p. 58). Applied to the aphasias (Meyer distinguished many interactions), he could conclude only that more painstaking anatomical data were needed, then added rather limply "But we certainly have gained some broad lines of hemisphere function" (Meyer, 1910, p. 243).

Joseph Collins (1866–1958)

The lack of real progress in solid knowledge of the faculty of speech was evident in a monograph by Joseph Collins, "Professor of Diseases of the Mind and Nervous System" at the New York Post-Graduate Medical School. *The Genesis and Dissolution of the Faculty of Speech: A Clinical and Psychological Study of Aphasia* (1898) was awarded the Alveranga Prize by the College of Physicians of Philadelphia and was a clear statement of the concept that "the zone of language" was not a map in the traditional (European) sense because the phylogeny and ontogeny were too varied to submit to rigid delineation (an echo of Hughlings Jackson's views). The zone was where processes essential to the speech synthesizing faculty – conception, comprehension, expression – occurred, and both topographically and morphologically was expediently close to the frontal convolutions, the intellectual area of the brain. Collins gave full weight to pathological and clinical experience and declared

> that in every form of genuine aphasia ... there is some perversion of idea externalization and of language. It is to Dejerine and [others] that we owe the proof of this statement ... and to them we cheerfully accord deserved praise for maintaining their contention in the face of very nearly universal opposition (Collins, 1898, p. 131).

Proof of Localization

Persuasive support of the close relation between specific brain regions and speech production came from the work of Wilder Penfield and his associates at the Montreal Neurological Institute. During neurosurgical operations (see Chapter 7) they electrically stimulated the surface of the exposed unanesthetized brains of patients who were talking. On-going speech was blocked by excitation of the parietal-temporal area, the inferior frontal area, and the supplemental motor area of the left hemisphere (Penfield & Roberts, 1959). Those experimental protocols were a continuation of Penfield's long quest to add to the knowledge of body representations on the neocortex. The early American introduction of "principles of localization" to the neurosurgical operating room are discussed in Chapter 4.

Ebb and Flow According to Geschwind

Until the 1920s, a general view of advancement in understanding aphasia held that after Broca there ensued a 60-year Dark Ages, dominated for the most part by German scholars with anatomical tunnel vision (Geschwind, 1964; 1966). That picture was drastically revised when it was shown that the classical figures were indeed aware of both physiologic and behavioral aspects of speech.

A Tidal Reversal

The most authoritative accounts of aphasiology's 20th-century history were written by the neurologist who had a major role in its shaping. In 1964, Norman Geschwind posited his disquietude regarding the "standard" version of the revolt against the classical anatomic beliefs, led by Marie in France, von Monakov in Switzerland, Goldstein in Germany, and Head in England: Had it been as revolutionary as depicted? Those "rebels" had stigmatized their predecessors as "naive and incompetent in the area of the higher functions" (Geschwind, 1974, p. 63). Geschwind and his colleagues at Boston Veterans Administration (VA) Hospital carefully read the original papers of the offending "indelible" figures and concluded: "The broadside accusations of carelessness, of inadequate examination, of unconcern with psychology were all unfounded, at least when the best of the classicists were involved" (Ibid., pp. 63–64).

To Geschwind the most interesting "reformer" was the German-American, Kurt Goldstein (1878–1965), "who has actually had much more profound effects on the thought about the higher functions, certainly in the United States and probably in England, than any of the others, even including Head" (Geschwind, 1974, p. 65). A one-time assistant to Edinger, neurologist-psychiatrist Goldstein developed a concept of holistic organismal approach to brain disease or trauma, an original idea that had less impact than warranted due to competition from advances in electrical techniques and neurosurgery (Denny-Brown, 1966). Some disorders of behavior described by Goldstein lent themselves, according to Geschwind, to clarification from the point of view of organismic reactions, others from an "atomistic" perspective; the error came from not considering both views to determine which is most appropriate to the patient.

Norman Geschwind (1926–1984)

A top student in all his high school subjects and protégé of his teachers, Geschwind's (Fig. 15.4) eagerness to learn and capacity for scholarship took him through Harvard college and medical school. In the middle of college he served in the infantry in Europe and Japan and discovered his interest in human behavior under stress, a subject which shaped his career (Damasio & Galaburda, 1985). After an internship at Boston City Hospital, he had 3 productive postgraduate years at London's National Hospital Queen Square (1952–1954), then returned to Boston City Hospital as Denny-Brown's chief resident. A 2-year vacation from the clinic was spent with F.O. Schmitt at MIT, working on the axonal membrane (see Chapter 6). Those years also harbored a quantitative study of language of aphasic patients in cooperation with

Fig. 15.4. In mid-20th century, Norman Geschwind singlehandedly revived the flagging American interest in the aphasias. He and associates discovered a neural basis for functional differentiations that related handedness to brain asymmetries.

Davis Howes, an experimental psychologist interested in language (Geschwind and Howes, 1962).

In 1958, Geschwind joined the neurology staff at the Boston VA Hospital affiliated with Boston University, and intensified his study of unusual aphasic syndromes (Geschwind & Kaplan, 1962). Their patient was the

> first modern case of disconnection syndrome at a time when the neurological literature had all but denied the fundamental role of the neocortical commissures, and neglected, or actively rejected, the notion of disconnection as useful in the conceptualization of higher brain function (Damasio & Galaburda, 1985, p. 501).

This was before the work of Sperry and his associates on W.J. (noted below) and culminated in *Disconnexion Syndromes in Animals and Man* (Geschwind, 1965), a compendium of the earlier literature as well as the author's own work and ideas. The monograph, issued in three parts, stressed the role of connections between brain systems responsible for particular behaviors, and is anonymously credited as launching the new subdiscipline of behavioral neurology (Morrell, 1985; Damasio & Galaburda, 1985). In 1968 Geschwind and Livitsky published their macroscopic findings on normal human brain asymmetry and related it to handedness and other conditions, thus providing the anatomic substrate for lateralization of function and initiating searches for other behavioral and anatomic asymmetries. Relatedly and more recently, the connection of preferred handedness with one or the other hemispheres was elucidated with the Wada test (see below): In normal left and right handers; 95 percent of the right-handed and, surprisingly, 60 percent to 70 percent of the left-handed possessed language in the left hemisphere. In addition, the left-handed had "a greater degree of bilateral representation of language and other cognitive functions than right handers" (Witelson, Bryden, & Bulman-Fleming, 1999, p. 855).

The decade Geschwind spent at Boston University-VA saw the organization and immediate success of the Aphasia Research Center, representing a multidisciplinary attack on many components of the disorder which attracted both funds and a wide range of researchers. Even after moving to Harvard Medical School, Geschwind remained active in the center. His tenure of the James Jackson Putnam chair was largely devoted to a brisk teaching schedule, but it also yielded an inspired and original concept: a relation between cerebral lateralization and the endocrine and immune systems (Geschwind & Behan, 1982). *Cerebral Lateralization* was published posthumously twice, in *Archives of Neurology* in 1985 and 2 years later by The MIT Press; the author considered it "his most mature, pathmaking, and lasting contribution" (Damasio & Galaburda, 1985, p. 502). That Norman Geschwind was one of the century's most farsighted contributors to the brain sciences is seen in the final sentence of the monograph: "Understanding the ultimate origins of asymmetry will require studying not only the foundations of biology but very possibly also those of physics itself" (Geschwind & Galaburda, 1987, p. 239).

NEUROLOGISTS AND FUNCTIONAL DIFFERENTIATION

By the end of the 19th century postgraduate schools for improving the training of men who already held the M.D. had gained solid acceptance (see p. 29) and were spreading across the country as physicians treating mental and organic brain diseases were building specialized practices and clinics. A case study of the emergence of a school in this pattern is illustrative of how the fortuitous juxtaposition of a specialized group promoted a climate conducive to creative progress. Whereas an integrative neuroscience emerged on the East Coast from traditional and well established schools and laboratories, the comparable development on the American Pacific Crest arose from struggling medical schools and no visible research laboratories (Flexner, 1910). The institutional setting in Southern California – its hospitals and medical schools – was derived from the motivation and energy of individuals. The pioneer specialists trained in the disorders of the nervous system first took up practice there before the First World War and by midcentury their cooperative efforts had reversed the early situation and a vibrant cluster of schools and institutes was in place, creating an atmosphere that encouraged innovation and experiment, with a spectacular outcome.

Pioneers in a School of Neurology[5]

Carl Wheeler Rand (1886–1973)
Born in Massachusetts, after Williams College, Rand (Fig. 15.5) attended Johns Hopkins where he came under the influence of Harvey Cushing and the challenge of brain disorders. After receiving his M.D. in 1912, he followed his mentor, who was making his own crucial move from Hopkins to Harvard, and completed his residency at Cushing's clinic in Boston. Coming to Los Angeles in 1915, he established a practice in general surgery, then served in a mobile operating unit in the First World War. By the time Rand returned to Los Angeles after military service, he had become an experienced surgeon and had made the decision to devote his time to neurosurgery. He accepted an appointment to establish the first neurosurgical service at the Los Angeles County Hospital and when the University of Southern California (USC)

[5] An early version of the following sections was published in the brochure, "History of the Human Brain" (Brewer, Lomax, Marshall, O'Neill, & Magoun, 1983).

Fig. 15.5. Carl Rand arrived in Southern California in 1915, filling a void in neurosurgery in that region and contributing to its "school" of neurology.

reopened a medical school in 1928, Rand served as clinical professor of neurosurgery until his retirement in 1954.

In addition to becoming the outstanding brain surgeon of his time in Los Angeles, Carl Rand was one of the founders of the local neurological society and edited the surgical section of its *Bulletin*. In 1961, the Society of Clinical Neurosurgeons held their annual meeting in Los Angeles to honor their distinguished colleague. Rand was not overly vocal or pushy, but had earned a reputation of always being willing to help or teach anything he could. Soon after Rand began his service at the County Hospital, Samuel Ingham, a neurologist-alienist, arrived and the cooperative relationship which developed between the two men was significant in its long-term effects on local neurological practice and research.

Samuel Delucenna Ingham (1875–1965)

The second pioneer, well trained with a specific interest in the nervous system, had been raised in homespun style by his widowed mother who ran the family foundry near Philadelphia. After graduating from the Medical Chirugical College in Philadelphia in 1899, as a fledgling general practitioner

Fig. 15.6. A refugee from Philadelphia's elitism, Samuel Ingham combined neurology and psychiatry in a productive Southern California career.

Ingham (Fig. 15.6) spent several years in Colorado, then returned to Philadelphia to complete his training. While there he looked around at the many eminent, established, and competent neurologists who were too young and decided that if he stayed he would be a second-rate physician no matter how well he did, for a good many years.[6] In 1920 Ingham took his young family to Southern California. There he mingled with neuropsychiatrists and secured appointments at two hospitals; at the reopening of the USC medical school he was appointed head of neurology and psychiatry in the Department of Medicine.

Ingham's contribution to early neurology in Southern California was two-fold. He held together the Los Angeles Neurological Society when it first brought into contact practitioners in the fields of neurology, neurosurgery, and psychiatry. In addition, he had a career-long interest in the aphasias – developing specialized techniques for diagnosis of subtle deficits in

[6]From the taped interview with his son, Harrington Ingham, 15 December 1981; Series NRL Code ING, Neuroscience History Archives, Brain Research Institute, University of California, Los Angeles.

communication and for assembling and documenting material from the large number of cases at the County Hospital.

The Second Generation

Among Rand's and Ingham's many interns and residents during the years, it was a matter of good fortune that two young men came along who would later join their mentors in organizing an already active but loosely knit group into the Los Angeles Neurological Society. With the prestige of Rand and Ingham in back of them, the two former students began the innovations in the 1930s which laid the groundwork for research and advances in later years.

Johannes Maagaard Nielsen (1890–1969)

The usefulness of Ingham's aphasia cases comes to light in Nielsen's monograph *Agnosia, Apraxia, Aphasia: Their Value in Cerebral Localization* (1936), a study of clinical signs in 240 patients with verification of the lesion in 40 of them. Following the Second World War, when the Veterans Administration in Washington organized specialized treatment and research centers at a small number of VA Hospitals throughout this country, that for aphasic disorders was established at the Long Beach VA Hospital, with Nielsen in charge.

Johannes Nielsen (Fig. 15.7; right) migrated with his parents from Denmark. He learned to work when young to help support his sisters and widowed mother. His grammar school principal, "apparently because of his iconoclastic attitude, advised his mother that Johannes was not high school material" and so he was apprenticed to a carpenter. By age 17, he was a foreman and wanted a change; a career in medicine challenged him. Said to be fluent in six languages and good in math, he developed a pattern of hard work and graduated from the University of Chicago Medical School in 1923. His first taste of the Los Angeles scene was as an intern at the County Hospital where he met Samuel Ingham, who channeled his interests to neurology. Nielson's postgraduate work at Battle Creek Sanitarium, Michigan was followed by a year with Otto Marburg in Vienna. The story is told that when Marburg asked him why he was not like the other frivolous American students, Nielsen replied that he was not really an American, he was a Dane.

The standing invitation of Ingham to join him in private practice was acted on in 1929 and they became associated in the practice of neurology and psychiatry in a compatible relationship. Nielsen's prodigious role in the maturation of neuroscience in Southern California included helping organize, with Eugene Ziskind, the Los Angeles Society of Neurology and Psychiatry; being a founder of the Society of Biological Psychiatry (president 1947 and 1948); and serving as president of the American Neurological Association in 1955. A true neuroscientist, he collaborated with George Thompson on *The*

Fig. 15.7. Johannes M. Nielsen (right) became the most prominent neurologist of his era in Southern California and director of the specialized diagnostic and treatment center of the U.S. Veterans Administration Hospital at Long Beach in the aftermath of the Second World War (see also Fig. 13.3, Chapter 13). Neuropathologist Cyril B. Courville (left) held concurrent appointments in neurology and in his field at Loma Linda School of Medicine and the Los Angeles County Hospital. His editing and organizational efforts fostered the cohesiveness of the early Southern California school of neurology. Photograph from Denny-Brown, 1975, p. 484.

Engrammes of Psychiatry (1947), dedicated "To the scientists in the neuronal basis of behavior."

Cyril Brian Courville (1900–1968)

One of Rand's early residents, Courville spent a year with Cushing, as had Rand, but as a voluntary assistant. Percival Bailey was then at the Peter Bent Brigham Hospital studying the histology of the brain tumors removed at Cushing's operations. Courville was attracted by this introduction to neuropathology and returned to a residency in that field at Los Angeles County Hospital. In 1924, he established and became director of the Santiago Ramón y Cajal Laboratory of Neuropathology there and later was founding president of the American Association of Neuropathologists. At his famous brain-cutting sessions, Courville always insisted on being told the history and examination of the patient whose brain was under study. Then he would tell where and what the lesions were before the brain was cut, usually correctly.

Concurrently, Courville (15.7; left) was professor of neurology at his Alma Mater, the Loma Linda School of Medicine. In 1936, with Ingham, Rand, and Nielsen, he helped found the *Bulletin of the Los Angeles Neurological Society*, and served as its managing editor for 30 years. Courville was notable for his historical interest in skull fractures and other cranial injuries and amassed an outstanding library and collection of skulls together with related weapons and protective armor. His studies ranged from prehistory through early cultures in both the Old World and the New.

A Climactic Attainment

This tradition of integrative activity in neurological sciences prepared a fertile ground for the crowning event of the early Los Angeles School of Neurology in the fall of 1981 with the award of the Nobel Prize in physiology or medicine to Roger Sperry. The pathway to fame commenced with the chance encounter of a patient admitted in status epilepticus by a surgical resident at Los Angeles's White Memorial Hospital in 1961. During his medical training, Joseph E. Bogen had had summer research experience with Magoun's group at the Long Beach VA Hospital and with Antoine van Harreveld at Caltech, where he met Ronald Myers and Roger Sperry. Bogen was able to persuade the patient and family, neurosurgeon Philip Vogel, and eventually psychobiologist Sperry to become part of a drastic investigation: What is a human being with a split brain and could we get an answer by surgically treating patients whose epilepsy was otherwise intractable? Sperry's graduate student, Michael Gazzaniga, put "W.J." through all the sophisticated psychological tests that should be carried out preoperatively. When Bogen wrote up the case history of W.J., he argued: "The work was done here, it ought to be in the *Neurological Bulletin*" and the editor, Courville, said "Absolutely."[7] In 1969, after the number of operated patients had grown to eight, Bogen published a series of three papers under the general title "The other side of the brain" all in the *Bulletin*.

Bogen's patient was not the first in America to undergo a commissurotomy; two decades earlier the procedure had been carried out on a small group of epileptic patients at the University of Rochester (Van Wagenen & Herren, 1940). Post-operative study of the patients by a psychiatrist and a psychologist (Smith & Atelaitis, 1942) gave essentially negative results, i.e., only minor behavioral or motor changes were found, apparently as a result of incomplete transections and inadequate psychological testing (Gazzaniga, 1970). The success of Vogel's transection of the corpus callosum was twofold: Spread of the epileptic seizure to the opposite hemisphere was

[7] Quoted from the interview June 6, 1981; Series NRL Code BOG, Neuroscience History Archives, Brain Research Institute, University of California, Los Angeles.

prevented so W.J. could live a normal life; and a series of psychological studies by Sperry and his associates was initiated that "could demonstrate convincingly that the right hemisphere has its own consciousness ... independent from high level verbal processing" (Damasio, 1982, p. 224).

Roger Wolcott Sperry (1913–1994)

Following his liberal arts college education, Sperry (Fig. 15.8) was a graduate student at the University of Chicago and received the Ph.D. in 1941. Like Walter Cannon a generation earlier, he avoided training abroad, went to Harvard, and after a year became a research associate at Harvard's Yerkes Laboratories of Primate Biology. In 1946 he returned to his Alma Mater, and in 1954 added to the luster of brain science in Southern California by appointment as Hixon Professor of Psychobiology at the California Institute of Technology (Caltech) in Pasadena.

Sperry's influential contributions to knowledge about the brain concerned two related problems. At the inception of his career, he was curious about what controls the organization of neural nets. Using frogs subjected to various combinations of optic nerve and chiasma transection with reversal of retinal visual fields, he and his associates showed that regeneration of the visual

Fig. 15.8. Roger W. Sperry was doubly qualified to share in 1981 the Nobel Prize in physiology or medicine by his early work on the organization of nerve growth and his later studies on cerebral hemispheric lateralization. Photographed ca. 1975.

pathways followed the original distribution even though such reconnection resulted in inappropriate behavior. Those findings, chiefly from visual stimuli but expanded to touch and pressure, challenged the belief prevalent in the 1940s that experience and learning governed the formation of a neural network. Sperry concluded that the outgoing processes from the neuron were guided to their targets by an innate "chemoaffinity," a prepatterned, "hard-wired" view of neuronal growth that was later questioned in turn.

While at the University of Chicago, Sperry commenced studies of cerebral organization using a split-brain approach. With graduate student Ronald E. Myers (1953), the effects of severing the corpus callosum in cats and monkeys was studied. Although in general the animals' behaviors seemed unchanged, there were subtle alterations in more specific behaviors revealed by adequate psychological testing. Those experiments were continued at Caltech and when ultimately carried out with humans they proved that neither cerebral hemisphere dominated the other. As expressed by Michael Gazzaniga (1981, p. 518), "For the first time in the history of brain science the specialized functions of each hemisphere could be positively demonstrated as a function of which hemisphere was asked to respond." The test findings on those patients caused a rethinking of the mind-brain question and consciousness by Sperry and others (see Chapter 13) and opened a new dimension of knowledge of the human communicative sciences. The limitations of the earlier concept of dominance by one or the other cerebral hemisphere were shattered by the complementary relationship revealed by the work of Sperry and associates (B. Milner, 1974a).

Differential Hemianesthesia

Alternative methods of study of functional differentiation of the cerebral hemispheres soon became available, and because of their relative simplicity and reversibility, they were applied in humans experimentally as well as diagnostically, and confirmed and extended the new information obtained after commissurotomy. The Wada test (Wada & Rasmussen, 1960) was the intracarotid injection of sodium amytal, a short-acting barbiturate, which temporarily inactivated the ipsilateral cerebral hemisphere. During the short period of anesthesia (1 to 10 minutes depending on dose), the asymmetry of the laterality for speech could be determined, a preoperative safeguard to protect speech functions, and the test was widely adopted as a routine procedure (B. Milner, 1974b). That the differences were present at birth was shown by Molfese and associates (1975). A second method, called dichotic-listening and devised in England, was tested by Doreen Kimura (1961) at the Montreal Neurological Institute on patients with epileptogenic lesions whose hemispheric dominance for speech had been verified by the Wada test. Two different verbal stimuli were presented simultaneously, one to each ear, and the patients were asked to report what they heard; the normal controls revealed a right-ear superiority for digits. Later tests with musical stimuli (Kimura, 1964) confirmed that the superiority had switched to the left ear, indicating that sound processing had moved to the right hemisphere. Those results were credited with "providing the first clear demonstration of dual functional asymmetry in the normal brain" (B. Milner, 1974b, p. 75).

LANGUAGE AND NEUROLINGUISTICS

During the uncertainties surrounding localization of speech and its disorders in the central nervous system, a related branch of inquiry opened from a different perspective, language and the subdiscipline of neurolinguistics. The teasing question of how and why speech developed, phylogenetically and ontogenetically, became a major topic of research and debate.

Evolutionary Origins

A Medley of Theories

A reliable sampling of then current suppositions regarding the evolutionary source of speech is found in a symposium organized by the New York Academy of Sciences in the year of the American bicentennial on *Origins and Evolution of Language and Speech* (Harnad, Steklis, & Lancaster, 1976). Held 2 years after a similar symposium under the auspices of the American Anthropological Association, the academy meeting had a more diverse roster of participants and concepts, but they shared a common theme: language development is an evolutionary phenomenon constricted to the human species.

Peter Marler, with a grounding in research on bird song and chimpanzee vocalization (see Chapter 14) and having in mind Lorenz's ideas of imprinting, proposed a phylogenetic assumption of a predisposed, selective responsiveness to survival-enhancing adjustment of templates, in both avian and human vocal learning. In other words, there was a plasticity of the auditory memory against which perceived sounds could be matched. The dependence of the adult repertoire on auditory feedback was one of the chief supports of that notion, but Marler felt it also "points up the lacunae in our knowledge of primate vocal perception and ontogeny … ." (Marler, 1976, p. 392).

A psychobiologist at Stanford University, Karl H. Pribram, propounded a cultural hypothesis of human language origin. Stressing the symbolic nature and sign aspect of lower animal communication, Pribram suggested they had become fused in the human experience. "Convergence of significant and symbolic processing at some subcortical locus or loci is, on the basis of subhuman evidence, a serious contender as an explanation of the emergence of language" (Pribram, 1976, p. 801). He continued, addressing the means to that end: "When the neural sign system becomes sufficiently powerful (i.e., has sufficient memory and coding capability), it can treat the tokens of expression (of others and of self) as signs, signifying social rather that physical situations" (Ibid., p. 802). This bimodal view of language development was seen as having occurred through cultural pressures.

The chapter by Ronald E. Myers, a neurologist who had gained unusual insight into subcortical mechanisms through his graduate work with Sperry (see above) stressed the functional properties of the human posterior cerebrum. Human speech

> evolved not from the vocal responses of lower primates, but ... *de novo* .... From a neurologic view, the evolution of speech must represent ... function[s] to analyze the information of the senses, to establish memories thereof, and to organize voluntary responses which proceed from these analyses or memories (Myers, 1976, p. 755).

On this physiology, he constructed a dichotomy in the uses of face and voice bound together by emotion.

The introduction of emotion into the conference deliberations was underscored by the limbic influence on human speech described by neurologist Bryan W. Robinson, clinician from Tallahassee, Florida. He recalled earlier work that showed the primary role of the limbic system in infrahuman vocalization and the growing belief that speech developed independently, and wondered if there were limbic involvement in human speech as well. In view of the "wide expanses of frontal and temporal cortex lying near Broca's area ... heavily connected to the limbic system" (Robinson, 1976, p. 765) and that system in turn, through the reticular activating and other systems having reinforcing access to the motivational and adaptive behaviors, Robinson concluded that human speech depended on two systems working together.

> The one system is phylogenetically older, is subordinate, and is supplementary to the other .... [V]estigial limbic speech is still possible through its brain stem connections. In rational and logical discourse, the neocortical system is predominant, but in times of emotional stress the limbic system reclaims its old primacy (Robinson, 1976, p. 767).

Discussion following the Myers and Robinson papers did not challenge their concepts of limbic involvement, curiously, but revolved, again, around fine details of localization, conditioning, and vocalization. Historically, perhaps the most interesting points were that Myers confessed to having undergone some professional ostracism as a result of his views and cautioned his colleagues not to overgeneralize their results.

Neurocommunications at MIT

The Boston area's reputation for excellence in the domains of mathematics and electrical engineering was anchored by the applied mathematician Claude Elwood Shannon (1916–2001). First as a graduate student, then associated with MIT from 1958 but still working at Bell Laboratories, he was surrounded by such figures as Warren McCulloch, Walter Pitts, Walter Rosenblith, Jerome Lettvin, and many others who were interested in neural networks. Shannon realized that Boolean logic with 0 and 1 symbols could be applied automatically by electrical switching circuits. His published paper, "A mathematical theory of communication" (1948), proved that with the binary system the nature of the message (sound, words, numbers) was unimportant, thus launching the digital revolution. By the end of the century the elementary interest in nerve nets had metamorphosed into a new discipline, neuroinformatics, with a wide range of subdisciplines. Human language was the specialty of linguist Noam Chomsky, a slightly younger contemporary of Shannon at MIT and an articulate and agitating force in the argument over the substrates of human language. Sniffing for truffles beneath the effusion of words, one discerns a belief that language is human through Darwinian natural selection of innate core modules. Chomsky's conception (summarized in 1984) of a genetically transmitted human "language organ" was prominent in American debates for three decades. Some adversaries argued that he had been seduced by MIT's digital computer ambience to champion a hypothetical universal grammar that humans possess and the other animals do not. The rules for Chomsky's "universal grammar" became more complex as more languages were studied and the digital computer programs failed to represent them (Lieberman, 1998). Instead of a "deep structure" related to meaning, "[t]he neural basis of human speech motor control and syntax appear to be linked together in a functional language system that produces and comprehends spoken language" (Ibid., p. 122). Lieberman went further: the functional language system conforms to neurophysiologic and neuroanatomic principles "that shape unique functional systems adapted to other behaviors that exist in other species" (Ibid., p. 132). A conciliatory note was sounded in an exchange of letters: "[T]he important thing now is that the way is open for linguists and geneticists to work together on the origin of linguistic competency, both in evolution and in individual development" (J.M. Smith, 1996, p. 41).

Small Steps Forward

An overview of progress in the various domains of speech research – language perception and production, acquisition and comprehension – would contain not only work already mentioned but other minor breakthroughs which should not go unnoted. The first demonstration of loci of language production in normal brain was by means of the EEG (McAdam & Whitaker,

1971). Maximum slow negative potentials occurred over Broca's area when the subjects spoke polysyllabic words and bilateral symmetric potentials when nonspeech gestures were made. A common system for language production and comprehension was identified by stimulation mapping during open-cranial surgery under local anesthesia at the University of Washington (Ojemann & Mateer, 1979). This refinement of earlier work by Penfield and others revealed a system surrounding a common motor pathway for speech which in turn was surrounded by a verbal-memory system; this work may be seen as an exemplar of the power of the interrelatedness approach to the discovery of knowledge.

The usefulness of speech errors or slips of the tongue as clues to substrates of communication was the central interest of Victoria Fromkin (1924–2000) and she became the most prominent contemporary researcher in that field (Gleason, 1982). Her teaching and research at the University of California, Los Angeles included coauthorship of an introductory textbook widely used in academic courses in linguistics (Fromkin & Rodman, 1978).

DEVELOPMENTAL PROBLEMS

Among the behavioral asymmetries that Norman Geschwind and his associates correlated with an anatomic substrate, the planum temporale, was the medical complex collectively known as developmental disabilities. Characterized by reversed interpretation of symbols, an estimated 2 percent to 20 percent of the American population had difficulty learning to read and/or write and spell; extreme variation in the degree of affliction and methods of teaching presented a major problem for schools from kindergarten to college.

Initial Recognition

Disturbances in speech have been known for centuries (O'Neill, 1980) and the causes were still mysterious when the magnitude of the problem struck home. In the United States serious study commenced in the 1920s at the University of Iowa with the quantitative measurements of Carl Seashore and work on stuttering by Lee Travis. The latter abandoned an electromyographic method and turned to the new electroencephalography (EEG) as an indicator of the brain's involvement in the communicative process (Travis & Knott, 1936; see Chapter 7). He was the earliest American to acknowledge the priority of Richard Caton and Hans Berger in recording brain waves, respectively, from the exposed cortex of animal brains and through the scalp (O'Leary & Goldring, 1976). Under Travis's leadership as head of the department of psychology, the Iowa group became internationally known for

Fig. 15.9. Specializing in neuropathology, Samuel T. Orton was an unusual psychiatrist for his period in that he carried out research in addition to his organizational, administrative, and clinical activities. He and his wife, June Lyday Orton, a psychiatric social worker, left an indelible imprint on the field of developmental disabilities. Photographs kindness of Dr. Bennett Shaywitz.

its active EEG program emphasizing communicative disorders.[8] The most significant contributions to speech from the University of Iowa, however, had been made earlier by a psychiatrist interested in research.

Samuel Torrey Orton (1879–1948)

With an M.D. from the University of Pennsylvania (1905) specializing in the neuropathology of mental disorders, and after serving in the First World War, Orton (Fig. 15.9) advanced to the position of scientific director at the Pennsylvania State Hospital for Mental Diseases. In 1919, he was appointed professor of psychiatry at the medical school of the University of Iowa with the mandate to build and direct the State Psychopathic Hospital. Joining the movement then underway to take psychiatry out of the state hospitals and into the community through out-patient clinics, a mobile mental hygiene unit was established. Under Orton's direction, the team located school children who were nonreaders and noticed that they had some of the clinical signs of aphasics but without the brain damage shown by adults.[9] These patients presented an

[8] Details of the advanced instrumentation developed there were described by O'Leary and Goldring (1976) and by Lindsley (1969).
[9] Relatedly, the rehabilitation of adult aphasic patients utilized many of the methods effective in children with reading difficulties.

unusual opportunity for systematic study and research. In the words of his widow, a psychiatric social worker,

> Dr. Orton was unusually well prepared for such investigations because he had been trained as a neuropathologist and at that time had studied more human brains at post mortem by laboratory methods than anyone else in the country (J.L. Orton, 1957, p. 2).

Samuel Orton's first report on "word blindness" in school children was to the American Neurological Association in 1925. He applied the term "strephosymbolia" to the disorientation of letters, words, and phrases characteristic of specific reading disability (S.T. Orton, 1928) and collaborated with Lee Travis in the latter's ongoing study of stuttering (Orton & Travis, 1929). He propounded (1928) a theory based on physiology rather than pathology to explain the problem as the patient's confusion over which was the dominant hemisphere in interpretation of sensory stimuli. In his view, lower levels of sensory elaboration were bilateral, but the higher level dealing with symbols and spoken words was unilateral.

The Ortons returned to the East Coast in 1928 and he became neuropathologist at the New York Neurological Institute and professor of neuropathology at Columbia University P&S in 1930. In quick succession, he delivered the James Arthur Lecture of the American Museum of Natural History and the Thomas W. Salmon Memorial Lectures of the New York Academy of Medicine, both during 1936. Published in several editions, they championed the normality of left-handedness. In a pattern similar to Norman Geschwind's, Samuel Orton's final decade was devoted to teaching and training those who would follow him. Among many others were Loretta Bender (Fig. 15.10) and Margaret Byrd Rawson (1899–2001; Fig. 15.11). During an unusually long practice and teaching career, Rawson advanced human communicative sciences by preparing a timely bibliography (1966) and in a longitudinal study documented the capacity of dyslexic boys to learn to compete in the "reading" world (Rawson, 1968).

During his career, the Rockefeller Foundation supported three projects directed by Orton. First, it continued the mobile unit mentioned above that had so successfully identified and treated nonreaders throughout Iowa. Second, the Language Research Project of the New York Neurological Institute provided opportunity for research in teaching methods and training. And the third program included studies of motor coordination of children with reading disabilities. As noted, his efforts to inform and educate the public culminated in the Salmon Lectures, published with the subtitle *A Presentation of Certain Types of Disorder in the Development of the Language Faculty* (1937). A measure of the appeal and success of Orton's broad yet specific approach to dyslexia was manifested a year after his death

Fig. 15.10. Psychologist Loretta Bender, as the first Percival Bailey Lecturer of the Illinois Neuropsychiatric Institute, Chicago, on October 24, 1960, held the rapt attention of Dr. Bailey. Photograph courtesy of Mrs. Bailey.

by the incorporation of The Orton Society with its *Bulletin* by individuals who believed in a comprehensive and orderly approach to each child's education. The society is now the International Dyslexic Association and continues to serve a large membership.

The Mohonk Conference

One federal response to the midcentury crescendo of the American public's concern about the prevalence of developmental disabilities was the funding of a large invitational conference held at one of the country's historic hostelries, the Mohonk Mountain House. Situated about 90 miles from New York City at the rocky crest of the Catskill Mountains, this Quaker-owned hotel has since 1883 quietly hosted a long series of conferences and family

Fig. 15.11. Margaret B. Rawson's seven decades of research, teaching, and writing about dyslexia and learning to read changed approaches and methodologies of studies of human communication. Photograph 1965 courtesy of the Beneficial-Hodson Library of Hood College, Frederick, Maryland.

vacations (W.H. Marshall, 1970). The conference proceedings, *Early Experience and Visual Information Processing in Perceptual and Reading Disorders* (Young & Lindsley, 1970), represented a consensus of the organizers, the Committee on Brain Sciences of the National Research Council-National Academy of Sciences (see Volume 2). The Mohonk Conference supplemented the work of the 3-year study of the Joint Commission on Child Mental Health and the National Advisory Committee on Dyslexia and Related Reading Disabilities. In the congenial mountain setting and informative tone, the meeting opened the eyes of most participants to the depth and complexity of the problem, from the neurophysiology of visual processes to classroom pedagogy and clinical diagnosis. Even nomenclature was identified as a barrier to interdisciplinary communication and a glossary was added to the published proceedings.

The conference's goal was to remedy the situation in which

> much new information ... about the mechanisms of the eye and the brain has not been brought to the attention of those concerned with the applied problems in which these mechanisms are involved. Furthermore, considerable effort has been expended in the search for the onset and development of various psychologic and behavioral functions, as well as their anatomic and physiologic precursors (Young & Lindsley, 1970, p. 2).

In the concluding panel discussion there was agreement that not enough data existed on the normal learning-to-read process because research on reading was in its infancy. Help to the classroom teacher faced with nonreaders held primary effective importance. And last, the meeting itself was "an example of the tremendous effort it takes to absorb the detailed knowledge, the vocabulary, and the concepts of related fields" (Robinson, 1970, p. 475). And reading is but one facet of the communicative sciences!

Samuel Orton's fervid endeavor to identify, define, and treat developmental dyslexia had rich returns in directions other than preparation of teachers – in the organization of centers of research. Clinics specializing in testing and treating children with reading disorders appeared often in conjunction with established neurologic or child study centers. One such group, selected from many flourishing programs aimed at obtaining new knowledge about this complex disability, is at Yale University. The Yale Center for the Study of Learning and Attention has mobilized large teams to address many aspects of dyslexia and related problems of attention, learning, and hyperactivity, taking advantage of functional neuroimaging techniques. Cooperative studies have been carried out with the well known Haskins Laboratories; relatedly, it was A.M. Liberman (1974) of the Haskins Laboratory who pointed out that speech perception is more complex than other perceptions and requires a higher degree of re-encoding, thus presenting more loci for potential disconnections.

The Yale Center's codirectors, Sally E. Shaywitz and Bennett A. Shaywitz received their medical degrees respectively from Albert Einstein College of Medicine and Washington University School of Medicine. In addition to the collaborative work mentioned, Sally's special interests have focused on endocrinologic influences and gender differences underlying developmental disabilities (S.E. Shaywitz, 1996). In addition to directing pediatric neurology at the medical school, Bennett's interests have centered on the component cognitive processes of the brain for reading and language (Pugh, B.A. Shaywitz, S.E. Shaywitz, et al., 1996). The Shaywitz's long-term contributions to unraveling the knot of learning disabilities were initiated during their association with the Connecticut Longitudinal Study of 400 representative children, an association which afforded the opportunity to follow normal cognitive development and examine specific epidemiologic influences (Shaywitz, Shaywitz, Fletcher, & Escobar, 1990).

Conclusion

After a slow start, the communicative sciences in North America more than made up for the initial neglect in the 20th century and continue to surge ahead. As a reminder that new knowledge has unfolded from information exchange based on numbers, this chapter concludes with reference to the thought-provoking essay of the American psychologist, George A. Miller. At end-century he summarized some informational concepts in *Psychological Review* in a paper titled "The magical number seven," noting the severe limitations of the amount of information humans can receive, process, and remember; the largely unknown recording procedures they have developed to deal with those limitations, and the usefulness of the theory of information to quantify and test the available and new concepts. On an optimistic note of imminent progress, he enumerated some of the sevens – the seas, the wonders of the world, the days of the week, and so on – and wrote: "… perhaps there is something deep and profound behind all these sevens. But I suspect that it is only a pernicious pythagorean coincidence" (G.A. Miller, 1994. p. 341). And so the human communicative sciences have come full circle to the mathematics of information exchange.

Chapter 16
Postscript[1]

In General

Before attempting a summary of North American neuroscience in the 20th century, it is instructive to review the discipline from a more global, less parochial perspective. The hurdles to be cleared in a synoptic view of neuroscience history are formidable. The first is the nature of neuroscience – a convergence of subdisciplines that encompass the myriad experiences of humankind. Second is the increase in rate of discoveries, fueled by the speed of technological advances. And perhaps most significant (and difficult to describe) is the intangible "feel" for science, the Zeitgeist that prevails at any given time. Yet in spite of the hurdles, the recency of neuroscience's entry into the biomedical sciences requires a certain defensive stance in the historical posture.

The Main Questions

A look at passed mileposts affords the opportunity to judge how well the branch fits the tree of human experience. Does neuroscience – represented by a hybrid noun invented by combining two[2] of the world's great languages – constitute a permanent addition to the scientific domain in spite of the lateness of its emergence? Or does it occupy only a niche in the biomedical ecology and is destined to fractionate into specialized subdisciplines? And last, does neuroscience offer the power to solve the riddle of the conscious brain-mind problem? Answers to these questions are not yet possible, but a look backward reveals some healthy signs of permanence.

Some Answers

The emergence of this broad and basic discipline rode on a ground swell powered by the growing comprehension that as two cerebral hemispheres are better than one, so the combined study of brain and behavior provided a methodology more revealing than that constricted to separate subdisciplines. Conjoining neurophysiology and neuroanatomy was a seamless transition, abiding as it did

[1] This is an added chapter by the editor.
[2] "Neuro" from the Greek and "science" from the Latin word meaning "knowledge."

in the postulate that form follows function (see Marshall & Magoun, 1998). Other disciplines soon discovered that they needed to know about the nervous system. For example, neurochemistry budded from biochemistry, neuropsychopharmacology from pharmacology, neuroendocrinology from its parent, and so on. The latest to join the tribe, and not the last, is neurogenetics, the twin of behavioral genetics. Founding of this field is credited to Seymour Benzer who initiated the genetic dissection of behavior in Drosophila (1973) and has demonstrated monoclonal antibody cross reactions between the fly and human brain (Miller & Benzer, 1983).

In Particular

The domain of neuroscience stretches from interpretations of discoveries from prehistory to the present and from first speculations about consciousness to modern gene expression, parallel, yet two-dimensional concepts of time and event. The fossil record of hominid development shows a disproportionately greater expansion of cerebral tissue than that of other organ systems, interpreted as due to the pressure of increasing behavioral complexity. The concepts gleaned from perceptions of the world by the ancient western writers were absorbed and significantly extended by Arabic physicians and then reintroduced in the West. Writings that have survived from antiquity are witness to the early philosopher-physicians' concerns with the mind (soul)-body problem and how the muscles worked. Their mechanistic concepts culminated in Descartes's hydraulic scheme, an idea replaced in the next century by Galvani's demonstration of "animal electricity." In turn, the vagaries of animal electricity were overcome through the aggressive promotion of exact chemistry and physics by that remarkable quadrumvirate of German physiologists – Helmholtz, Dubois-Reymond, Brücke, and Ludwig – who initiated the movement that eventually freed biology from philosophy.

Prodroma

During the 19th century the virtual formation of western neuroscience as a discipline commenced with a gradual recognition of a direct connection between behavior and the brain. Examples of the awakening selected from a host of possibilities include: coincidence of speech impairment with a lesion in the left frontal lobe of the brain (Broca, Wernicke); mental faculties localized at specific cerebral sites (Gall); movement in contralateral muscles after stimulation of specific points on the brain surface (Fritz and Hitzig, Ferrier); directional flow of impulses both within and across neurons shown to be contiguous, not reticulated (Golgi, Ramón y Cajal); hierarchies of behavioral modes with anatomic levels of the central nervous system (Hughlings Jackson).

Carl Wundt's laboratory became the mecca of psychologists worldwide, offering a new thinking about the human psyche through a combination of experimentation with sensual introspection. He had trained with Helmholtz and DuBois-Reymond, enjoyed institutional support, and deigned to keep close touch with his students; he also wrote prolifically. At a time when American biomedical science was immature, the urge for advanced training in Europe was irresistible. Significant to the development of neuroscience in America in this pattern were periods spent abroad by psychologists McKeen Cattell and Stanley Hall and anatomist Clarence Luther Herrick, whose younger brother and student, C. Judson Herrick, established a distinguished school of comparative neurology at the new University of Chicago. The younger Herrick elucidated phylogenetically the intimate relationship between behavior and brain, as James Papez demonstrated neurologically at Cornell University. Clarence Herrick's early introduction of the term "psychobiology" symbolized the confluence of the two streams of knowledge and completed its separation from philosophy.

North American Contributions

Continuing into the 20th century, physiology caught up with anatomy in the advance of neuroscience. To highlight the movement in America, again a brilliant array of discoveries embedded in technologic advances such as stereotaxy, radioisotopes, and microimaging is available. In somewhat chronologic order they include the following: quantification of memory and learning by maze running in lower animals (John Watson; Robert Yerkes); electrophysiological studies of nerve impulse conduction (Adrian, Gasser, Erlanger, Bishop); the commanding role of the hypothalamus and neurohypophysis in processes essential to survival (Cushing, Fröhlich, Hinsey, Ranson); the cybernetic concept of feedback and feed forward (McCulloch, Walter Pitts, Weiner); and electrolyte fluxes across axonal membranes (Cole, Huxley, Hodgkin). And finally, the 1950s highlighted the nature of human sexual behavior (Masters and Johnson); demonstration of genetic influences on animal behavior as well as in physical characteristics (Lorenz, Benzer); and the conceptualization of the reticular activating system in mammals (Moruzzi and Magoun).

Support Systems

By midcentury fundamental neuroscience was a global concept of research and its novel approach to brain-behavior solutions had absorbed a wide spectrum of molecular components. Teams of investigators replaced the activity of the isolated investigator in fruitful research programs. With a surge in public and private funding for education as well as for investigation, there developed a critical mass of restive personnel and a need for assessment of the status quo. The earliest inquiry of an interdisciplinary-multidisciplinary science of brain and behavior was led by neuroembryologist Paul Weiss and sponsored by the U.S. National Academy of Sciences. Apprehensive that students of neurobiology were not adequately prepared to pursue advanced research in the field, Weiss and colleagues (1952) produced a comprehensive bibliography and a census of investigators working in neurobiology. In another direction, the psychologists were being listed selectively (and starred) in McKeen Cattell's *American Men* (later *and Women*) *of Science* (1906). Robert Yerkes launched *Psychological Abstracts* in 1916, thus enabling those early psychologists to identify and bond with their peers in academia.

Committees, listings, and research reports were not enough to catalyze a movement, especially when disciplinary lines continued to retard the exchange of information. Turning to more personal encounters (for dinner and discussions of common interests), neurophysiologist Ralph Gerard brought together those physiologists working on the peripheral nervous system. Thus "axonology" was born amid exciting and exuberant exchanges about their research and it continued to dominate neurophysiology until the axonologists disbanded in 1942.

After the Second World War funding agencies were interested in sponsoring symposia around central themes with increasingly diverse fields represented by the participants. Two such series under the general direction of Frank Fremont-Smith of The Josiah Macy, Jr., Foundation had an unusually wide influence on the direction of neuroscience in America as well as internationally. They were the conferences organized and carefully tended by Warren McCulloch, illuminating the concept of cybernetics, and the series "Brain and Behavior," chaired by Horace Magoun, and bringing together diverse neural interests and backgrounds. The ongoing Nebraska Symposia on Motivation have included, among many other themes, analysis of the discovery by Olds and Milner (1954) of the behavioral effects of operant control of stimulation of sites deep in the cerebrum. An additional series of meetings, with extensive federal funding, was crafted by Francis Schmitt at MIT. The Neurosciences Research Program's core of associate members met several times a year to discuss leading-edge research with invited experts and also organized three longer summer programs with published proceedings.

A Survey's Impact

As a direct result of the ringing success of the Moscow Colloquium in 1958, and with UNESCO sponsorship, the International Brain Research Organization (IBRO) was established with six speciality categories of membership. That organization represented the first gathering of a ground swell that crested eventually in the formation of the Society for Neuroscience and conferred substantive permanence to the discipline. The first project undertaken by IBRO was a world survey of personnel and facilities for brain research. When it came time to set in motion the U.S. component, the U.S. National Academy of Sciences was approached with funding from the National Institute of Mental Health. The director of the National Research Council (NRC) Division of

Medicine, biochemist Keith Cannan, combined the survey with a request from the National Library of Medicine to review its coverage in the behavioral sciences and convened the Committee on Brain Sciences to supervise in parallel those two projects. It was not long before the committee's agenda, under the chairmanship of Neal E. Miller, generated discussions of a new professional society. Ralph Gerard was tapped and skillfully orchestrated assessments of need and feasibility from representative neuroscientists in 20 national regions selected on the basis of the IBRO survey. Buoyed by an overwhelming positive response, an executive group led by Edward Perl prepared bylaws and in late 1969 the Society for Neuroscience was incorporated in the District of Columbia.

In addition to the energy generated at the crest of the wave, the new organization had several attributes crucial to its subsequent success. A "grass roots" element was built in from the start in the form of local autonomous chapters. These tended to smoothly absorb the many small and specialized regional groups that had been meeting in reaction to the less satisfactory huge meetings of FASEB and AAAS. A second essential was the prompt establishment of the *Neuroscience Newsletter*, the inhouse publication which maintained interest and involvement in organizational and disciplinary activities. Third, the good will of the National Academy of Sciences, due to the fact that almost all the planners were members, yielded free office space, secretarial, and legal assistance for the first year of operation. Fourth, the bylaws required that the slates for nominations of officers shall represent both geographic and speciality membership demographics. The participation of medical specialists was a goal of the society, as evidenced by the series of workshops and symposia at the annual meetings.

It is clear that neuroscience sprang from a convergence of three streams of knowledge: neural, behavioral, and communicative. Each has its medical components and their combination yields powerful approaches to solving problems. Today the drive for human and veterinarian good health dictates the new directions the discipline is taking in neuroengineering and neurogenetics. Curiously, the end of the century has seen revival of the interest in consciousness, by philosophers (Searle, Suppes) and neuroscientists (Libet) alike. Publications popular and abstruse, seminars, and academic departments abound in a frontal attack on this elusive domain of life. This trend augurs well for continued probing in spite of – or perhaps because of – behaviors constantly shown to be more complex as new knowledge of the central nervous system becomes more detailed.

The author of this volume, Horace Winchell Magoun, saw neuroscience from a long perspective. Never happy with microscopic details (personal communication), yet quick to perceive slight differences, he looked for signs of the greater sweep in small events. He witnessed a succession of happenings in the field, as the lead U.S. delegate to the Moscow Colloquium, chair of the brain history panel of IBRO, a member of the NRC Committee on Brain Sciences, prepared the first directory of educational programs in neuroscience, was the prime mover in establishing an institute, and served on federal and private advisory committees. All this in addition to recognizing and analyzing a fundamental process in mind/brain coordinating mechanisms. In sum, Professor Magoun's many contributions, like his viewpoint, were instrumental in securing a permanent spot for neuroscience in the biomedical world. We hope his view of history may illustrate some of the ways this came about.

REFERENCES

Abel, J.J. 1899. Über den Blutdtruckerregenden der Nebenniere das Epinephrin. *Hoppe-Seylers Z. Physiol. Chem.*, 28: 318–362.

Abel, J.J. 1903. On epinephrin and its compounds, with special reference to epinephrin hydrate. *Am. J. Pharmacol.*, 75: 301–325.

Aberle, S.D. and G.W. Corner. 1953. *Twenty-five Years of Sex Research: History of the National Research Council Committee for Research in Problems of Sex, 1922–1947.* Philadelphia: W.B. Saunders.

Ackerknecht, E.H. and H.V. Vallois. 1956. *Franz Joseph Gall, Inventor of Phrenology and His Collection.* Wisconsin Studies in Medical History No. 1. Trans. by C. St Léon. Madison: University of Wisconsin Medical School.

Adler, H.E. 1977. The vicissitudes of fechnerian psychophysics in America. Pages 21–32 in R.W. Rieber and K. Salzinger, eds., *The Roots of American Psychology: Historical Influences and Implications for the Future.* Ann. New York Acad., 291. New York: The New York Academy of Sciences.

Adrian, E.D. 1934. Electrical activity of the nervous system. *Arch. Neurol. Psychiat.*, 32: 1125–1136.

Adrian, E.D. 1966. Thomas Graham Brown. *Biog. Mem. Fellows R. Soc.*, 12: 23–33.

Adrian, E.D. and D.W. Bronk. 1928. The discharge of impulses in motor nerve fibres. Part I. Impulses in single fibres of the phrenic nerve. *J. Physiol.* (London), 66: 81–101.

Adrian, E.D. and D.W. Bronk. 1929. The discharge of impulses in motor nerve fibres. Part II. The frequency of discharge in reflex and voluntary contractions. *J. Physiol.* (London), 67: 119–151.

Adrian, E.D. and B.H.C. Matthews. 1934. The Berger rhythm: potential changes from the occipital lobes in man. *Brain*, 57: 355–385.

Aldrich, T.B. 1901. A preliminary report on the active principle of the suprarenal gland. *Am. J. Physiol.*, 5: 457–461.

Alexander, R.S. 1946. Tonic and reflex functions of medullary sympathetic cardio-vascular centers. *J. Neurophysiol.*, 9: 205–217.

Allen, B.M. 1916. The results of exterpation of the anterior lobe of the hypophysics and of the thyroid of *Rana pipiens* larvae. *Science*, 44: 755–758.

Allen, B.M. 1918. The results of thyroid removal in the larvae of *Rana pipiens*. *J. Exp. Zool.*, 24: 499–519.

Allen, B.M. 1924. Brain development in anurar larvae after thyroid or pituitary gland removal. *Endocrinology*, 8: 639–651.

Allen, G.W. 1967. *William James. A biography.* New York: Viking Press.

Allen, W.F. 1932. Formatio reticularis and reticulospinal tracts, their visceral functions and possible relationships to tonicity. *J. Wash. Acad. Sci.*, 22: 490–495.

Allis, E.P. 1897. The cranial muscles and cranail and first spinal nerves in Amia calva. *J. Morphol.*, 12: 487–806.

Alpers, B.J. 1975. Charles Karsner Mills 1845–1931. Pages 80–84 in D. Denny-Brown, ed., *Centennial Anniversary Volume of the American Neurological Association 1875–1975.* New York: Springer Publishing Co.

Anderson, E. 1969. Earlier ideas of hypothalamic function, including irrelevant concepts. Pages 1–12 in W. Haymaker, E. Anderson, and W.J.H. Nauta, eds., *The Hypothalamus.* Springfield, Illinois: C.C. Thomas.

Angell, J.R. 1916. A reconsideration of James' theory of emotion in the light of recent criticisms. *Psychol. Rev.*, 25: 251–261.

Anonymous. 1906. [Report of demonstration]. *Brit. Med. J.*, 2, part 1: 633.

Anonymous. 1984. Family reunion. The Talk of the Town. *The New Yorker.* 16 July, 1984.

Arey, L.B. 1943. Stephen Walter Ranson 1880–1942. *Anat. Rec.*, 86: 610–618.

Armstrong, C.M. and F. Bezanilla. 1973. Currents related to movement of the gating particles of the sodium channels. *Nature*, 242: 459–461.

Arnold, A.P. 1975. The effects of castration and androgen replacement on song, courtship, and aggression in zebra finches (*Poephila guttata*). *J. Exp. Zool.*, 191: 309–326.

Arnold, A.P. 1982. Neural control of passerine song. Pages 75–94 in D.E. Kroodsma, E.H. Miller, and H. Ouellet, eds., *Acoustic Communication in Birds. Vol. I: Production, Perception, and Design Features of Sounds.* San Diego: Academic Press.

Arnold, A.P., J. Wade, W. Grisham, E.C. Jacobs, and A.T. Campagnoni. 1996. Sexual differentiation of the brain in songbirds. *Dev. Neurosci.*, 18: 124–136.

Aronson, L.R. and G.K. Noble. 1945. The sexual behavior of Anura. II. Neural mechanisms controlling mating in the male leopard frog. *Bull. Am. Mus. Nat. Hist.*, 86: 83–140.

Aronson, L.R. and J.W. Papez. 1934. Thalamic nuclei of Pithecus (Macacus) rhesus. II. Dorsal thalamus. *Arch. Neurol. Psychiat.*, 32: 27–44.

Arvanitaki, A. 1942. Effects evoked in an axon by the activity of a contiguous one. *J. Neurophysiol.*, 5: 85–108.

Arvanitaki, A. and N. Chalizonitis. 1949. Prototypes d'interactions neuroniques et transmissions synaptiques. Données bioéléctriques de préparations cellulaires. *Arch. Sci. Physiol.*, 3: 547–566.

Asanuma, H. and I. Rosen. 1972. Functional role of afferant inputs to the monkey motor cortex. *Brain Res.*, 40: 3–5.

Aserinsky, E. and N. Kleitman. 1953. Regularly appearing periods of eye motility and concomitant phenomena, during sleep. *Science*, 118: 273–274.

Astwood, E.B. and H.L. Fevold. 1939. Action of progesterone on the gonadotropic activity of the pituitary. *Am. J. Physiol.*, 127: 192–198.

Atlas, D. and W.R. Ingram. 1937. Topography of the brain stem of the rhesus monkey with special reference to the diencephalon. *J. Comp. Neurol.*, 66: 263–289.

Axelrod, J. 1975. Biochemical and pharmacological approaches in the study of sympathetic nerves. Pages 191–208 in F.G. Worden, J.P. Swazey, and G. Adelman, eds., *The Neurosciences: Paths of Discovery.* Cambridge, Massachusetts: MIT Press.

Bacq, Z.M. 1934. La pharmacologie du système nerveux autonome, et particulièrement du sympathique d'après la théorie neurohumorale. *Ann. Physiol.*, 10: 467–553.

Bacq, Z.M. 1975. Walter B. Cannon's contribution to the theory of chemical mediation of the nerve impulse. Pages 68–83 in C. McC. Brooks, K. Koizumi, and J.O. Pinkston, eds., *The Life and Contributions of Walter Bradford Cannon 1871–1945*. Brooklyn: State University of New York Downstate Medical Center.

Bagley, C. and C.P. Richter. 1924. Electrically excitable region of the forebrain of the alligator. *Arch. Neurol. Psychiat.*, 11: 257–263.

Bailey, P. 1956. The Academic Lecture. The great psychiatric revolution. *Am. J. Psychiat.*, 113: 387–406.

Bailey, P. 1958. Silas Weir Mitchell, February 15, 1829–January 4, 1914. *Natl. Acad. Sci. Biog. Mem.*, 32: 334–353.

Bailey, P. 1965. *Sigmund the Unserene, A Tragedy in Three Acts*. Springfield, Illinois: C.C. Thomas.

Bailey, P. 1969. *Up From Little Egypt*. Chicago: The Buckskin Press.

Bailey, P. and F. Bremer. 1921. Experimental diabetes insipidus. *Arch. Intern. Med.*, 28: 773–803.

Bailey, P. and H. Cushing. 1926. *A Classification of the Tumors of the Glioma Group on a Histogenetic Basis with a Correlated Study of Prognosis*. Philadelphia: J.B. Lippincott Co.

Bailey, P. 1975. Moses Allen Starr 1854–1932. Pages 112–113 in D. Denny-Brown, ed., *Centennial Anniversary Volume of the American Neurological Association 1875–1975*. New York: Springer Publishing Co.

Balda, R.P., I.M. Pepperberg, and A.C. Kamil, eds. 1998. *Animal Cognition in Nature: the Convergence of Psychology and Biology in Laboratory and Field*. San Diego: Academic Press.

Bard, P. 1929. The central representation of the sympathetic nervous system as indicated by certain physiologic observations. *Arch. Neurol. Psychiat.*, 22: 230–246.

Bard, P. 1934. On emotional expression after decortication with some remarks on certain theoretical views. Parts I and II. *Psychol. Rev.*, 41: 309–329; 424–449.

Bard, P. 1939. Central nervous mechanisms for emotional behavior patterns in animals. *Res. Pub. Assoc. Res. Nerv. Ment. Dis.*, 19: 190–218.

Bard, P. 1950. Central nervous mechanisms for the expression of anger in animals. Chapter 18 in M.L. Reymert, ed., *Feelings and Emotions, the Mooseheart Symposium in Connection with the University of Chicago*. New York: McGraw-Hill Book Co.

Bard, P. 1973. The ontogenesis of one physiologist. *Annu. Rev. Physiol.*, 35: 1–16.

Bard, P. and V.B. Mountcastle. 1948. Some forebrain mechanisms involved in expression of rage with special reference to suppression of angry behavior. *Res. Pub. Assoc. Res. Nerv. Ment. Dis.*, 27: 362–399.

Bard. P. and D. McK. Riock. 1937. A study of four cats deprived of neocortex and additional portions of the forebrain. *Bull. Johns Hopkins. Hosp.*, 60: 73–147.

Barker, L.F. 1899. *The Nervous System and Its Constituent Neurones*. New York: Appleton.

Barker, L.F. 1942. *Time and the Physician. The Autobiography of Lewellys F. Barker.* New York: G.P. Putnam's Sons.

Barrett, E.S. 1999. Aggression impulsivity: Neurobiological corelates. Pages 35–37 in G. Adelman and B.H. Smith, eds., *Encyclopedia of Neuroscience*, vol. 1, 2nd ed. Amsterdam: Elsevier Science B.V.

Barrett, R., H.H. Merritt, and A. Wolf. 1967. Depression of consciousness as a result of cerebral lesions. *Res. Pub. Assoc. Res. Nerv. Men. Dis.*, 45: 241–272.

Bartholow, R. 1874. Experimental investigations into the functions of the human brain, *Am. J. Med. Sci., New Series*, 67: 305–313.
Bartley, S.H. and E.B. Newman. 1930. Recording cerebral action currents. *Science*, 71: 587.
Bazett, H.C. and W. Penfield. 1922. A study of the Sherrington decerebrate animal in the chronic as well as the acute condition. *Brain*, 45(II): 185–264.
Beach, F.A. 1937. The neural basis of innate behavior. I. Effects of cortical lesions upon the maternal behavior pattern in the rat. *J. Comp. Psychol.*, 24: 393–434.
Beach, F.A. 1942. Central nervous mechanisms involved in the reproductive behavior of vertebrates *Psychol. Bull.*, 39: 200–226.
Beach, F.A. 1944. Effects of injury to the cerebral cortex upon sexually-receptive behavior in the female rat. *Psychosomat. Med.*, 6: 40–48.
Beach, F.A. 1961. Karl Spencer Lashley, June 7, 1890–August 7, 1958. *Biog. Mem.* Washington, DC: National Academy of Sciences (U.S.).
Beach, F.A., ed. 1965. *Sex and Behavior.* New York: John Wiley & Sons.
Beach, F.A., D.O. Hebb, C.T. Morgan, and H.W. Nissen, eds. 1960. *The Neuropsychology of Lashley.* New York: McGraw-Hill.
Beadle, D.J., D. Hicks, and C. Middleton. 1982. Fine structure of Periplaneta americana neurons in long-term culture. *J. Neurocytol.*, 11: 611–626.
Beadle, D.J., G. Lees, and S.B. Kater, eds. 1988. *Cell Culture Approaches to Invertebrate Neuroscience.* London: Academic Press.
Beaumont, W. 1833. *Experiments and Observations on the Gastric Juice and the Physiology of Digestion.* Plattsburg, New York: F.P. Sller. A facsimile was published in 1929 by the Harvard University Press.
Beck, J.R. 1872. How spermatozoa enter the uterus, *Am. J. Obst. Dis. Women Children.*, Nov., p. 440.
Beer, C. 1986. Reviews. Words and Chimps. [Review of D. Premack, Gavagai...] *Nat. Hist.*, 5: 86–89.
Beevor, C.E. and V. Horsley. 1890. A record of the results obtained by electrical excitation of the so-called motor cortex and internal capsule in an orang-outang (Simia satyrus). *Phil. Trans. R. Soc. Lond. (B)*, 181: 129–158.
Ben-David, J. and R. Collins. 1966. Social factors in the origins of a new science: the case of psychology. *Am. Sociol. Rev.*, 31: 451–465.
Benison, S., A.C. Barger, and E.L. Wolfe. 1987. *Walter B. Cannon. The Life and Times of a Young Scientist.* Cambridge. Massachussetts: Harvard University Press.
Bennett, L.L. 1975. Endocrinology and Herbert M. Evans. Pages 247–272 in C.H. Li, *Hormonal Proteins and Peptides.*, v. 3. New York: Academia Press.
Berger, H. 1929. Über das Electrenkephalogramm des Menschen. *Arch. Psychiat. Nervenkr.*, 87: 527–570.
Berkley, H.J. 1894. The nerve elements of the pituitary gland. *Johns Hopkins Hosp. Rep.*, 4: 285–295.
Bianchi, L. 1895. The functions of the frontal lobes. Translated from the original MS. by A. de Watteville. *Brain*, 18: 497–522.
Bissinger, H.G. 1997. *A Prayer for the City.* New York: Random House.
Block, N. 1981. Review of Jaynes, *Cognition Brain Theor.*, 4: 81–83.
Blustein, B.E. 1991. *Preserve Your Love for Science. Life of William A. Hammond, American Neurologist.* Cambridge: Cambridge University Press.
Bogen, J.E. 1969a. The other side of the brain. I. Dysgraphia and dyscopia following cerebral commissurotomy. *Bull. Los Angeles Neurol. Soc.*, 34: 73–105.

Bogen, J.E. 1969b. The other side of the brain. II. An appositional mind. *Bull. Los Angeles Neurol. Soc.*, 34: 135–162.
Bogen, J.E. and G.M. Bogen. 1969. The other side of the brain. III. The corpus callosum and creativity. *Bull. Los Angeles Neurol. Soc.*, 34: 191–220.
Bogen, J.E. and P.J. Vogel. 1962. Cerebral commissurotomy in man. Preliminary case report. *Bull. Los Angeles Neurol. Soc.*, 27: 169–172.
Boring, E.G. 1929. *A History of Experimental Psychology.* New York: Century.
Boring, E.G. 1950. The influence of evolutionary theory upon American psychological thought. Pages 267–298 in S. Persons, ed., *Evolutionary Thought in America*, 1950. New Haven: Yale University Press.
Boring, E.G. 1953. A history of introspection. *Psychol. Bull.*, 50: 169–186.
Boring, E.G. 1958. Karl M. Dallenbach, *Am. J. Psychol.*, 71: 1–40.
Bowers, J.Z. 1972. *Western Medicine in a Chinese Palace: Peking Union Medical College, 1917–1951.* New York: Josiah Macy, Jr. Foundation.
Bowers, J.Z, and E.F. Purcell, eds. 1976. *Advances in American Medicine: Essays on the Bicentennial*, 2 vol. New York: Josiah Macy, Jr., Foundation.
Brady, J.V. and W.J.H. Nauta. 1953. Subcortical mechanisms in emotional behavior: affective changes following septal forebrain lesions in the albino rat. *J. Comp. Physiol. Psychol.*, 46: 339–346.
Brady, J.V. and W.J.H. Nauta. 1955. Subcortical mechanisms in emotional behavior: the duration of affective changes following septal and habenular lesions in the albino rat. *J. Comp. Physiol. Psychol.*, 48: 412–420.
Brady, J.V. and W.J.H. Nauta, eds. 1972. *Principles, Practices, and Positions in Neuropsychiatric Research. A Volume in Honor of Dr. David McKenzie Rioch.* New York: Pergamon Press.
Brazier, M.A.B. 1961. *A History of the Electrical Activity of the Brain.* New York: Macmillan.
Brecher, E.M. 1969. *The Sex Researchers.* Boston: Little, Brown.
Bremer, F., R.S. Dow, and G. Moruzzi. 1939. Physiological analysis of the general cortex in reptiles and birds. *J. Neurophysiol.*, 2: 473–487.
Brewer, F.A., E.M.R. Lomax, L.H. Marshall, Y.V. O'Neill, and H.W. Magoun. 1983. *History of the Human Brain.* Los Angeles: Brain Research Institute, University of California, Los Angeles.
Brobeck, J.R., O.E. Reynolds, and T.A. Appel, eds. 1987. *History of the American Physiological Society: the First Century, 1887–1987.* Bethesda, Maryland: The American Physiological Society.
Broca, P. 1861. Remarques sur le siège de la faculté du langage articulé, suivies d'une observation d'aphémie (perte de la parole). *Bull. Soc. Anat. Paris*, 36: 330–357. Trans. by G. von Bonin, pages 49–72 in G. von Bonin, 1960. *Some Papers on the Cerebral Cortex.* Springfield, Illinois: C.C Thomas.
Broca, P. 1878. Anatomie comparée des convolutions cérébrales. Le grand lobe limbique et la scissure limbique dans le série des mammifières. *Rev. Authropol. Paris, 2nd series*, I: 385–498.
Brock, L.G., J.S. Coombs, and J.C. Eccles. 1951. Action potentials of motoneurones with intracellular electrode. *Proc. Univ. Otago Med. Sch.*, 29: 14–15.
Bronk, D.W. 1929a. The action of strychnine on sensory end organs in muscle and skin of the frog. *J. Physiol.* (London), 67: 17–25.
Bronk, D.W. 1929b. Fatigue of the sense organs in muscle. *J. Physiol.* (London), 67: 270–281.

Brookhart, J.M. and F.L. Dey. 1941. Reduction of sexual behavior in male guinea pigs by hypothalamic lesions. *Am. J. Physiol.*, 133: 551–554.
Brookhart, J.M., F.L. Dey, and S.W. Ranson. 1940. Failure of ovarian hormones to cause mating reactions in spayed guinea pigs with hypothalamic lesions. *Proc. Soc. Exp. Biol. Med.*, 44: 61–64.
Brookhart, J.M., F.L. Dey, and S.W. Ranson. 1941. The abolition of mating behavior by hypothalamic lesions in guinea pigs. *Endocrinology*, 28: 561–585.
Brooks, C. McC. 1986. Taped interview, Neuroscience History Archives, Oval History Project Series CON Code BKS, University of California, Los Angeles.
Brooks, C. McC. 1988. The history of thought concerning the hypothalamus and its functions. *Brain Res. Bull.*, 20: 657–667.
Brooks, C. McC., K. Koizumi, and J.D. Pinkston, eds. 1975. *The Life and Contributions of Walter Bradford Canvon 1871–1945, His Influence on the Development of Physiology in the Twentieth Century*. Albany, New York: State University of New York Press.
Brouwer, B. 1933. Centrifugal influence on centripetal systems in the brain. *J. Nerv. Ment. Dis.*, 77: 621–627.
Brown, J.W. and H.A. Matzke. 1989. John Gibbons Roofe 1899–1988. *Anat. Rec.*, 225(2): 67A–68A.
Brown, S. and E.A. Schäfer. 1888. An investigation into the functions of the occipital and temporal lobes of the monkey's brain. *Phil. Trans. R. Soc. Lond. B.*, 179: 303–327.
Brown, T.G. 1913–1914. On postural and non-postural activities of the mid-brain. *Proc. R. Soc. B.* (London), 87: 145–163.
Brown-Séquard, C.E. 1877. The dual character of the brain. *The Toner Lectures, Lecture II. Smithsonian Miscellaneous Collections*. Washington, DC: Smithsonian Institution.
Bucy, P.C. 1933. Electrical excitability and cyto-architecture of the premotor cortex in monkeys. *Arch. Neural. Psychiat.*, 30: 1205–1225.
Bucy, P.C. and H. Klüver. 1940. Anatomic changes secondary to temporal lobectomy. *Arch. Neurol. Psychiat.*, 44: 1142–1146.
Bueker, E.D. 1948. Implantation of tumors in the hind limb field of the embryonic chick and the developmental response of the lumbosacral nervous system. *Anat. Rec.*, 102: 369–389.
Bull, N. 1962. *The Body and Its Mind. An Introduction to Attitude Psychology.* New York: Las Americas Publ. Co.
Bullock, T.H. 1970. The reliability of neurons. The Jacques Loeb Memorial Lecture. *J. Gen. Physiol.*, 55: 563–584. Introductory remarks by Detlev W. Bronk.
Bullock, T.H. 1984. Comparative neuroscience holds promise for quiet revolutions. *Science*, 225: 473–478.
Bullock, T.H. 1995. Are the main grades of brains different principally in numbers of connections or also in quality? Pages 439–448 in O. Brcidbach and W. Kutsch, eds., *The Nervous Systems of Invertebrates: an Evolutionary and Comparative Approach.* Basel: Birkhäuser Verlag.
Bullock, T.H. and G.A. Horridge. 1965. *Structure and Function in the Nervous Systems of Invertebrates.* San Francisco: W.A. Freeman.
Bunge, M.B., R.P. Bunge, and G.D. Pappas. 1962. Electron microscopic demonstration of connections between glia and myelin sheaths in the developing mammalian central nervous system. *J. Cell. Biol.*, 12: 448–454.

Bunge, R.P. 1968. Glial cells and the central myelin scheath, *Physiol. Rev.*, 48: 197–251.

Burgus, R., T. Dunn, D. Disiderio, and R. Guillemin. 1969. Structure moléculaire du facteur hypothalamique hypophysiotrope TRF d'origine ovine: mise en evidence par spectrométrie de masse de la séquence PCA-His-Pro-NH_2. *C.R. Acad. Sci.*, (Paris), 269: 1870–1873

Burnham, J.C. 1971. The early twentieth century: prestige and maturity. Pages 252–259 in J.C. Burnham, ed., *Science in America: Historical Selection.* New York: Holf, Reinhart, and Winston.

Burr, A.R. 1929. *Weir Mitchell: His Life and Letters.* New York: Duffield & Company.

Cable, G., R.P. Balda, and W.R. Willis. 1983. The physics of leaping animals and the evolution of preflight. *Am. Nat.*, 121: 455–476.

Caldwell, M.C. and D.K. Caldwell, 1965. Individualized whistle contours in bottlenosed dolphins (*Tursiops truncatus*). *Nature*, 207: 434–435.

Caldwell, M.C., D.K. Caldwell, and P. Tyack. 1990. Review of the signature-whistle hypothesis for the Atlantic bottlenose dolphin. Pages 199–234 in S. Leatherwood and R.R. Reeves, eds., *The Bottlenose Dolphin.* San Diego: Academic Press.

Camus, J. and G. Roussy. 1920. Experimental researches on the pituitary body. Diabetes insipidus, glycosuria and those dystrophies considered as hypophyseal in origin. *Endocrinology*, 4: 507–522.

Canavan, M.M. 1925. *Elmer Ernest Southard and His Parents. A Brain Study.* Cambridge, Mass: Privately printed by the University Press.

Cannon, W.B. 1909. The influence of emotional states on the functions of the alimentary canal. *Am. J. Med. Sci.*, 137: 480–487.

Cannon, W.B. 1914a. Some characteristics of antivivisection literature. *Sci. Am. Suppl.*, 78: 58.

Cannon, W.B. 1914b. The interrelations of emotions as suggested by recent physiological researches. *Am. J. Psychol.*, 25: 256–282.

Cannon, W.B. 1915. *Bodily Changes in Pain, Hunger, Fear and Rage.* New York: Appleton.

Cannon, W.B. 1918. The physiological basis of thirst. Croonian Lecture. *Proc. R. Soc. B.* (London), 99: 283–301.

Cannon, W.B. 1925. Some general features of endocrine influence on metabolism. *Trans. Cong. Am. Physicians & Surgeons*, 13: 31–53. Also 1926, *Am. J. Med. Sci.*, 71: 1–20.

Cannon, W.B. 1927. The James-Lange Theory of emotions: a critical examination and an alternative theory. *Am. J. Psychol.*, 39: 106–124.

Cannon, W.B. 1929. Organization for physiological homeostasis. *Physiol. Rev.*, 9: 399–431.

Cannon, W.B. 1931. Again the James-Lange and the thalamic theories of emotion. *Psychol. Rev.*, 38: 281–295.

Cannon, W.B. 1932. *The Wisdom of the Body.* New York: W.W. Norton.

Cannon, W.B. 1945. *The Way of an Investigator. A Scientist's Experience in Medical Research.* New York: Hafner.

Cannon, W.B. and Z.M. Bacq. 1931. Studies on the conditions of activity in endocrine organs. XXVI. A hormone produced by sympathetic action on smooth muscle. With the assistance of R.M. Moore. *Am. J. Physiol.*, 96: 392–412.

Cannon, W.B. and S.W. Britton. 1925. Studies on the conditions of activity in endocrine glands. XV. Pseudaffective medulliadrenal secretion. *Am. J. Physiol.*, 72: 283–294.

Cannon, W.B. and A. Rosenblueth. 1933. Studies on conditions of activity in endocrine organs. xxx. Sympathin E and Sympathin I. *Am. J. Physiol.*, 104: 557–574.
Carlson, A.J. 1938. Silas Weir Mitchell, 1829–1914. *Science*, 87: 474–478.
Carpenter, C.R. 1932. Relation of the male avian gonad to responses pertinent to reproductive phenomena. *Psychol. Bull.*, 29: 509–527.
Catchpole, C.K. and P.J.B. Slater. 1995. *Bird Song: Biological Themes and Variations.* Cambridge: Cambridge University Press.
Cattell, J. McK. 1895. Correspondence. Experimental psychology in America. *Science*, 2: 627–628.
Cattell, J. McK. 1921. In memory of Wilhelm Wuudt by his American students. *Psychol. Rev.*, 28: 155–159.
Cattell, J. McK. 1928. Early psychological laboratories. *Science*, 67: 543–548.
Cattell, J.M. 1929. Psychology in America. *Science*, 70: 335–347.
Chang, H.-T. 1984. Pilgrimage to Yale. *The Physiologist*, 27(6): 390–392.
Channing, W. 1894. Some remarks on the address delivered to the American Medico-Psychological Association by S. Weir Mitchell, M.D., May 16, 1894. *Am. J. Insanity*, 51: 171–181.
Chase, M.H., ed. 1972. *The Sleeping Brain.* Proceeding of the Symposia of the First International Congress of the Association for the Psycho-physiological Study of Sleep. Bruges, Belgium June 19–24, 1971. Los Angeles: Brain Research Institute, Univeristy of California, Los Angeles.
Chase, M.W. and C.C. Hunt. 1995. Herbert Spencer Gasser 1888–1968. *Biog. Mem. Natl. Acad. Sci. USA*, 67: 1–33.
Chase, R. 1979. The mentalist hypothesis and invertebrate neurobiology. *Persp. Biol. Med.*, 23: 103–117.
Chen, M.P., R.K.S. Lim, S.C. Wang, and C.L. Yi. 1937. On question of myelencephalic sympathetic centre in medulla. *Chinese J. Physiol.*, 11: 355–366; 367–384; 385–407.
Cherry, C. 1966. *On Human Communication: A Review, a Survey and a Criticism.* 2nd ed. Cambridge, Massachusettes: The MIT Press.
Chomsky, N. 1984. The formal nature of language. Pages 397–442 in E.H. Lenneberg, *Biological Foundations of Language*, Appendix A. Malabar, Florida: Robert E. Krieger Publishing Co.
Clark, G. 1939. The use of the Horsley-Clarke instrument on the rat. *Science*, 90: 92.
Clark, G., H.W. Magoun, and S.W. Ranson. 1939. Hypothalamic regulation of body temperature. *J. Neurophysiol.*, 2: 61–80.
Clarke, E. and L.S. Jacyna. 1987. *Nineteenth-Century Origins of Neuroscience Concepts.* Berkeley/Los Angeles: University of California Press.
Clemente, C.D. 1969. Cortical synchronization and the onset of sleep. Pages 80–88 in A. Kales, ed., *Sleep Physiology and Pathology. A Symposium.* Philadelphia: J.B. Lippincott.
Clemente, C.D. and M.B. Sterman. 1967. Basal forebrain mechanisms for internal inhibition and sleep. Pages 127–147 in S.S. Kety, E.V. Evarts, and H.L. Williams, eds., *Sleep and Altered States of Consciousness, Res. Pub. Assoc. Res. Nerv. Ment. Dis.*, v. 45. Baltimore: Williams and Wilkins.
Cobb, S. 1952. On the nature and locus of mind. *Arch. Neurol. Psychiat.*, 67: 172–177.
Coghill, G.E. 1902. The cranial nerves of Amblystoma tigrinum. *J. Comp. Neurol.*, 12: 205–289.
Coghill, G.E. 1929a. *Anatomy and the Problem of Behavior.* New York: MacMillan.

Coghill, G.E. 1929b. The early development of behavior in Amblyophia and in Man. *Arch. Neurol. Psychiat.*, 21: 989–1009.

Cole, K.S. 1949. Dynamic electrical characteristics of the squid giant axon membrane. *Arch. Sci. Physiol.*, 3: 253–258.

Cole, K.S. 1968. *Membranes, Ions and Inpulses: a Chapter of Classical biophysics.* Berkeley: University of California Press.

Cole, K.S. 1979. Interview series CON code COL, Oral History Project, Neuroscience History Archives, Brain Research Institute, University of California, Los Angeles.

Cole, K.S. and H.J. Curtis. 1939. Electric impedance of the squid giant axon during activity. *J. Gen. Physiol.*, 22: 649–670.

Cole, K.S. and A.L. Hodgkin, 1939. Membrane and protoplasm resistance in the squid giant axon. *J. Gen. Physiol.*, 22: 671–687.

Cole, M.F. and M. Cole. 1971. *Pierre Marie's Papers on Speech Disorders.* New York: Hafner.

Collins, J. 1898. *The Genesis and Dissolution of the Faculty of Speech. A Clinical and Psychological Study of Aphasia.* New York: Macmillan.

Collip, J.B., E.M. Anderson, and D.L. Thomson. 1933. The adrenotropic hormone of the anterior pituitary lobe. *Lancet*, 2: 347–348.

Conklin, E.G. 1939. Henry Herbert Donaldson 1857–1938. *Biog. Mem., Natl. Acad. Sci. USA*, 20: 229–243.

Cooper, J.R. 1987. Neuropharmacology. Pages 841–842 in G. Adelman and B.S. Smith, eds., *Encyclopedia of Neuroscience*, vol. 2. Boston: Birkhäuser.

Cooper, S. and J.C. Eccles. 1929. Isometric muscle twitch. *J. Physiol.*, London, iii–v.

Corner, G.W. 1928. Physiology of the corpus luteum. I. The effect of very early ablation of the corpus luteum upon embryos and uterus. *Am. J. Physiol.*, 86: 74–81.

Corner, G.W. 1945. Development, organization, and breakdown of the corpus luteum in the rhesus monkey. *Contrib. Embryo.*, 31: 117–146.

Corner, G.W. 1958. *Anatomist at Large: An Autobiography and Selected Essays.* New York: Basic Books.

Corner, G.W. and W.M. Allen. 1929. Physiology of the corpus luteum. II. Production of the special uterine reaction (progestational proliferation) by extracts of the corpus luteum. *Am. J. Physiol.*, 88: 326–339.

Cowan, W.M., D.H. Harter, and E.R. Kandel. 2000. The emergence of modern neuroscience: Some implications for neurology and psychiatry. *Annu. Rev. Neurosci.*, 23: 343–391.

Craigie, E.H. 1925. *An Introduction to the Finer Anatomy of the Central Nervous System Based upon That of the Albino Rat.* Toronto: University of Toronto Press.

Crick, F. 1994. *The Astonishing Hypothesis: The Scientific Search for the Soul.* New York: Charles Scribner's Sons.

Crile, G.W. 1950. *The Origin and Nature of the Emotions: Miscellaneous Papers.* Edited by A.F. Rowland. Philadelphia: Saunders.

Crocker, E.C. 1945. *Flavor.* New York: McGraw-Hill.

Crosby, E.C. 1917. The forebrain of Alligator mississippiensis. *J. Comp. Neurol.*, 27: 325–402.

Crosby, E.C. 1960. Charles Judson Henrick October 6, 1866–January 29, 1960. *J. Comp. Neurol.* 115: 3–8.

Crouch, R.L. 1934. The nuclear configuration of the hypothalamus and the subthalamus in Macacus rhesus. *J. Comp. Neurol.*, 59: 431–485.

Cushing, H. 1906. Sexual infantilism with optic atrophy in cases of tumor affecting the hypophysis cerebri. *J. Nerv. Ment. Dis.*, 33: 704–716.

Cushing, H. 1909. A note upon the faradic stimulation of the postcentral gyrus in conscious patients. *Brain*, 32: 44–53.

Cushing, H. 1912. *The Pituitary Body and Its Disorders; Clinical States Produced by Disorders of the Hypophysis Cerebri*. Philadelphia: Lippincott.

Cushing, H. 1930. Neurohypophysial mechanisms from a clinical standpoint. *The Lancet*, July 19, 1930: 119–127; July 26, 1930: 175–184.

Dale, H.H. 1936–1937. Transmission of nervous effects by acetylcholine. *The Harvey Lectures*, 32: 229–245.

Dale, H.H. 1937–1938. Du Bois-Reymond and chemical transmission. *J. Physiol.* (London), 91: 4P.

Dallenback, K.M. 1955. Phrenology versus psychoanalysis. *Am. J. Psychol.*, 68: 511–525.

Dalton, J.C. 1885. *Topographical Anatomy of the Human Brain*, 3 vol. Philadelphia: Brothers.

Damasio, A.R. 1982. Reflecting on the work of R.W. Sperry. *Trends Neurosci.*, 5: 222–224.

Damasio, A.R. and A.M. Galaburda. 1985. Norman Geschwind. *Arch. Neurol.*, 42: 500–504.

Damasio, A.R. and N. Geschwind. 1984. The neural basis of language. *Annu. Rev. Neurosci.*, 7: 127–147.

Dana, C.L. 1921. The anatomic seat of the emotions: a discussion of the James-Lange theory. *Arch. Neurol. Psychiat.*, 6: 634–639.

Darwin, C. 1872. *The Expression of the Emotions in Man and Animal.* London: J. Murray.

Dave, A.S., A.C. Yu, and D. Margoliash. 1998. Behavioral state modulation of auditory activity in a vocal motor system. *Science*, 282: 2250–2254.

Davenport, H.W. 1975. Walter B. Cannon's contribution to gastroenterology. Pages 3–25 in C. McC. Brooks, K. Koizumi, and J.D. Pinkston, eds., *The Life and Contributions of Walter Bradford Cannon 1871–1945*. Albany, New York: State University of New York Press.

Davenport, H.W. 1982. Epinephrin(e). *The Physiologist*, 25: 76–82.

Davis, H. 1968. Early notes by Charles Henry of HD's account (5 April 1968) of early EEG at Harvard. Revised by H. Davis 18 June 1968. Unpublished typescript.

Dejerine, J. and M. Vialet. 1893. Contribution à l'étude de la localization anatomique de la cécité verbale pure. *C.N. Soc. Bio. Paris*, 45: 790–791.

Delgado, J.M.R., W.W. Roberts, and N.E. Miller, 1954. Learning motivated by electrical stimulation of the brain. *Am. J. Physiol.*, 179: 587–593.

Dement, W. 1958. The occurrance of low voltage, fast, electroencephalogram patterns during behavioural sleep in the cat. *Electroenceph. Clin. Neurophysiol.*, 10: 291–296.

Dennett, D.C. 1998. *Brainchildren. Essays on Designing Minds.* Cambridge. Massachussetts: MIT Press.

Denny-Brown, D. 1966. The organismic (holistic) approach: The neurological impact of Kurt Goldstein. *Neuropsycholica*, 4: 293–297.

Denny-Brown, D., ed. 1975. *Centennial Anniversary Volume of the American Neurological Association 1875–1875.* New York: Springer Publishing Co.

DeVoogd, T.J. 1991. Endocrine modulation of the development and adult function of the avian song system. *Psychoneuroendocrinology*, 16: 41–66.

Diamond, M.C., A.B. Scheibel, G.M. Murphy, Jr., and T. Harve. 1985. On the brain of a scientist: Albert Einstein. *Exp. Neurol.*, 88: 198–204.

Dickinson, R.L. 1933. *Human Sex Anatomy.* Baltimore: Williams & Wilkins.

Dickinson, R.L. and L.S. Bryant. 1931. *Control of Conception; an Illustrated Medical Manual.* Baltimore: Williams & Wilkins.

Donaldson, H.H. 1891. Anatomical observations on the brain and several sense-organs of the blind deaf-mute, Laura Dewey Bridgman. *Am. J. Psychol.*, 3: 293–342; 4: 248–294.

Donaldson, H.H. 1892. The extent of the visual cortex in man, as deduced from the study of Laura Bridgman's brain. *Am. J. Psychol.*, 4: 503–513.

Donaldson, H.H. 1895. *The Growth of the Brain., a Study of the Nervous System in Relation to Education.* London: Scott.

Donaldson, H.H. 1900. The functional significance of the size and shape of the neurone. *J. Nerv. Ment. Dis.*, 27: 526–527.

Donaldson, H.H. 1908. A comparison of the albino rat with man in respect to the growth of the brain and of the spiral cord. *J. Comp. Neurol.*, 18: 345–392.

Donaldson, H.H. 1925. Research at the Wistar Institute, 1905–1925. *Bulletin*, 6: 43–51.

Donaldson, H.H. 1928. A study of the brains of three scholars: Granville Stanley Hall, Sir William Osler, and Edward Sylvester Morse. *J. Comp. Neurol.*, 46: 1–95.

Doupe, A.J. and M. Konishi. 1991. Tone-selective auditory circuits in the vocal control system of the zebra finch. *Proc. Natl. Acad. Sci. USA.* 88(24): 1339–1343.

Du Vigneaud, V., C. Ressler, J.M. Swan, C.W. Roberts, P.G. Katsoyannis, and S. Gordon. 1953. The synthesis of an octapeptide amide with the hormonal activity of oxytoxin. *J. Am. Chem. Soc.*, 75: 4879–4880.

Duncan, I. 1927. *My Life.* New York: Boni and Liveright.

Dunlap, K. 1912. Discussion. The case against introspection. *Psychol. Rev.*, 19: 404–412.

Dunlap, K., ed. 1922. *The Emotions.* Baltimore: Williams & Wilkins.

Dunlap, K. 1930. The National Institute of Psychology. *Science*, 71: 130–131.

Dunlap, K. 1930/1961. Knight Dunlap. Pages 35–61 in C. Murchison, ed., *A History of Psychology in Autobiography*, vol. 2. New York: Russell and Russell.

Dusser de Barenne, J.G. 1916. Experimental researches on sensory localizations in the cerebral cortex. *Q. J. Exp. Physiol.*, 9: 355–390.

Dusser de Barenne, J.G. 1924. Experimental researches on sensory localization in the cerebral cortex of the monkey (*Macacus*). *Proc. R. Soc. B.* (London), 96: 272–291.

Dusser de Barenne, J.G. 1935. Central levels of sensory integration. Pages 274–288 in C.A. Patten, A.M. Frantz, and C.C. Hare, eds., *Sensation: Its Mechanisms and Disturbances.* Res. Pub. Assoc. Res. Nerv. Ment. Dis., v. 15.

Dusser de Barenne, J.G. 1937. Letter to Marion Hines, October 11, 1937. Cushing/Whitney Medical Library, Yale University.

Dusser de Barenne, J.G. and W.S. McCulloch. 1938. The direct functional interrelation of sensory cortex and optic thalamus. *J. Neurophysiol.*, 1: 176–186.

Dutt, R.H. and L.E. Caside. 1948. Alteration of the estral cycle in sheep by use of progesterone and its effect upon subsequent ovulation and fertility. *Endocrinology*, 43: 208–217.

Earnest, E. 1950. *S. Weir Mitchell, Novelist and Physician.* Philadelphia: University of Pennsylvania Press.

Eccles, J.C., ed. 1966. *Brain and Conscious Experience.* Study Week September 28 to October 4, 1964, of the Pontificia Academia Scientiarum. New York: Springer-Verlag.

Eccles, J.C. 1970. Alexander Forbes and his achievement in electrophysiology, *Persp. Biol. Med.*, 13: 388–404.

Eccles, J.C. and W.C. Gibson. 1979. *Sherrington, His Life and Thought.* New York: Springer International.

Eccles, J.C., P. Fatt, and S. Landgren. 1957. Central pathway for direct inhibitory action of impulses in largest afferent nerve fibres to muscle. *J. Neurophysiol.*, 19: 75–98.

Edelman, G.M. 1992. *Bright Air, Brilliant Fire. On the Matter of the Mind.* New York: Basic Books.

Edinger, L. 1877. Die Endigung der Hautnerven bei Pterotrachen. *Arch. Mikr. Anat.*, 14: 171–179.

Edinger, L. 1882. Rückenmark und Gehirn in einem Falle von angeborenem Mangel eines Vorderarms. *Virch. Arch.*, 89: 46–63.

Edinger, L. 1885. *Zehn Vorlesungen über den Bau der nervösen Centralorgane des Menschen und der Thiere*, 5th ed. Leipzig: Vogel.

Edinger, L. 1899. *The anatomy of the Central Nervous System of Man and of Vertebrates in General.* Trans. from the Fifth German Edition by W.S. Hall, P.L. Holland, and E.P. Carlton. Philadelphia: F.A. Davis Co.

Edinger, T. 1955. Hearing and smell in cetacean history. *Monatsch. Psychiat. Neurol.*, 129: 37–58.

Emlen, S.T. 1972. An experimental analysis of the parameters of bird song eliciting species recognition. *Behavior*, 41: 120–171.

Epstein, Jason. 1999. Always time to kill. *The New York Review*, 46(17): 57–64.

Erlanger, J. and H.S. Gasser, with the collaboration... of G.H. Bishop. 1924. The compound nature of the action current of nerve as disclosed by the cathode ray oscilloscope. *Am. J. Physiol.*, 70: 624–666.

Erlanger, J. and H.S. Gasser. 1937. *Electrical Signs of Nervous Activity.* Philadelphia: University of Pennsylvania Press.

Erlanger, J., G.H. Bishop, and H.S. Gasser. 1926. The action potential waves transmitted between the sciatic nerve and its spinal roots. *Am. J. Physiol.* 78: 574–591.

Evans, E.C. III and K.S. Norris. 1988. On the evolution of acoustic communication systems in vertebrates. Part I. Historical aspects. Part II. Cognitive aspects. Pages 655–681 in P.E Nachtigall and P.W.B. Moore, eds., *Animal Sonar: Processes and Performance.* New York: Plenum Press.

Evans, H. McL. 1925. The function of the anterior hypophysis. *Harvey Lect.*, 19: 212–235.

Evarts, E.V. 1967. Unit activity in sleep and wakefulness. Pages 545–556 in Quarton, G.C., T. Melnechuk, and F.O. Schmitt, eds., *The neurosciences–A Study Program.* New York: Rockfeller University Press.

Eve, A.S. 1939. *Rutherford.* New York: Macmillan.

Eysenck, H.J., ed. 1960. *Behavior Therapy and the Neuroses: Readings in Modern Methods of Treatment Derived from Learning Theory.* New York: Macmillan.

Falls, J.B. 1963. Properties of bird song eliciting responses from territorial males. *Proc. XIII Int. Ornithol. Congr.*, 259–271.

Falls, J.B. 1992. Playback: a historical perspective. Pages 11–33 in P.K. McGregor, ed., *Playback and Studies of Animal Communication*. New York: Plenum Press.

Feindel, W. 1958. Discussion. Pages 669–670 in H.H. Jasper, L.D. Proctor, R.S. Knighton, W.C. Noshay, and R.T. Costello, eds., *Reticular Formation of the Brain*. Boston: Little, Brown and Co.

Fenn, W.O. 1963. *History of the American Physiological Society: The Third Quarter Century, 1937–1962*. Washington, DC: The American Physiological Society.

Ferrier, D. 1873. Experimental researches in cerebral physiology and pathology. *West Riding Lunatic Asylum Med. Rep.*, 3: 30–96.

Ferrier, D. 1876. *Functions of the Brain*. New York: Putnam.

Finger, S. 1999. Margaret Kennard on sparing and recovery of function: a tribute on the 100th anniversary of her birth. *J. Hist. Neurosci.*, 8: 269–285.

Finger, S. and D.G. Stein. 1982. *Brain Damage and Recovery*. New York: Academic Press.

Fisher, C., W. Ingram, and S.W. Ranson. 1938. *Diabetes Insipidus and the Neurohormonal Control of Water Balance.* Ann Arbor, Michigan: Edwards Bros., Inc. [litho-printed].

Fisher, C., H.W. Magoun, and S.W. Ranson. 1938. Dystocia in diabetes insipidus. *Am. J. Obstet. Gynecol.*, 36: 3–11.

Fitzgerald, J.E. and W.F. Windle. 1942. Some observations on early human fetal movements. *J. Comp. Neurol.*, 76: 159–167.

Flechsig, P. 1901. Developmental (myelogenetic) localisation of the cerebral cortex in the human subject. *Lancet*, ii: 1027–1029.

Flexner, A. 1910. *Medical Education in the United States and Canada. A Report...* Reprinted 1972. New York: Arno Press & The New York Times.

Flexner, A. 1914. *Prostitution in Europe.* New York: Bureau of Social Hygiene.

Flourens, P. 1843. De la phrénologie et des études vraies sur le cerveau. Paris: Garniers Frères.

Flourens, P. 1846. *Phrenology examined.* Translated by D.L. Meigs. Philadelphia: Hogan and Thompson. Reprinted 1978 in D.N. Robinson, ed., *Significant Contributions to the History of Psychology, 1750–1920.* Washington, DC.: University Publications of America.

Forbes, A. 1922. The interpretation of spinal reflexes in terms of recent knowledge of nerve conduction. *Physiol. Rev.*, 2: 361–414.

Forbes, A. and C. Thacher. 1920. Amplification of action currents with the electron tube in recording with the string galvanometer. *Am. J. Physiol.*, 52: 409–471.

Forbes, A., C.J. Campbell, and H.B. Williams. 1924. Electrical records of afferent nerve impulses from muscular receptors. *Am. J. Physiol.*, 69: 238–303.

Forel, A. 1937. *Out of My Life and Work.* New York: Norton.

Fouts, R. 1972. The use of guidance in teaching sign language to a chimpanzee. *J. Comp. Physiol. Psychol.*, 80: 515–522.

Fowler, O.S. 1870. *Sexual Science. Including Manhood, Womanhood, and Their Mutual Interrelations…As Taught by Phrenology*. Philadelphia: National Publishing Co.

Fowler, O.S. 1889. *Amativeness: Embracing the Evils and Remedies of Excessive and Perverted Sexuality, Including Warning and Advice to the Married and Single.* New York: Fowler and Wells Co., Publishers. New & revised ed. Reprinted in Rosenberg, C. and C. Smith Rosenberg. eds, 1974. *Sex and Science*. New York: Arno Press.

Frank, R.G., Jr., L.H. Marshall, and H.W. Magoun. 1976. The neurosciences. Pages 552–613 in J.Z. Bowers and E.F. Purcell, eds., *Advances in American Medicine: Essays at the Bicentennial*, 2 v. New York: Josiah Macy, Jr. Foundation.

Franz, S.I. 1907. On the functions of the cerebrum. The frontal lobes. *Arch. Psychol.*, 2: 4–64.

Franz, S.I. 1915. Symptomatological differences associated with similar cerebral lesions in the insane. Pages 1–79 in *Psychol. Monogr.* 19(1). Princeton, New Jersey: Psychological Review Co.

Franz, S.I. 1923. *Nervous and Mental Re-education.* New York: The Macmillan Company.

Franz, S.I. 1933. *Studies in Cerebral Function, I–IX.* Berkeley: University of California Press.

Franz, S.I. 1961. Shepherd Ivory Franz. Pages 89–113 in C. Muschison, ed., *A History of Psychology in Autobiography*, Vol. II. New York: Russell & Russell.

Freeman, W.J. and J.W. Watts. 1942. *Psychosurgery in the Treatment of Mental Disorders and Intractable Pain.* Springfield, Illinois: C.C Thomas.

French, J.D. 1952. Brain lesions associated with prolonged unconsciousness. *Arch. Neurol. Psychiat.*, 68: 727–740.

French, J.D., D.B. Lindsley, and H.W. Magoun. 1984. *An American Contribution to Neuroscience: The Brain Research Institute, UCLA 1959–1984.* Los Angeles: Brain Research Institute. University of California, Los Angeles.

Fields, W.S. and W.H. Sweet, eds. 1975. *Neural Basis of Violence and Aggression.* St. Louis: Warren H. Green.

Freud, S. 1891. *Zur Auffassung der Aphasien.* Leipzig: Franz Dentiche.

Fritsch, G. and E. Hitzig. 1870. Über die elektrische Erregbarkeit des Grosshirns. *Arch. Anat. Physiol.* (Leipzig), 37: 300–332.

Fröhlich, A. 1901. Ein Fall von Tumor der Hypophysis cerebri ohne Akronmegalie. *Wien. Klin. Rundsch.* 15: 883–886.

Fromkin, V. and R. Rodman. 1978. *An Introduction to Language*, 2nd ed. New York: Holt, Rinehart, and Winston.

Fulton, J.F. 1926. *Muscular Contraction and the Reflex Control of Movement.* Baltimore: Williams & Wilkins Co.

Fulton, J.F. 1938. *Physiology of the Nervous System.* London: Oxford University Press.

Fulton, J.F. 1946. *Harvey Cushing. A Biography.* Springfield, Illinois: C.C Thomas.

Fulton, J.F. 1951. *Frontal Lobotomy and Affective Behavior. A Neurophysiological Analysis.* New York: Norton.

Fulton, J.F. and R.W. Gerard. 1940. J.G. Dusser de Barenne 1885–1940. *J. Neurophysiol.*, 3: 282–292.

Fulton, J.F. and A.D. Keller. 1932. *The sign of Babinski. A Study of the Evolution of Cortical Dominance in Primates.* Springfield, Illinois: C.C Thomas.

Fulton, J.F. and M.A. Kennard. 1934. A study of flaccid and spastie paralyses produced by lesions of the cerebral cortex in primates. Pages 158–210 in *Localization of Function in the Cerebral Cortex. Res. Pub. Assoc. Nerv. Ment. Dis.*, 13.

Fye, W.B. 1987. *Development of American Physiology.* Baltimore: Johns Hopkins Press.

Gage, S.H. 1911. Retirement of Professor Burt Green Wilder. *Anat. Rec.*, 5: 359–362.

Gall, F.J. and G. Spurzheim. 1810–1819. *Anatomie et physiologie du système nerveux en généŕal, et du cerveau en particulier.* Paris: F. Schoell.

Gall, F.J., Vimont, and Broussais. 1838. *On the functions of the cerebellum*, trans. from the French by George Combe .... Edinburgh: MacLachlan & Stewart.

Gardener, H.H. 1893. Sex in Brain. Pages 95–125 in *Facts and Fictions in Life* by H.H. Gardener. Chicago: Charles H. Kerr and Co.

Gardener, R.A. and B.T. Gardener. 1969. Teaching sign language to a chimpanzee. *Science*, 165: 644–672.

Garner, R.L. 1892. *The Speech of Monkeys*. London: Heinemann.

Garner, R.L. 1896. Gorillas & Chimpanzees. London: Osgood, McIlvaine & Co.

Garstang, M., D. Lorom, R. Raspet, and M. Lindeque. 1995. Atmospheric controls on elephant communication. *J. Exp. Biol.*, 198: 939–951.

Garvey, C.R. 1929. List of American psychology laboratories. *Psychol. Bull.*, 26: 652–660.

Gasser, H.S. 1952. Discussion of B. Frankenhaeuser, The hypothesis of saltatory condition. *Cold Spring Harb. Sym. Q. Biol.*, 17: 32–36.

Gasser, H.S. 1964. Herbert Spencer Gasser 1888–1963 [an Autobiography]. *Exp. Neurol.*, Suppl. 1: iv–vii, 1–38.

Gasser, H.S. and J. Erlanger. 1922. A study of the action currents of nerve with the cathoderay oscillograph. *Am. J. Physiol.*, 62: 496–524.

Gasser, H.S. and H.S. Newcomer. 1921. Physiological action currents in the phrenic nerve. An application of the thermionic vacuum table to nerve physiology. *Am. J. Physiol.*, 57: 1–26.

Gazzaniga, M.S. 1970. *The Bisected Brain*. New York: Appleton-Centry-Crofts.

Gazzaniga, M.S. 1981. 1981 Nobel Prize for physiology or medicine. *Science*, 214: 517–518.

Gerard, R.W. 1970. Warren Sturgis McCulloch: rebel genius. *Trans. Am. Neurol. Assoc.* 95: 344–345.

Gerard, R.W. 1975. The minute experiment and the large picture. Pages 456–474 in F.G. Worden, J.P. Swazey, and G. Adelman, eds., *The Neurosciences: Paths of Discovery*. Cambridge, Massachusetts: The MIT Press.

Gerard, R.W. and W.H. Marshall. 1933. Nerve conduction velocity and equilibration. *Am. J. Physiol.* 104: 575–585.

Gerard, R.W., W.H. Marshall, and L.J. Saul. 1934. Brain action potentials. *Am. J. Physiol.* 109: 38–39.

Gerard, R.W., W.H. Marshall, and L.J. Saul. 1936. Electrical activity of the cat's brain. *Arch. Neurol. Psychiat.*, 36: 675–735.

Geren, B.B. 1954. The formation from the Schwann cell surface of myelin in the peripheral nerves of chick embryos. *Exp. Cell Res.*, 7: 558–562.

Geschwind, N. 1964. The paradoxical position of Kurt Goldstein in the history of aphasia. *Cortex*, 1: 214–224. Reprinted in N. Geschwind, *Selected Papers on Language and the Brain*, pp. 62–72. Bastow: D. Reidel Publishing Co., 1974.

Geschwind, N. 1965. Disconnexion syndromes in animals and man. *Brain*, 88: 237–294; 585–644.

Geschwind, N. 1966. Carl Wernicke, the Breslau school, and the history of aphasia. Pages 1–16 in E.C. Carterette, ed., *Brain Function*, Vol. III. Proceedings of the Third Conference, November 1963: Speech, Language, and Communication. Berkeley and Los Angeles: University of California Press.

Geschwind, N. and P. Behan. 1982. Left-handedness: association with immune disease, migraine, and developmental learning disorder. *Proc. Natl. Acad. Sci. USA*, 79: 5097–5100.

Geschwind, N. and A.M. Galaburda. 1987. *Cerebral Lateralization: Biological Mechanisms, Associations, and Pathology*. Cambridge, Massachusetts: The MIT Press.

Geschwind, N. and E. Kaplan. 1962. A human cerebral disconnection syndrome. A preliminary report. *Neurology*, 12: 675–685.

Geschwind, N. and W. Levitsky. 1968. Human brain: left-right asymmetrics in temporal speech region. *Science*, 161: 186–187.

Gibbs, F.A. and E.L. Gibbs. 1936. The convulsion threshold of various parts of the cat's brain. *Arch. Neurol. Psychiat.*, 35: 109–116.

Gibbs, F.A., E.L. Gibbs, and W.G. Lennox. 1938. Cerebral dysrhythmias of epilepsy. *Arch. Neurol. Psychiat.*, 39: 298–314.

Gibbons, A. 1991. Research News. Déjà vu all over again: chimp-language wars. *Science*, 251: 1561–1562.

Gladwell, M. 2000. John Rock's error. *The New Yorker*, Mar. 13, 2000: 52–63.

Gleason, J.B. 1982. Linguistic Slips. Review of V.A. Fromkin, ed., *Errors in Linguistic Performance: Slips of the Tongue, Ear, Pen, and Hand.* New York: Academic Press.

Goddard, H.H. 1919. *Psychology of the Normal and Subnormal.* New York: Dodd, Mead.

Goldby, F. and H.J. Gamble. 1957. The reptilian cerebral hemispheres. *Biol. Rev.*, 32: 383–420.

Goldenberg, M., M. Faber, E.J. Alston, and E.C. Chargaff. 1946. Evidence for the occurrence of nor-epinephrine in the adrenal gland. *Science*, 109: 534–535.

Goldman, S.A. and F. Nottebohm. 1983. Neuronal production, migration, and differentiation in a vocal control nucleus of the adult female canary brain. *Proc. Nat. Acad. Sci. USA*, 80: 2390–2394.

Goldstein, K. 1948. *Language and Language Disturbances. Aphasic Symptom Complexes and Their Significance for Medicine and Theory of Language.* New York: Grune & Stratton.

Goltz, F.L. 1881. [Report on Goltz's decerebrate dogs] *Trans. Int. Med. Cong.*, London, Kolckmann, I, 218–243.

Goodall, J. 1986. *The Chimpanzees of Gombe: Patterns of Behavior.* Cambridge, Massachusetts: Harvard University Press.

Goodman, E.S. 1980. Margaret F. Washburn (1871–1939): first woman Ph.D. in psychology. *Psychol. Women Q.*, 5: 69–80.

Gorski, R.A. and R.E. Whalen. 1966. *Brain and Gonadal Function.* Brain and Behavior. v.3, UCLA Forum in Medical Science. Berkeley and Los Angeles: University of California Press.

Gould, S.J. 1978–1979. Wide hats and narrow minds. *Nat. Hist.*, 88: 34–40.

Gould, S.J. 1999. Editorial: Take another look. *Science*, 286: 899.

Grafstein, B. 1971. Transneuronal transfer of radioactivity in the central nervous system. *Science*, 172: 177–179.

Grass, A.M. 1980. Interview 11 November 1980, Series CON Code GRA. Neuroscience History Archives, Brain Research Institute, University of California, Los Angeles.

Grass, A.M. 1984. *The Electroencephalic Heritage.* Quincy, Massachusetts: Grass Instrument Co.

Green, J.D. 1964. The Hippocampus. *Physiol. Rev.* 44: 561–608.

Green, J.D. and W.R. Adey. 1956. Electrophysiological studies of hippocampal connections and excitability. *Electroencephal. Chin. Neurophysiol.*, 8: 245–263.

Green, J.D. and J. Harris. 1946. A note on the blood supply and nerve supply of the hypophysis cerebri. *J. Anat.* 80: 247 (Proc.).

Green, J.D., C.D. Clemente and J. de Groot. 1957. Rhinencephalic lesions and behavior in cats; an analysis of the Klüver-Bucy syndrome with particular reference to normal and abnormal sexual behavior. *J. Comp. Neurol.*, 108: 505–545.

Greenblatt, M. and H.C. Solomon, eds. 1953. *Frontal Lobes and Schizophrenia; Second Lobotomy Project of Boston Psychopathic Hospital.* New York: Springer Publishing Co.

Greenblatt, M., R. Arnot, and H.C. Solomon, eds. 1950. *Studies in Lobotomy.* New York: Grune & Stratton.

Greenwalt, C.H. 1968. *Bird Song: Acoustics and Physiology.* Washington, DC: Smithsonian Institution Press.

Greep, R.O. 1974. History of research on anterior hypophysial hormones. Pages 1–27 in E. Knobil and W.H. Sawyer, eds., *Handbook of Physiology, Section 7: Endocrinology, Vol. IV, the Pituitary Gland and its Neuroendocrinological Control, Part 2.* Washington, DC: American Physiological Society.

Griffin, D.R. 1953. Bat sounds under natural conditions, with evidence for the echolocation of insect prey. *J. Exp. Zool.*, 123: 435–466.

Griffin, D.R. 1984. *Animal Thinking.* Cambridge, Massachusetts: Harvard University Press.

Griffin, D.R. 1985. Recollections of an experimental naturalist. Pages 120–142 in D.A. Dewsbury, ed., *Leaders in the Study of Animal Behavior.* Lewisburg, Pennsylvania: Buchnell University Press.

Griffin, D.R. and R. Galambos. 1941. The sensory basis of obstacle avoidance by flying bats. *J. Exp. Zool.*, 86: 481 506.

Grinnell, A.D. and P. Brown. 1978. Long-latency "subthreshold" collicular responses to the constant-frequency components emitted by a bat. *Science*, 202: 996–999.

Grob, G.N. 1998. Presidential Address. Psychiatry's Holy Grail: the search for the mechanisms of mental diseases. *Bull. Hist. Med.*, 72: 189–219.

Grünbaum, A.S.F. and C.S. Sherrington. 1904. Observations on the physiology of the cerebral cortex of the anthropoid apes. *Proc. R. Soc. B.* (London), 72: 152–155.

Guillemin, R. 1978. Peptides in the brain. The new endocrinology of the neuron. Pages 160–193 in *Les Prix Nobel.* Stockholm: Almquith & Wiksell International.

Guillemin, R. 1979. New endocrinology of the brain. *Persp. Bio. Med.*, 22: S74–S80.

Guillemin, R. and J.E. Gerich. 1976. Somatostatin: physiological and clinical significance. *Annu. Rev. Med.*, 27: 379–388.

Gurdjian, E.S. 1925. The olfactory connections in the albino rat, with special reference to the stria medullaris and the anterior commissure. *J. Comp. Neurol.*, 38: 127–163.

Gurdjian, E.S. 1927. The diencephalon of the albino rat. Studies on the brain of the rat, no. 2. *J. Comp. Neurol.*, 1–114.

Gurdjian, E.S. 1928. The corpus striatum of the albino rat. Studies on the brain of the rat, no. 3. *J. Comp. Neurol.*, 45: 249–281.

Haines, D.E. 1991. The contributors to volume 1 of *The Journal of Comparative Neurology*: C.L. Herrick, C.H. Turner, H.R. Pemberton, B.G. Wilder, F.W. Langdon, C.J. Herrick, C. von Kupffer, O.S. Strong, and J.B. Stowell. *J. Comp. Neurol.*, 314: 9–33.

Haines, D.E. 1995. Spitzka and Spitzka on the brains of the assassins of presidents. *J. Hist. Neurosci.*, 4: 236–266.

Hall, G.S. 1891. Review of James' Principles of Psychology. *Am. J. Psychol.*, 3: 578–591.

Hall, G.S. 1895. Discussion and Correspondence. Experimental psychology in America. *Science*, 2: 734–735.

Hall, G.S. 1912. *Founders of Modern Psychology*. New York: Appleton.

Hall, G.S. 1921. In memory of Wilhelm Wundt by his American students. *Psychol. Rev.*, 28: 154–155.

Hall, G.S. 1923. *Life and Confessions of a Psychologist.* New York: D. Appleton & Co.

Hall, K. 1999. Snails and tails or sugar and spice. [Review of "The Two Sexes" by E.E. Maccoby.] *Science*, 285: 1681–1682.

Halstead, W.C. 1947. *Brain and Intelligence. A Quantitative Study of the Frontal Lobes*. Chicago: University of Chicago Press.

Hamburger, V. 1934. The effects of wing bud extirpation on the development of the central nervous system in chick embryos. *J. Exp. Zool.*, 68: 449–494.

Hamilton, A. 1901. The division of differentiated cells in the central nervous system of the white rat. *J. Comp. Neurol.*, 11: 297–320.

Hamilton, G.V. 1911. A study of trial and error reactions in mammals. *J. Anim. Behav.*, 1: 33–66.

Hamilton, G.V. 1914. A study of sexual tendencies in monkeys and baboons. *J. Anim. Behav.*, 4: 295–318.

Hamilton, G.V. 1929. *A Research in Marriage.* New York: Garland.

Hamilton, G.V. and K. Macgowan. 1929. *What is Wrong with Marriage.* New York: Albert & Charles Boni.

Hammond, W.A. 1871. *A Treatise on the Diseases of the Nervous System.* New York: Appleton.

Hardy, J.D., H.G. Wolff, and H. Goodell. 1952. *Pain Sensations and Reactions.* Baltimore: Williams & Wilkins.

Harrell, W. and R. Harrison. 1938. The rise and fall of behaviorism. *J. Gen. Psychol.*, 18: 367–421.

Harlow, H.F. and R. Stagner. 1932. Psychology of feelings and emotions. I. Theory of feelings. *Psychol. Rev.*, 39: 570–589.

Harlow, J.M. 1848/49. Passage of an iron rod through the head. *Boston Med. Surg. J.* 39: 389–393.

Harlow, J.M. 1869. Recovery from the passage of an iron bar through the head. *Publ. Mass. Med. Soc. Boston*, 2: 327–346.

Harris, G.W. 1948. Electrical stimulation of the hypothalamus and the mechanism of neural control of the adenohypophysis. *J. Physiol.* (London), 107: 418–429.

Harrison, R.G. 1907. Observations on the living developing nerve fiber. *Anat. Rec.*, 1: 116–118.

Harrison, R.G. 1908. Embryonic transplantation and development of the nervous system. *Anat. Rec.*, 2: 385–410.

Hartridge, H. 1920. The avoidance of objects by bats in their flight. *J. Physiol.* (London), 54: 54–57.
Hartung, W.H. 1931. Epinephrine and related compounds: influence of structure on physiological activity. *Chem. Rev.*, 9: 389–465.
Hatai, S. 1901. The finer structure of the spinal ganglion cells in the white rat. *J. Comp. Neurol.*, 11: 1–24. On the presence of the centrosome in certain nerve cells of the white rat. Ibid.: 25–36.
Hatai. S. 1902. Number and size of the ganglion cells and dorsal root fibers in the white rat at different ages. *J. Comp. Neurol.*, 12: 107–123. Studies on the finer structures of neurones in the white rat. Ibid.: 199–204.
Hatai, S. 1907. On the zoological position of the rat. *Biol. Bull.*, 12: 266–273.
Hayes, K. 1951. *The Ape In Our Home*. New York: Harper.
Haymaker, W. 1970a. William Alexander Hammond (1828–1900). Pages 445–449 in W. Haymaker and F. Schiller, eds., *The Founders of Neurology*, 2nd ed. Springfield, Illinois: C.C. Thomas.
Haymaker, W. 1970b. James Papez (1883–1958). Pages 143–147 in W. Haymaker and F. Schiller, eds., *The Founders of Neurology*, 2nd ed. Springfield, Illinois: C.C. Thomas.
Haymaker, W. and F. Schiller, eds. 1970. *The Founders of Neurology*, 2nd ed. Springfield, Illinois: C.C Thomas.
Haymaker, W., E. Anderson, and W.J.H. Nauta, eds. 1969. *The Hypothalamus*. Springfield. Illinois: C.C Thomas.
Head, H. 1920. Ahasia: an historical review. *Brain*, 43: 390–411.
Head, H. and G. Holmes. 1911. Sensory disturbance from cerebral lesions. *Brain*, 34: 102–254.
Hécaen, H. 1977. *Évolution des Connaissances et des Doctrines sur les Localisations Cérébrales*. Paris: Desclees de Brouwer.
Heims, S.J. 1911. *The Cybernetics Group*. Cambridge, Massachusetts: The MIT Press.
Henneman, E., G. Somjen, and D.O. Carpenter. 1965. Functional significance of cell size in spinal motoneurons. *J. Neurophysiol.*, 28: 560–580.
Herrick, C.J. 1894. Studies from the neurological laboratory of Denison University. XI. The cranial nerves of *Amblystoma punctaturn*. *J. Comp. Neurol.*, 4: 193–207.
Herrick, C.J. 1899a. The cranial and first spinal nerves of Menidia; a contribution upon the nerve components of the bony fishes. *Arch. Neurol. Psychopathol.*, 2: 21–319.
Herrick, C.J. 1899b. The cranial and first spinal nerves of Menidia; a contribution upon the nerve components of the bony fishes. *J. Comp. Neurol.*, 9: 155–457.
Herrick, C.J. 1913. Some reflections on the origin and significance of the cerebral cortex. *J. Anim. Behav.*, 3: 222–236.
Herrick, C.J. 1926. *Brains of Rats and Men. A Survey of the Origin of the Cerebral Cortex*. Chicago: University of Chicago Press.
Herrick, C.J. 1933a. The amphibian forebrain. vi. Necturus. *J. Comp. Neurol.*, 58: 1–288.
Herrick, C.J. 1933b. The functions of the olfactory parts of the cerebral cortex. *Proc. Natl. Acad. Sci. USA*, 19: 7–14.
Herrick, C.J. 1941. The founder and the early history of the Journal. *J. Comp. Neurol.*, 74: 25–38.

Herrick, C.J. 1948. *The Brain of the Tiger Salamander, Ambystoma Tigrinum.* Chicago: University of Chicago Press.
Herrick, C.J. 1949. *George Ellett Coghill, Naturalist and Philosopher.* Chicago: University of Chicago Press.
Herrick, C.J. 1951. Introduction. Pages ix–xiii in N. Bull, *The Attitude Theory of Emotion.* New York: Nerv. and Ment. Dis. Mono., 81.
Herrick, C.J. 1954. One hundred volumes of the Journal of Comparative Neurology. *J. Comp. Neurol.*, 100: 717–756.
Herrick, C.J. 1955. Clarence Luther Herrick, pioneer naturalist, teacher and psychobiologist. *Am. Philos. Soc. Trans. Ser.2*, 45: 1–85.
Herrick, C.J. 1956. *The Evolution of Human Nature.* Austin: University of Texas Press.
Herrick, C.J. 1959. Elizabeth Caroline Crosby. *J. Comp. Neurol.*, 112: 13–17.
Herrick, C.J. 1961. Nervous mechanisms in behavior. *Fed. Proc.*, 20: 628–631.
Herrick, C.J. and G.H. Bishop. 1958. A comparative survey of the spinal lemniscus system. Pages 353–360 in H.H. Jasper, L.D. Procter, R.S. Knighton, W.C. Noshay and R.T. Costello, eds., *Reticular Formation of the Brain.* Boston: Little, Brown.
Herrick, C.L. and C.H. Turner. 1895. Second Report of the State Zoologist Including a Synopsis of the Entomostraca of Minnisota. *Zoological Series II. Geological and Natural History Survey of Minnesota.*
Herrnstein, R.J. 1969. Behaviorism. Pages 51–67 in D.L. Krantz, ed., *Schools of Psychology.* New York: Appleton, Century, Crofts.
Heywood, A. and H.A. Vortriede. 1905. Minor Studies from the Psychological Laboratory of Vassar College. Communicated by M.F. Washburn. I. Some experiments on the associative power of smells. *Am. J. Psychol.*, 16: 537–541.
Hilgard, E.R. 1939. Book Reviews. Skinner, B.F. The Behavior of Organisms. *Psychol. Bull.*, 36: 121–125.
Himwich, H.E. and L.H. Nahum. 1929. Respiratory quotient of brain. *Am. J. Physiol.*, 90: 389–390.
Himwich, W.A. 1980. Interview Series CON Code HIM, 8 November 1980. Neuroscience History Archives, University of California, Los Angeles.
Hines, M. 1936. The anterior border of the monkey's (Macaca mulatta) motor cortex and the production of spasticity. *Am. J. Physiol.*, 116: 76.
Hines, M. 1937. The "motor" cortex. *Johns Hopkins Hosp. Bull.*, 60: 313–336.
Hinsey, J.C. 1943. Stephen Walter Ranson, M.D. 1880–1942. *Arch. Neurol. Psychiat.*, 49: 457–463.
Hinsey, J.C. 1961. Ingredients in medicine research – The story of a method. *The Pharos*, 24: 13–23.
Hinsey, J.C. 1980. Interview Series CON Code HIN, Neuroscience History Archives, Brain Research Institute, University of California, Los Angeles.
Hinsey, J.C. 1981. A memoire of my years with Stephen Walter Ranson, 1924–1930. Pages 67–77 in W.F. Windle, ed., *Stephen Walter Ranson, Ground-Breaking Neuroscientist.* Los Angeles, California: Brain Research Institute, University of California, Los Angeles.
Hinsey, J.C. and J.E. Markee. 1933. Pregnancy following bilateral section of the cervical sympathetic trunk in the rabbit. *Proc. Soc. Exp. Biol. Med.*, 31: 270–271.
Hinsey, J.C., S.W. Ranson, and R.F. McNattin. 1930. The role of the hypothalamus and mesencephalon in locomotion. *Arch. Neurol. Psychiat.*, 23: 1–42.

Hitzig, E. 1874. *Untersuchungen über das Gehirn. Abhandlungen physiologischen und pathologischen Inhalts.* Berlin: Hirschwald.
Hoagland, H. 1974. *The Road to Yesterday.* Worchester, Massachusetts: Privately printed.
Hoagland, H. 1975. Reflections. Pages 146–157 in J. Meites, B.T. Donovan, and S.M. McCann, eds., *Pioneers in Neuroendocrinology.* New York: Plenum Press.
Hobson, J.A. and R.W. McCarley. 1977. *Neuronal Activity in Sleep. 1969–1974. An Annotated Bibliography.* Los Angeles: Brain Information Service/BRI Publications Office, University of California, Los Angeles.
Hoff, H.E. 1962. John Fulton's contribution to neurophysiology. *J. Hist. Med. Allied Sci.*, 17: 16–71.
Hoff, H.E., C.G. Breckenridge, and W.A. Spencer. 1952. Suprasegmental integration of cardiac innervation. *Am. J. Physiol.*, 171: 178–188.
Hollingworth, H.L. 1931. Effect and affect in learning. *Psychol. Rev.*, 38: 153–160.
Holmes, O.W. 1870. Mechanism in Thought and Morals. Reprinted in J.C. Burnham, ed., *Science in America: Historical Selections*, pp. 186–204. New York: Holt, Reinhart, and Winston.
Hooker, D. 1952. *The Prenatal Origin of Behavior.* Porter Lectures. Lawrence: University of Kansas Press.
Horsley, V.A. and R.H. Clarke. 1908. The structure and functions of the cerebellum examined by a new method. *Brain*, 31: 45–124.
Horsley, V.A. and E.A. Schäfer. 1888. A record of experiments upon the functions of the cerebral cortex. *Phil. Trans. R. Soc. Lond. (B)*, 179: 1–45.
Hoskins, E.R. and M. Morris. 1916–1917. On thyroidectomy in amphibia. *Anat. Rec.*, 11: 363.
Houssay, B.A. 1918. The pituitary body and polyuria. *Endorcrinology*, 2: 94–97.
Houssay, B.A., A. Biasotti, and R. Saminartino. 1935. Modifications fonctionnelles de l'hypophyse après les lesions infundibulo-tubériennes chez le crapaud. *C.R. Soc. Biol.*, 120: 725–727.
Howell, W.H. 1898. The physiological effects of extracts of the hypophysis cerebri and infundibular body. *J. Exp. Med.*, 3: 245–258.
Hoyle, G., ed. 1977. *Identified Neurons and Behavior of Arthropods.* New York: Plenum Press.
Huber, G.C. and E.C. Crosby. 1929. The nuclei and fiber paths of the avian diencephalon, with consideration of telencephalic and certain mesencephalic centers and connections. *J. Comp. Neurol.*, 48: 1–225.
Hull, C.L. 1942. Conditioning: outline of a systematic theory of learning. Pages 61–95 in *The 41st Yearbook of the National Society for the Study of Education, Part II. The Psychology of Learning.* Bloomington: Public School Publishing Co.
Humphrey, T. 1951. The caudal extent of the descending nerve during the period of early human fetal activity (8 to 8.5 weeks of menstrual age). *Anat. Rec.*, 109: 306–307.
Huxley, Andrew. 1995. Electrical activity of nerve: the background up to 1952. Pages 3–10 in S.G. Waxman, J.D. Kocsis, and P.K. Stys, eds., *The Axon.* New York: Oxford University Press.
Huxley, T.H. 1863. *Evidence as to Man's Place in Nature.* London: Williams Norgate.

Ingram, W.R. 1975. A personal neuroscientific development with remarks on other events and people. Pages 175–191 in J. Meites, B.T. Donovan, and S.M. McCann, eds., *Pioneers in Neuroendocrinology*. New York: Plenum Press.
Ingram, W.R. and S.W. Ranson. 1932. Postural reactions in cats following destruction of both red nuclei. *Proc. Soc. Exp. Biol. Med.*, 29: 1089.
Ingram, W.R., F.I. Hannett and S.W. Ranson. 1932. The topography of the nuclei of the diencephalon of the cat. *J.Comp. Neurol.* 55: 333–394.
Jacobs, M.S., P.J. Morgane, and W.L. McFarland. 1971. The anatomy of the brain of the bottlenose dolphin (Tursiots truncatus). Rhinic lobe (rhinencephanlon). I. The paleocortex. *J. Comp. Neurol.*, 141: 205–271.
Jacobsen, C.F. 1931. A study of cerebral functioning in learning. The frontal lobes. *J. Comp. Neurol.*, 52: 271–340.
Jacobsen, C.F. 1936. The functions of the frontal association areas in monkeys. Pages 1–60 in Studies of cerebral function in primates. *Comp. Psychol. Monogr.*, 12–13.
James, W. 1899. *The Principles of Psychology*, 2 v. New York: Henry Holt and Co.
James, W. 1904. Does "consciousness" exist? *J. Phil. Psychol. Sci. Methods.*, 1: 477–491.
James, W. 1967. What is an emotion? Pages 11–30 in facsimile of 1922 edition of *Mind*, 9: 188–205. Reprinted New York: Hafner Publishing. Co.
Jasper, H.H. and L. Carmichael. 1935. Electrical potentials from the intact human brain. *Science*, 81: 51–53.
Jaynes, J. 1976. *On the Origin of Consciousness in the Breakdown of the Bicameral Mind.* Princeton: Princeton University Press.
Jennings, H.S. 1906. *Behavior of the Lower Organisms*. New York: Macmillar Co.
Johnson, R.A. 1995. Thomas Jefferson's letters to friends about a new book by Pierre Flourens. *Kopf Carrier*, 41: 1–3.
Johnston, J.B. 1902. The brain of Acipenser. A contribution to the morphology of the vertebrate brain. *Zoologische Jahrbucher für Anatomie und Ontologie der Thiere*, 15: 59–260.
Johnston, J.B. 1906. *The Nervous System of Vertebrates*. Philadelphia: P. Blakiston's Son & Co.
Johnston, J.B. 1913. The morphology of the septum, hippocampus, and pallial commissures in reptiles and mammals. *J. Comp. Neurol.*, 23: 371–478.
Johnston, J.B. 1916. Evidence of a motor pallium in the forebrain of reptiles. *J. Comp. Neurol.*, 26: 475–479.
Jonas, G. 1972. Profiles. Visceral learning. I and II. *The New Yorker*, Aug. 19, pp. 34–57; Aug. 26, pp. 30–57.
Jones, E. 1961. *The Life and Work of Sigmund Freud.* Edited and abridged by L. Trilling and S. Marcus. New York: Basic Books.
Judd, C.H. 1961. Charles H. Judd. Pages 207–235 in C. Murchison, ed., *A History of Psychology in Autobiography*, v. 2. New York: Russell & Russell.
Jung, R. 1975. Some European neuroscientists: a personal tribute. Pages 476–511 in F.G. Worden, J.P. Swazey, and G. Adelman, eds., *The Neurosciences: Paths of Discovery*. Cambridge, Massachusetts: The MIT Press.
Kales, A., ed. 1969. *Sleep Physiology & Pathology. A Symposium.* Philadelphia: J.B. Lippincott Co.
Kandel, E.R. 1969. The organization of sub-populations in the abdominal ganglion of *Aplysia*. Pages 71–111 in Brazier, M.A.B., ed., *The Interneuron*. UCLA Forum in Medical Sciences #11. Berkeley and Los Angeles: University of California Press.

Kappers, C.U.A. 1921. Die Vergleichende Anatomie des Nervensystems Wirbeltiere und des Menschen. 2 v. Haarlem: Bohn.
Kappers, C.U.A., G.C. Huber, and E.C. Crosby. 1936. *The Comparative Anatomy of the Nervous System of Vertebrates, Including Man*, 2 vol. New York: Macmillan.
Katz, I.C. and M.E. Gurney. 1981. Auditory responses in the Zebra Finch's motor system for song. *Brain Res.*, 211: 192–197.
Keegan, J.J. 1920. The Indian Brain. *Am. J. Physical Anthropol.*, 3: 25–62.
Keen, W.W. 1891. Five cases of cerebral surgery. I and II. For epilepsy following trauma. III. For insanity following trauma. IV. For cerebral tumor. V. For defective development. *Am. J. Med. Sci.*, 102: 219–238.
Keller, A.D., W. Noble and J.W. Hamilton. 1936. Effects of anatomical separation of the hypophysis from the hypothalamus in the dog. *Am. J. Physiol.*, 117: 467–473.
Kellogg, R. 1928. The History of whales. Their adaptation to life in water. *Q. Rev. Biol.*, 3: 174–208.
Kellogg, W.N., R. Kohler, and H.N. Morris. 1953. Porpoise sounds as sonar signals. *Science*, 117: 239–243.
Kellogg, W.N. 1960. Auditory scanning in the dolphin. *Psychol. Rec.*, 10: 25–27.
Kellogg, W.N. 1961. *Porpoises and Sonar*. Chicago: University of Chicago Press.
Kennard, M.A. 1938. Reorganization of motor function in the cerebral cortex of monkeys deprived of motor and premotor areas in infancy. *J. Neurophysiol.*, 1: 477–496.
Kennedy, D. 1975. Behavior to neurobiology: a zoologist's approaches to nervous systems. Pages 47–62 in F.G. Worden, J.P. Swazey, and G. Adelman, eds., *The Neurosciences: Paths of Discovery*. Cambridge, Massachusetts: The MIT Press.
Kies, M.W. 1985. Species-specificity and localization of encephalitogenic sites in myelin basic protein. *Springer Seminars in Immunopathology*, 8: 295–303.
Kimura, D. 1961. Cerebral dominance and the perception of verbal stimuli. *Can. J. Psychol.*, 15: 166–171.
Kimura, D. 1964. Left-right differences in the perception of melodies. *Q.J. Exp. Psychol.*, 16: 355–358.
Kinnaman, A.J. 1902. Mental life of two *Macacus rhesus* monkeys in captivity. I. *Am. J. Psychol.*, 13: 98–148, II, 173–218.
Kinsey, A.C., W.B. Pomeroy, and C.E. Martin. 1948. *Sexual Behavior in the Human Male*. Philadelphia: W.B. Saunders Co.
Kinsey, A.C., 1953. *Sexual Behavior in the Human Female*: by the Staff of the Institute for Sex Research, Indiana Univeristy. Philadelphia: W.B. Saunders Co.
Kleitman, N. 1969. Basic sleep-rest cycle in relation to sleep and wakefulness. Pages 33–38 in A. Kales, ed., *Sleep Physiology and Pathology. A Symposium*. Philadelphia: J.B. Lippincott.
Kline, L.W. 1898–1899. Methods in animal psychology. *Am. J. Psychol.*, 10: 256–279.
Klüver, H. 1951. Functional differences between the occipital and temporal lobes. With special reference to the interrelations of behavior and extracerebral mechanisms. Pages 147–199 in L.A. Jeffries, ed., *Cerebral Mechanisms and Behavior*, the Hixon Symposium 1948. New York: Wiley.
Klüver, H. 1955. Stephen Polyak 1889–1955: *J. Comp. Neurol.*, 103: 1–9.
Klüver, H. 1958. "The temporal lobe syndrome" produced by bilateral ablations. *Ciba Foundation Symposium on the Neurological Basis of Behavior*, pp. 175–182.

Klüver, H. 1965. The neurobiology of normal and abnormal perception. Pages 1–40 in P.H. Hoch and J. Zubin, eds., *Psychopathology of Perception*. New York: Grune and Stratton.

Klüver, H. and P.C. Bucy. 1937. "Psychic blindness" and other symptoms following bilateral temporal lobectomy in Rhesus monkeys. *Am. J. Physiol.*, 119: 352–353.

Klüver, H. and P.C. Bucy. 1939. A preliminary analysis of the functions of the temporal lobes in monkeys. *Trans. Am. Neurol. Assoc.*, 65: 170–175.

Kohler, R.E. 1982. *From Medical Chemistry to Biochemistry: the Making of a Biomedical Discipline*. Cambridge: Cambridge University Press.

Konishi, M. 1985. Birdsong: from behavior to neuron. Annu. Rev. Neurosci., 8: 125–170.

Koppányi, T. and J.F. Pearcy. 1925. Comparative studies on the excitability of the forebrain. *Am. J. Physiol.*, 71: 339–343.

Krieg, W.J.S. 1975. *Stereotaxy*. Evanston, Illinois: Brain Books.

Kroodsma, D.E. 1976. Reproductive development in a female songbird: differential stimulation by quality of male song. *Science*, 192: 574–575.

Krücke, W., H. Spatz, K. Goldstein, M. Flesch-Thebesius, and A. Lauche. 1959. *Ludwig Edinger 1855–1918*. Wiesbaden: Steiner.

Kruger, L. 1959. The thalamus of the dolphin (*Tursiops truncatus*) and comparison with other mammals. *J. Comp. Neurol.*, 111: 133–194.

Kubie, L.S. 1953. Some implications for psychoanalysis of modern concepts of organization of brain. *Psychoanalyt. Q.*, 22: 21–68.

Ladd, G.T. 1887. Elements of Physiological Psychology. Reviewed by G.S. Hall. *Am. J. Psychol.* 1: 159–164.

Ladd, G.T. 1895. Correspondence. Experimental psychology in America. *Science*, 2: 626–627.

Lambert, E.F., B.F. Skinner, and A. Forbes. 1933. Some conditions affecting intensity and duration thresholds in motor nerve, with reference to chronaxie of subordination. *Am. J. Physiol.*, 106: 721–737.

Lange, C.G. 1967. *The Emotions. A psychophysiological study.* Trans. I.A. Haupt from the authorized German translation by H. Kurella, 1887. Reprinted. New York: Hafner Publishing. Co.

Langley, J.N. 1905. On the reaction of cells and nerve-endings to certain poisons, chiefly as regards the reaction of striated muscle to nicotine and curari. *J. Physiol.* (London), 33: 374–413.

Langley, L.L., ed. 1975. *Homeostasis: Origins of the Concept.* Stroudburg, Pennsylvania: Dowder, Hutelusion & Ross.

Langworthy, O.R. 1931. Factors determining the differentiation of the cerebral cortex in sea-living mammals (the cetacea). A study of the brain of the porpoise, Tursiops, truncatus. *Brain*, 54: 225–236.

Langworthy, O.R. 1932. A description of the central nervous system of the porpoise (Tursiops truncatus). *J. Comp. Neurol.*, 54: 437–488.

Langworthy, O.R. and C. Bagley, Jr. 1926. Forebrain and midbrain of alligator with experimental transections of brain stem; study of electrically excitable regions. *Arch. Neurol. Psychiat.*, 16: 154–166.

Laqueur, T.W. 1999. [Review of Wijingaard, M.v.d., "Reinventing the Sexes: the Biomedical Construction of Feminity and Masculinity."] *Bull. Hist. Med.*, 73: 546–547.

Lashley, K.S. 1929. *Brain Mechanisms and Intelligence.* Chicago: University of Chicago Press.

Lashley, K.S. 1933. Integrative functions of the cerebral cortex. *Physiol. Rev.*, 13: 1–42.

Lashley, K.S. 1938. The thalamus and emotion. *Psychol. Rev.*, 45: 42–61.

Lashley, K.S. 1950. In search of the engram. *Symp. Soc. Exp. Biol.*, 4: 454–482.

Lashley, K.S. 1952. Neuropsychology. Pages 18–23 in *Survey of Neurobiology.* Washington, DC: National Academy of Sciences-National Research Council.

Leake, L.D. and R.J. Walker. 1980. *Invertebrate Neuropharmacology.* New York: Wiley.

Lehrman, D.S. 1953. A critique of Konrad Lorenz's theory of instinctive behavior. *Q. Rev. Biol.*, 28: 337–363.

Lenneberg, E.H. 1967. *Biological Foundations of Language.* New York: Wiley. Reprinted 1984.

Lentz, T.L. 1968. *Primitive Nervous Systems.* New Haven: Yale University Press.

Leonard, S.L. 1939. Induction of singing in female canaries by injections of male hormone. *Proc. Soc. Exp. Bird. Med.*, 41: 229–230.

Lettvin, J.Y. 1989. Strychnine neuronography. Pages 50–58 in R. McCulloch, ed., *Collected Works of Warren S. McCulloch,* 4 v. Salinas, California: Intersystems Publication.

Lettvin, J.Y., H.R. Maturana, W.S. McCulloch, and W.H. Pitts. 1959. What the frog's eye tells the frog's brain. *Proc, Inst. Radio Eng.*, 47: 1940–1951.

Leuret, F. et L.P. Gratiolet. 1839. *Anatomie comparée du système nerveux considéré dans ses rapports avec l'intelligence*, v. 1. Paris: Baillière.

Levi-Montalcini, R. 1975. NGF: An uncharted route. Pages 244–265 in F.G. Worden, J.P. Swazey, and G. Adelman, eds., *The Neurosciences: Paths of Discovery.* Cambridge, Massachusetts: The MIT Press.

Levin, P.M. 1936. The efferent fibers of the frontal lobe. *J. Comp. Neurol.*, 63: 369–420.

Leyton, A.S.F. and C.S. Sherrington. 1917. Observations on the excitable cortex of the chimpanzee, orang-utan, and gorilla. *Q.J. Exp. Physiol.*, 11: 135–222.

Liberman, A.M. 1974. The specialization of the language hemisphere. Pages 43–56 in F.O. Schmitt and F.G. Worden, eds., *The Neurosciences Third Study Program.* Cambridge, Massachusetts: The MIT Press.

Libet, B. 1994. A testable field theory of mind-brain interaction. *J. Conscious. Stu.*, 1: 119–126.

Libet, B. 1996. Neurol processes in the production of conscious experience. Pages 96–117 in M. Velmans, ed., *The Science of Consciousness. Psychol. Neuropsych. Clin. Rev.* London: Routledge.

Libet, B., W.W. Alberts, F.W. Wright, L. Delattre, G. Levin, and B. Feinstein. 1964. Production of threshold levels of conscious sensation by electrical stimulation of human somatosensory cortex. *J. Neurophysiol.*, 27: 546–578.

Lieberman, P. 1998. *Eve Spoke: Human Language and Human Evolution.* New York: Norton.

Lilly, J.C. 1958. Some considerations regarding basic mechanisms of positive and negative types of motivations. *Am. J. Psychiat.*, 115: 498–507.

Lilly, J.C. 1961. *Man and Dolphin.* Garden City, New York: Doubleday.

Lilly, J.C. 1962. Consideration of the relation of brain size to capability for language activity as illustrated by Homo sapiens and Tursiops truncatus (bottlenose dolphin). *Electroenceph. Clin. Neurophysiol.*, 14: 424.

Lilly, J.C. 1965. Vocal mimicry in *Tursiots*. Ability to match numbers and durations of human vocal bursts. *Science*, 147: 300–301.

Lilly, J.C. 1978. *Communication Between Man and Dolphin: The Possibilities of Talking with Other Species*. New York: Crown Publishers, Inc.

Lilly, J.C. and A.M. Miller. 1961a. Sounds emitted by the bottlenose dolphin. *Science*, 133: 1689–1693.

Lilly, J.C. and A.M. Miller. 1961b. Vocal exchanges between dolphins. *Science*, 134: 1873–1876.

Lilly, J.C. and A.M. Miller. 1962. Operant conditioning of the bottlenose dolphin with electrical stimulation of the brain. *J. Comp. Physiol. Psychol.*, 55: 73–79.

Lindsley, D.B. 1938. Electrical potentials of the brain in children and adults. *J. Gen. Psychol.*, 19: 285–306.

Lindsley, D.B. 1969. Average evoked potentials–achievements, failures and prospects. 1–43 in E. Donchin and D.B. Lindsley, eds., *Average Evoked Potentials. Methods, Results, and Evoluations*. Washington, DC: National Aeronautics and Space Administration.

Lindsley, D.B., J. Bowden, and H.W. Magoun. 1949. Effect upon the EEG of acute injury to the brain stem activating system. *Electroeuceph. Clin. Neurophysiol.* 1: 475–486.

Ling, G. and R.W. Gerard. 1949. The normal resting potential of frog sartorius muscle. *J. Cell Comp. Physiol.*, 34: 383–396.

Livingston, R.B. and J.F. Fulton. 1948. "Cortical instability": a study of frequency effects. *Fed. Proc.*, 7: 74.

Lloyd, D.P.C. 1941. A direct central inhibitory action of dromically conducted impulses. *J. Neurophysiol.*, 4: 184–190.

Lloyd, D.P.C. 1946a. Facilitation and inhibition of spinal moto-neurons. *J. Neurophysiol.*, 9: 421–438.

Lloyd, D.P.C. 1946b. Integrative pattern of excitation and inhibition in two-neuron reflex arcs. *J. Neurophysiol.*, 9: 439–444.

Lockard, I. 1992. *Desk Reference for Neuroscience*, 2nd ed. New York: Springer-Verlag.

Locke, J. 1700/1964. An essay concerning human understanding. Abridged and edited by A.S. Pringle-Pattison. Oxford: Clarendon Press.

Loeb, J. 1900. *Comparative Physiology of the Brain and Comparative Psychology*. New York: G.P. Putnam's & Sons.

Loeb, J. 1905. Heliotropism. Pages 1–88 in *Studies in General Physiology*, 2 v. Chicago: University of Chicago Press.

Loeser, W. 1905. A study of the functions of different parts of the frog's brain. *J. Comp. Neurol. Psychol.*, 15: 355–373.

Loewi, O. 1921. Über humorale Übertragbarkeit der Herznervenwirkung. I. Mitteilung *Pflügers Arch.*, 189: 239–242.

Lomax, E. 1977. The Laura Spelman Rockefeller Memorial: Some of its contributions to early research in child development. *J. Hist. Behav. Sci.*, 13: 283–293.

Loomis, A.L., E.N. Harvey, and G.A. Hobart. 1935. Potential rhythyms of the cerebral cortex during sleep. *Science*, 81: 597–598.

Lotze, H. 1851. *Allgemeine Physiologie des Körperlichen Lebens*. Leipzig: Weidman.

Lotze, H. 1852. *Medicinsche Psychologie: oder, Physiologie der Seele*. Leipzig: Weidmann.

Lotze, H. 1885. *Outlines of Psychology. Dictations from Lectures.* Trans. with a Chapter on the Anatomy of the Brain by C.L. Herrick. Minneapolis: S.M. Williams.
Luse, S.A. 1956. Formation of myelin in the central nervous system of mice and rats, as studied with electron microscope. *J. Biophys. Biochem. Cytol.*, 2: 777–783.
Lundberg, A. and C.G. Phillips. 1973. T. Graham Brown's film on locomotion in the decerebrate cat. *J. Physiol.* (London), 231: 90P–91P.
Lunt, G.G. and R.W. Olsen, eds. 1988. *Comparative Invertebrate Neurochemistry.* Ithaca, New York: Cornell University Press.
Maccoby, E.E. 1998. *The Two Sexes: Growing up Apart, Coming Together.* Cambridge, Massachusetts: Harvard University Press.
MacLean, P.D. 1949. Psychosomatic disease and the "visceral brain." Recent developments bearing on the Papez Theory of emotion. *Psychosom. Med.*, 11: 338–353.
MacLean, P.D. 1952. Some psychiatric implications of physiological studies on frontotemporal portion of limbic system (visceral brain). *Electroenceph. Clin. Neurophysiol.*, 4: 407–418.
MacLean, P.D. 1954. Studies in limbic system (visceral brain) and their bearing on psychosomatic problems. Pages 101–125 in E.D. Wittkower and R.A. Cleghorn, eds., *Recent Developments in Psychosomatic Medicine.* Philadelphia: J.B. Lippincott.
MacLean, P.D. 1967. The brain in relation to empathy and medical education. *J. Nerv. Ment. Dis.*, 144: 374–382.
MacLean, P.D. 1973. A triune concept of the brain and behavior. Pages 1–66 in *Hincks Memorial Lectures*, 1969. T.J. Boag and D. Campbell, eds. Toronto: University of Toronto Press.
MacLean, P.D. 1978. Challenges of the Papez heritage. Page 1–15 in K. Livingston and O. Hornkiewicz, eds., *Limbic Mechanisms.* New York: Plenum.
MacLean, P.D. 1990. *The Triune Brain in Evolution. Role in Paleocerebral Functions.* New York: Plenum Press.
MacLean, P.D. and J.M.R. Delgado. 1953. Electrical and chemical stimulation of frontotemporal portion of limbic system in the waking animal. *Electroenceph. Clin. Neurophysiol.*, 5: 91–100.
Magnus, R. and E.A. Schäfer. 1901. The action of pituitary extracts upon the kidney. *J. Physiol.* (London), 27: ix–x.
Magoun, H.W. 1942. Stephen Walter Ranson 1880–1942, An appreciation. *Q. Bull. Northwestern University Med. Sch.*, 16: 302–304.
Magoun, H.W. 1959. Preface to the Section on Neurophysiology. Page xi in J. Field, ed., *Handbook of Physiology. Section 1: Neurophysiology.* Vol. I. Washington, DC: American Physiological Society.
Magoun, H.W. 1960a. Evolutionary concepts of brain function following Darwin and Spencer. Pages 187–209 in S. Tax, ed., *Evolution After Darwin.* Chicago: University of Chicago Press.
Magoun, H.W. 1960b. Background and interests of the participants. Pages 27–28 in M.A.B. Brazier, ed., *The Central Nervous System and Behavior*. Transactions of the Third Conference … New York: Josiah Macy Jr. Foundation.
Magoun, H.W. 1966. Introduction. Pages 1–2 in R.A. Gorski and R.E. Whalen, eds., *Brain and Behavior, v. III. Brain and Gonadal Function.* Proceedings of the Third Conference, 1963, UCLA Forum in Medical Sciences.

Magoun, H.W. 1979. Interview Series BRI Code MAG. with L.H. Marshall. Oral History Project, Neuroscience History Archives, Brain Research Institute, University of California, Los Angeles.

Magoun, H.W. 1975. The role of research institutes in the advancement of neuroscience: Ranson's Institute of Neurology, 1928–1942. Pages 515–527 in F.G. Worden, J.P. Swazey, and G. Adelman, eds., *The Neurosciences: Paths of Discovery.* Cambridge, Massachusetts: The MIT Press.

Magoun, H.W. 1981a. Revival of the Horsley-Clarlse instrument, in studies of posture and locomotion, at Ranson's Institute of Neurology. Pages 103–122 in W.F. Windle, ed., *Stephen Walter Ranson: Ground-Breaking Neuroscientist.* Los Angeles: The Brain Research Institute, University of California, Los Angeles.

Magoun, H.W. 1981b. John B. Watson and the study of human sexual behavior. *J. Sex Res.*, 17: 368–378.

Magoun, H.W. 1985. The Northwestern connection with the reticular formation. *Surg. Neurol.*, 24: 250–252.

Magoun, H.W. and C.D. Clemente. 1976. Percival Bailey: 1892–1973. *Anat. Rec.*, 186: 235–237.

Magoun, H.W. and C. Fisher. 1980. Walter R. Ingram at Ranson's Institute of Neurology, 1930–1936. *Persp. Biol. Med.* 24: 31–56.

Magoun, H.W., W.K. Hare and S.W. Ranson. 1935. Electrical stimulation of the interior of the cerebellum in the monkey. *Am. J. Physiol.*, 112: 329–339.

Magoun, H.W. and R. Rhines. 1947. *Spasticity: The Stretch Reflex and Extra-Pyramidal Systems.* Springfield, Illinois: C.C Thomas.

Makepeace, A.W., G.L. Weinstein, and M.H. Friedman. 1937. The effect of progestrin and progesterone on ovulation in the rabbit. *Am. J. Physiol.*, 119: 512–516.

Mall, F.P. 1909. On several anatomical characters of the human brain, said to vary according to race and sex, with especial reference to the weight of the frontal lobe. *Am. J. Anat.*, 9: 1–32.

Mall, F.P. 1913. Plea for an institute of human embryology. *J. Am. Med. Assoc.*, 60: 1599–1601.

Malmo, R.B. 1942. Interference factors in delayed response in monkeys after removal of frontal lobes. *J. Neurophysiol.*, 5: 295–308.

Marie, P. 1886. Sur deux cas d'acromegalie. *Rev. Med.* (Paris) 6: 297–333.

Marler, P. 1970. Birdsong and speech development: could there be parallels? *Am. Sci.*, 58: 669–673.

Marler, P. 1976. An ethological theory of the origin of vocal learning. Pages 386–395 in S.R. Harnad, H.D. Steklis, and J. Lancaster, eds., *Origins and Evolution of Language and Speech.* New York: The New York Academy of Sciences.

Marmont, G. 1949. Studies on the axon membrane. I. A new method. *J. Cell. Comp. Physiol.*, 34: 351–382.

Marshall, L.H. 1983. The fecundity of aggregates: the axonologists at Washington University, 1922–1942. *Persp. Biol. Med.*, 26: 613–636.

Marshall, L.H. 1987. An annotated interview with Giuseppe Moruzzi, 1910–1986. *Exp. Neurol.*, 97: 225–242.

Marshall, L.H. and H.W. Magoun. 1998. *Discoveries in the Human Brain: Neuroscience Prehistory, Anatomy, and Function.* Totowa, New Jersey: Humana Press.

Marshall, L.H. and M.M. Patterson, comp. 1991. Stereotaxie Atlases and Related References. Privately published by David Kopf Instruments, Tujunga, CA 91043.

Marshall, W.H. 1950. Relation of dehydration of brain to spreading depression of Leão. *Electroncephal. Clin. Neurophysiol.*, 2: 177–185.

Marshall, W.H. 1970. Foreword. Pages v–vii in F.A. Young and D.B. Lindsley, eds. *Early Experience and Visual Information Processing in Perceptual and Reading Disorders*. Washington, DC: National Academy of Sciences.

Marshall, W.H. and S.A. Talbot. 1942. Recent evidence for neural mechanisms in vision leading to a general theory of sensory activity. *Biol. Sym. [Cold Spring Harbor]*, 7: 117–164.

Marshall, W.H., C.N. Woolsey, and P. Bard. 1937. Cortical representation of tactile sensibility as indicated by cortical potentials. *Science*, 85: 388–390.

Marshall, W.H., C.N. Woolsey, and P. Bard. 1941. Observations on cortical somatic sensory mechanisms of cat and monkey. *J. Neurophysiol.*, 4: 1–24.

Marx, O. 1984. The history of the biological basis of language. Pages 443–469 in E.H. Lenneberg, *Biological Foundations of Language,* Appendix B. Malabar, Florida: Robert E. Krieger Publishing Co.

Masters, W.H. and V.E. Johnson. 1966. *Human Sexual Response*. Boston: Little, Brown.

McAdam, D.W. and H.A. Whitaker. 1971. Langage production: electroencephalographic localization in the normal human brain. *Science*, 172: 499–502.

McBride, A.F. and D.O. Hebb. 1948. Behavior of the captive bottlenose dolphin, *Tursiops truncatus*. *J. Comp. Physiol. Psychol.*, 41: 111–123.

McClelland, D.C. 1955. Comments on Doctor Olds' paper. Pages 139–142 in M.R. Jones, ed., *Nebraska Symposium on Motivation 1955*. Lincoln: University of Nebraska Press.

McConnell, J.B. 1977. *Understanding Human Behavior*. New York: Holt, Rinebart, & Winston.

McCulloch, W.S. 1940. Joannes Gregorius Dusser de Barenne (1885–1940). *Yale J. Biol. Med.*, 12: 742–746.

McCulloch, W.S. 1944a. Cortico-cortical connections. Pages 211–242 in P.C. Bucy, ed., *The Precentral Motor Cortex*. Urbana, Illinois: University of Illinois Press.

McCulloch, W.S. 1944b. The functional organization of the cerebral cortex. *Physiol. Rev.*, 24: 390–407.

McCulloch, W.S. and W.H. Pitts. 1943. A logical calculus of the ideas immanent in nervous activity. *Bull. Math. Biophys.*, 5: 115–133.

McGill, T.E., D.A. Dewsbury, and B.D. Sachs, eds. 1978. *Sex and Behavior: Status and Prospects*. New York: Plenum Press.

McHenry, L.C., Jr. 1969. *Garrison's History of Neurology, Revised and Enlarged with a Bibliography of Classical, Original and Standard Works in Neurology*. Springfield, C.C Thomas.

McIlwain, H. 1966. *Biochemistry and the Central Neurons System*, 3rd ed. Boston: Little, Brown and Co.

Mello, C., F. Nottebohm, and D. Clayton. 1995. Repeated exposure to one song leads to rapid and persistent decline in and immediate early gene's response to the song in zebra finch telencephalon. *J. Neurosci.*, 15: 6919–6925.

Mello, C.V., D.S. Vicario, and D.F. Clayton. 1992. Song presentation induces gene expression in the songbird forebrain. *Proc. Nat. Acad. Sci. USA*, 89: 6818–6822.

Meltzer, H.Y. 1987. *Psychopharmachology: the Third Generation of Progress*. New York: Raven Press.

Menzel, C.R. 1999. Unprompted recall and reporting of hidden objects by a chimpamzee (Pan Troglodytes) after extended delays. *J. Comp. Psychol.*, 113: 426–434.

Mettler, F.A. 1935–1936. Corticofugal fiber connections of Macaca mulatta. *J. Comp. Neurol.*, 61: 1–37, 509–542; 62: 263–291; 63: 25–47.

Mettler, F.A., ed. 1949. *Selective Partial Ablation of the Frontal Cortex: a Correlative Study of Its Effects on Human Psychotic Subjects.* New York: Columbia Greystone Associates.

Mettler, F.A. 1958. James Wenceslas Papez. *Anat. Rec.*, 131: 279–282.

Mettler, F.A. 1975. Joseph Collins 1866–1930. Pages 117–121 in D. Denny-Brown, ed., *Centennial Anniversary Volume of the American Neurological Association 1875–1975.* New York: Springer Publishing Co.

Meyer, A. 1910. The present status of aphasia and apraxia. *Harvey Lect.*, 5: 228–250.

Meyer, A. 1941. The contemporary setting of the pioneer. *J. Comp. Neurol.*, 74: 1–24.

Meyer, A. 1970. Karl Friedrich Burdach and his place in the history of neuroanatomy. *J. Neurol. Neurosurg. Psychiat.*, 33: 553–561.

Meyer, A.C. 1971. *Historical Aspects of Cerebral Anatomy.* London: Oxford University Press.

Meyer, M.E. 1933. That whale among the fishes – the theory of emotions. *Psychol. Rev.*, 40: 292–300.

Meyers, R.E. 1976. Comparative neurology of vocalization and speech: proof of a dichotomy. Pages 745–757 in S.R. Harnad, H.D. Steklis, and J. Lancaster, eds., *Origins and Evolution of Language and Speech.* New York: The New York Academy of Science.

Miller, C.A. and S. Benzer. 1983. Monoclonal antibody cross-reactions between Drosophila and human brain. *Proc. Nat. Acad. Sci. USA*, 80: 7641–7645.

Miller, G.A. 1994. The magical number seven: Some limits on our capacity for processing information. *Psychol. Rev.*, 101: 343–352. Partial reprinting of *Psychol. Rev.*, 63: 81–97 (1956).

Miller, N.E. 1963. Some reflections on the law of effect produce a new alternative to drive reduction. *Nebraska Symposium on Motivation.* Lincoln, Nebraska: University of Nebraska Press.

Miller, N.E. 1980. Introduction: Brain stimulation reward and theories of reinforcement. Pages 1–7 in A. Routtenbery, ed., *Biology of Reinforcement: Facets of Brain-Stimulation Reward.* New York: Academic Press.

Miller, N.E. and A. Carmona. 1967. Modification of a visceral response, salivation in thirsty dogs, by instrumental training with water reward. *J. Comp. Physiol. Psychol.*, 63: 1–6.

Mills, C.K. 1898. *The Nervous System and Its Diseases: A Practical Treatise on Neurology for the Use of Physicians and Students.* Philadelphia: Lippincott.

Mills, C.K. 1924. Some recollections of the early meetings and personnel of the American Neurological Association.... Reprinted 1975. Pages 25–26 in D. Denny-Brown, ed., *Centennial Anniversary Volume of the American Neurological Association 1875–1975.* New York: Springer Publishing Company.

Mills, C.K. and W.G. Spiller. 1907. Symptomatology of lesions of the lenticular zone with some discussion of the pathology of aphasia. *J. Nerv. Ment. Dis.*, 34: 558–588.

Milner, B. 1974a. Introduction. Pages 3–4 in F.O. Schmitt and F.G. Worden, eds., *The Neurosciences Third Study Program.* Cambridge: The MIT Press.

Milner, B. 1974b. Hemispheric specialization: scope and limits. Pages 75–89 in F.O. Schmitt and F.G. Worden, eds., *The Neurosciences Third Study Program*. Cambridge: The MIT Press.

Milner, P.M. 1977. James Olds 1922–1976. *Neurosci. Res. Prog. Bull.*, 15: 139–140.

Mitchell, S.W. 1860. Researches upon the neuron of the rattlesnake, with an investigation of the anatomy and physiology of the organs concerned. *Smithonian Contributions to Knowledge*, 12(6): 1–145.

Mitchell, S.W. 1869. Researches on the physiology of the cerebellum. *Am. J. Med. Sci.*, 57: 320–338.

Mitchell, S.W. 1872. *Injuries of Nerves and Their Consequences*. Philadelphia: J.B. Lippincott & Co.

Mitchell, S.W. 1890. Memoir of John Call Dalton, 1825–1889. *Biog. Mem. Natl. Acad. Sci.*, USA, 3: 177–185.

Mitchell, S.W. 1894. Address before the fiftieth annual meeting of the American Medico-Psychological Association, *J. Nerv. Ment. Dis.*, 21: 413–437.

Mitchell, S.W. and E.T. Reichert. 1886. Researches upon the venons of poisonous serpents. *Smithsonian Contributions to Knowledge*, 26(1): 1–186.

Moffett, S.B. 1996. *Nervous System Regeneration in the Invertebrates*. Berlin: Springer.

Molfese, D.L., R.B. Freeman, and D.S. Palermo. 1975. The ontogeny of brain lateralization for speech and nonspeech stimuli. *Brain Lang.*, 2: 356–368.

Money, J. and A.A. Ehrhardt. 1972. *Man and Woman, Boy and Girl: Differentiation and Dimorphism of Gender Identity from Conception to Maturity*. Baltimore: Johns Hopkins University Press.

Moore, R.Y. and V.B. Eichler. 1972. Loss of circadian adrenal conticosterone rhythm following suprachiasmatic lesions in the rat. *Brain Res.*, 42: 201–206.

Morell, F. 1985. Norman Geschwind 1926–1984: an appreciation. *Neurology*, 35: 660–661.

Morell, P. 1984. A correlative synopsis of the leucodystrophies. *Neuropediatrics, Suppl*: 62–65.

Morgan, C.L. 1898. *An Introduction to Comparative Psychology*. London: W. Scott.

Morgan, L.H. 1868. *The American Beaver and His Works*. Philadelphia: J.B. Lippincott & Co.

Morgan, L.H. 1877. *Ancient Society*. New York: Henry Holt & Co. Reissued 1878.

Morgane, P.J., M.S. Jacobs, and W.L. McFarland. 1980. The anatomy of the brain of the bottlenose dolphin (Tursiops truncatus): surface configurations of the telencephalon of the bottlenose dolphin with comparative anatomical observations in four other cetacean species. *Brain Res. Bull.*, 5: Suppl.

Morison, R.S. and E.W. Dempsey. 1942. A study of thalamo-cortical relations. *Am. J. Physiol.*, 135: 381–292.

Moruzzi, G. 1960. Synchronizing influences of the brain stem and the inhibitory mechanisms underlying the production of sleep by sensory stimulation. Pages 231–256 in H.H. Jasper and G.D. Smirnov, eds., The Moscow Colloquium on Electroencephalography of Higher Neurons Activity. *Electroenceph. Clin. Neurophysiol.*, Suppl. 13, 1960.

Moruzzi, G. and H.W. Magoun. 1949. Brain stem reticular formation and activation of the EEG. *Electroenceph. Clin. Neurophysiol.*, 1: 455–473.

Mountcastle, V.B. 1977. Philip Bard. 1898–1977. *John Hopkins Med. J.*, 141: 296–298.

Murchison, C.A. ed. 1930. *A History of Psychology in Autobiography*. The International University Series in Psychology. Worchester, Massachusetts: Clark University Press.

Myers, R.E. 1976. Comparative neurology of vocalization and speech: proof of a dichotomy. Pages 745–757 in S.R. Harnad, H.D. Steklis, and J. Lancaster, eds., *Origins and Evolution of Language and Speech.* New York: The New York Academy of Sciences.

Myers, R.E. and R.W. Sperry. 1953. Interocular transfer of a visual form discrimination habit in cats after section of the optic chiasma and corpus callosum. *Anat. Rec.*, 115: 351–352.

Myers, R.E. and R.W. Sperry. 1956. Contralateral mnemonic effects with ipsilateral sensory inflow. *Fed. Proc.*, 15: 134.

Nachmansohn, D. 1946. The chemical mechanism of nerve activity. Pages 395–428 in D. Nachmansohn and others, The Physico-Chemical Mechanism of Nerve Activity. *Ann. New York Acad. Sci.*, 47: 375–602.

Nachmansohn, D. 1959. *Chemical and Molecular Basis of Nerve Activity.* New York: Academic Press.

Nagel, E.L., P.J. Morgane, and W.L. McFarland. 1964. Anesthesia for the bottlenose dolphin, Tursiots truncatus. *Science*, 146: 1591–1593.

Neher, E. and B. Sakmann. 1976. Single-channel current recorded from membrane of denervated frog muscle fibres. *Nature*, 260: 799–802.

Neider, C., ed. 1959. *The Autobiography of Mark Twain.* New York: Harper and Brothers.

Neumann, M.A. and R. Cohn. 1953. Incidence of Alzheimer's disease in a large mental hospital. *Arch. Neurol. Psychiat.*, 69: 615–636.

New Encyclopedia Britannica. 15th ed. 1998. "Introspection." Micropoedia. Ready Reference. Chicago: Encyclopedia Britannica, Inc.

Newman, E.B., F.F. Perkins, and R.H. Wheeler. 1930. Cannon's theory of emotion: a critique. *Psychol. Rev.*, 37: 305–326.

Nielsen, J.M. with the assistance of J.P. FitzGibbon. 1946. *Agnosia, Apraxia, Aphasia: Their Value in Cerebral Locations.* 2nd. ed. Los Angeles, Los Angeles Neurological Society; 2nd ed., New York: Hoeber.

Nielsen, J.M. and G.N. Thompson. 1947. *The Engrammes of Psychiatry.* Springfield, Illinois: C.C Thomas.

Noguchi, H. 1907. *Snake Venons: an Investigation of Venomous Snakes with Special Reference to the Phenomena of Their Venoms.* Washington: The Carnegie Institution of Washington (Pub. #111).

Norris, K.S., B. Würsig, R.S. Wells, and M. Würsig. 1994. *The Hawaiian Spinner Dolphin.* Berkeley: University of California Press.

Nottebohm, F. 1969. The "critical period" for song learning. *Ibis*, 111: 386–387.

Nottebohm, F. 1971. Neural lateralization of vocal control in a passerine bird. I. Song. *J. Exp. Zool.*, 177: 229–261.

Nottebohm, F. 1972. Neural lateralization of vocal control in a passerine bird. II. Subsong, calls, and a theory of learning. *J. Exp. Zool.*, 179: 35–49.

Nottebohm, F. 1980. Testoserone triggers growth of brain vocal control nuclei in adult female Canaries. *Brain Res.*, 189: 429–436.

Nottebohm, F. 1991. Reassessing the mechanisms and origins of vocal learning in birds. *Trends Neurosci.*, 14: 206–211.

Nottebohm, F., T.M. Stokes, and C.M. Leonard. 1976. Central control of song in the Canary, *Serinus Canarius. J. Comp. Neurol.*, 165: 457–486.

Novak, S.J. 1991. Pursuing scientific psychiatry. A 50 year journey. *NARSAD Research Newsletter*, May 1991: 1–5.

Ochs, S. and R.M. Wortle. 1978. Axoplasmic transport in normal and pathological systems. Pages 251–264 in S.G. Waxman, ed., *Physiology and Pathobiology of Axons*. New York: Raven Press.

Ojemann, G. and C. Mateer. 1979. Human language cortex: localization of memory, syntax and sequential motor-phoneme identification systems. *Science*, 205: 1401–1403.

Olds, J.A. 1954. A neural model for sign gestalt theory. *Psychol. Rev.*, 61: 59–72.

Olds, J.A. 1972. Neuropsychological studies of motivation. Pages 363–383 in W.R. Adey, *Brain Mechanisms and the Control of Behaviour*. New York: Crane, Russak.

Olds, J. and P. Milner. 1954. Positive reinforcement produced by electrical stimulation of septal areas and other regions of rat brain. *J. Comp. Physiol. Psychol.* 47: 419–427.

O'Leary, J.L. 1974. James Farquhar Fulton 1899–1960. *Surg. Neurol.*, 2: 143–145.

O'Leary, J.L. and G.H. Bishop. 1969. C.J. Herrick, scholar and humanist: a memorial essay written for his centenary. *Persp. Biol. Med.* 12: 492–513.

O'Leary, J.L. and S. Goldring. 1976. *Science and Epilepsy: Neuroscience Gains in Epilepsy Research*. New York: Raven Press.

Oliver, G. and E.A. Schäfer. 1895. The physiological effects of extracts from the suprarenal capsules. *J. Physiol.* (London), 18: 230–276.

O'Neill, Y.V. 1980. *Speech and Speech Disorders in Western Tought before 1600*. Westport, Connecticut: Greenwood Press.

Oppenheim, R.W. 1978. G.E. Coghill (1872–1941): Pioneer neuroembryologist and developmental psychobiologist. *Persp. Biol. Med.*, 22: 45–64.

Orbach, J., ed. 1982. *Neuropsychology After Lashley. Fifty Years Since the Publication of* Brain Mechanisms and Intelligence. Hillsdale, New Jersey: Lawrence Erlbaum Associates.

Orton, J.L. 1957. The Orton Story. *Bull. Orton Soc.*, 13: 1–8.

Orton, S.T. 1925. "Word blindness" in school children. *Arch. Neurol. Psychiat.*, 14: 581–615.

Orton, S.T. 1928. A physiological theory of reading disability and stuttering in children. *New Engl. J. Med.*, 199: 1046–1052.

Orton, S.T. 1928. Specific reading disability – strephosymbolia. *J. Am. Med. Assoc.*, 99: 1095–1099. Reprinted *Bull. Orton Soc.*, 13: 9–17, 1962.

Orton, S.T. 1937. *Reading, Writing and Speech Problems in Children. A Presentation of Certain Types of Disorders in the Development of the Language Faculty.* New York: W.W. Norton & Co.

Orton, S.T. and L.E. Travis. 1929. Studies in stuttering: IV. Studies of action currents in stutterers. *Arch. Neurol. Psychiat.*, 21: 61–68.

Osborn, Ht. F. 1888. A contribution to the internal structure of the amphibian brain. *J. Morphol.*, 2: 51–96.

Page, Irvine H. 1937. *Chemistry of the Brain*. Springfield, Illinois: C.C Thomas.

Papez, J.W. 1927. The brain of Helen H. Gardener (Alice Chenoweth Day). *Am. J. Physical Anthropol.*, 11: 29–87.

Papez, J.W. 1929. The brain of Burt Green Wilder 1841–1925. *J. Comp. Neurol.*, 47: 285–342.

Papez, J.W. 1937. A proposed mechanism of emotion. *Arch. Neurol. Psychiat.*, 38: 725–743.

Papez, J.W. 1956. Central reticular path to intralaminar and reticular nuclei of the thalamus for activating EEG related to consciousness. *Electroencephal. Clin. Neurophysiol.*, 8: 117–128.

Papez, J.W. 1958. The visceral brain, its components and connections. Pages 591–605 in H.H. Jasper, L.D. Proctor, R.S. Knighton, W.C. Noshay, and R.T. Costello, eds., *Reticular Formation of the Brain*. Boston: Little, Brown and Co.

Papez, J.W. and L.R. Aronson. 1934. Thalamic nuclei of Pithecus (Macacus) rhesus. I. Ventral thalamus. *Arch. Neurol. Psychiat.*, 32: 1–26.

Parascandola, J. 1992. *The Development of American Pharmachology: John J. Abel and the Shaping of a Discipline.* Baltimore: Johns Hopkins Press.

Park, R. 1889. Surgery of the brain based on the principles of cerebral localization. *Trans. Cong. Am. Phys. Surg.*, 1: 285–328.

Park, R. 1892. Clinical contributions to the subject of brain surgery. *Med. News*, 61: 617–621; 648–650.

Park, R. 1913. Conclusions drawn from a quarter century's work in brain surgery. *New York State J. Med.*, 13: 303–309.

Parker, A.J. 1878. Simian character in negro brains. *Proc. Acad. Nat. Sci. Philadelphia*, 30: 339–340.

Parker, G.H. 1907. Behavior of the lower organismss. *Science*, 26: 548–549.

Parker, S.T. 1990. Origins of comparative developmental evolutionary studies of primate mental abilities. Pages 3–63 in S.T. Parker and K.R. Gibbon, *"Language" and Intelligence in Monkeys and Apes: Comparative Development Perspectives.* Cambridge. England: Cambridge University Press.

Parker, S.T. and M.L. McKinney. 1999. *Origins of Intelligence: the Evolution of Cognitive Development in Monkeys, Apes, and Humans.* Baltimore: Johns Hopkins Press.

Patterson, F.G. 1981. Letters. Ape language. *Science*, 211: 87–88.

Patterson, F. and E. Linden. 1981. *The Education of Koko.* New York: Holt, Rinehart & Winston.

Patterson, M.M., comp. 1994. Stereotaxic Atlases and Related References. 1994 Supplement. Privately published by David Kopf Instruments, Tujunga, CA 91043.

Patton, H.D. 1994. David P.C. Lloyd September 23, 1911–April 20, 1985. *Biog. Mem.*, 65: 197–209. Washington, DC: U.S. National Academy of Sciences.

Pauly, P.J. 1979. Psychology at Hopkins: Its rise and fall and rise and fall and.... *Johns Hopkins Mag.*, 30(6): 36–41.

Pauly, P.J. 1980. Jacques Loeb and the Control of Life: an Experimental Biologist in Germany and America, 1895–1924. Dissertation, John Hopkins University.

Pauly, P.J. 1981. The Loeb-Jennings debate and the science of animal behavior. *J. Hist. Behav. Sci.*, 17: 504–515.

Pauly, P.J. 1987. *Controlling Life: Jacques Loeb and the Engineering Ideal in Biology.* (Monographs on the History and Philosophy of Biology.) New York: Oxford University Press.

Pavlov, I.P. 1928. *Lectures on Conditioned Reflexes.* Trans. by W.H. Gantt. New York: International Publishers.

Payne, K. 1998. *Silent Thunder: In the Presence of Elephants*. New York: Simon & Schuster.

Payne, K.B., W.R. Langbaur, Jr., and E.M. Thomas. 1986. Infrasonic calls of the Asian elephant (*Elephas maximu*). *Behav. Ecol. Sociobiol.*, 18: 297–301.

Payne, R.B. 1981. Song learning and social interaction in indigo buntings. *Anim. Behav.*, 29: 688–697.

Pearl, R. 1905. Biometrical studies in man. I. Variation and correlation in brain weight. *Biometrika*, 4: 13–104.

Pearson, R., ed. 1992. *Shockley on Eugenics and Race*. Washington, DC: Scott-Townsend Publishers.

Peek, F.W. 1972. An experimental study of the territorial function of vocal and visual display in the male red-winged blackbird. *Anim. Behav.*, 20: 112–118.

Penfield, W. 1936–1937. The cerebral cortex and consciousness. *Harvey Lect.*, 32: 35–69.

Penfield, W. 1938. I. The cerebal cortex and consciousness. *Arch. Neurol. Psychiat.*, 40: 417–422.

Penfield, W. 1958. Designated descussion. Pages 739–741 in H.H. Jasper, L.D. Protor, R.S. Knighton, W.C. Noshay, and R.T. Costello, eds., *Reticular Formation of the Brain*. Henry Ford Hospital International Symposium. Boston: Little, Brown & Co.

Penfield, W. 1967. *The Difficult Art of Giving. The Epic of Alan Gregg*. Boston: Little, Brown & Co.

Penfield, W. 1975. *The Mystery of the Mind, a Critical Study of Consciousness and the Human Brain*. Princeton: Princeton University Press.

Penfield, W. 1977. *No Man Alone: A Neurosurgeon's Life*. Boston: Little, Brown and Co.

Penfield, W. and H.H. Jasper. 1943. Electroencephalograms in post-traumatic epilepsy: pre-operative and post-operative studies. *Am. J. Psychiat.*, 100: 356–377.

Penfield, W. and H. Jasper. 1954. *Epilepsy and the Functional Anatomy of the Human Brain*. Boston: Little, Brown, and Co.

Penfield, W. and T. Rasmussen. 1949. Vocalization and arrest of speech. *Arch. Neurol. Psychiat.*, 61: 21–27.

Penfield, W. and T. Rasmussen. 1950. *The Cerebral Cortex of Man. A Clinical Study of Localization of Function*. New York: Macmillan.

Penfield, W. and L. Roberts. 1959. *Speech and Brain Mechanisms*. Princeton. New Jersey: Princeton University Press.

Pepperberg, I.F. 1999. *The Alex Studies: Cognitive and Communicative Abilities of grey Parrots*. Cambridge, Massachusetts: Harvard University Press.

Perry, R.B. 1926. *General Theory of Value; Its Meaning and Basic Principles Construed in Terms of Interest*. London: Longmans, Green.

Perry, R.B. 1954. *The Thought and Character of William James. Briefer Version*. New York: George Braziller, Publisher.

Pettigrew, J.D. 1988. Microbat vision and echolocation in an evolutionary context. Pages 645–650 in P.E. Nachtigall and P.W.B. Moore, eds., *Animal Sonar: Processes and Performance*. NATO Advanced Study Institute series A, Life Sciences v. 156. New York: Plenum Press.

Pfeiffer, C.A. 1936. Sexual differences of the hypophyses and their determination by the gonads. *Am. J. Anat.*, 58: 195–226.

Phoenix, C.H., R.W. Goy, A.A. Gerall, and W.C. Young. 1959. Organization action of prenatally administered teststerone propionate on the tissues mediating mating behavior in the female gninea pig. *Endocrinology*, 65: 369–382.
Pierce, G.W. and D.R. Griffin. 1938. Experimental determination of supersonic notes emitted by bats. *J. Mammal.*, 19: 454–455.
Piersol, G.A. 1911. A self-made naturalist. *Anat. Rec.*, 5: 71–86.
Pincus, G. and M.C. Chang. 1948. The effects of progesterone and related compounds on ovulation and early development in the rabbit. *Acta Physiol. Latinoamericano*, 3: 177–183.
Polyak, S. 1932. *The Main Afferent Fiber Systems of the Cerebral Cortex in Primates*. Berkeley: University of California Press.
Polyak, S. 1941. *The Retina*. Chicago: the University of Chicago Press.
Polyak, S., 1957. *The Vertebrate Visual System.* Edited by Heinrich Klüver. Chicago: the University of Chicago Press.
Pomeroy, W.B. 1972. *Dr. Kinsey and the Institute for Sex Research.* New York: Harper & Row.
Poole, J.H., K.B. Payne, W.R. Langbauer, Jr., and C.J. Moss. 1988. The social contexts of some very low frequency calls of African elephants. *Behav. Ecol. Sociobiol.*, 22: 385–392.
Popa, G. and U. Fielding. 1930. A portal circulation from the pituitary to the hypothalamic region. *J. Anat.*, 65: 88–91.
Pope, A., ed. 1983. *Human Brain Dissection.* Sponsored by National Institute of Neurological and Communicative Disorders and Stroke; National Institute of Mental Health. Bethesda, Maryland: U.S. Department of Health and Human Services, Public Health Service, National Institutes of Health.
Pope, A., W.F. Caveness, and K.E. Livingston. 1952. Architectonic distribution of acetylcholinesterase of psychotic and nonpsychotic patients. *Arch. Neurol. Psychiat*, 68: 425–443.
Pope, A., M.B. Lees, and G. Hauser. 1980. Obituary. Jordi Folch-Pi 1911–1979. *J. Neurochem.*, 35: 1–3.
Porter, R.W., J.V. Brady, M.A. Conrad, J.W. Mason, R. Galambos, and D.McK. Rioch. 1958. Some experimental observation on gastrointestinal lesions in behaviorally conditioned monkeys. *Psychosom. Med.*, 20: 379–394.
Postman, L. 1947. The history and present status of the law of effect. *Psychol. Bull.*, 44: 489–563.
Premack, D. 1971. Language in chimpanzee? *Science*, 172: 808–822.
Premack, D. 1986. *Gavagai! or the Future History of the Animal Language Controversy.* Cambridge, Massachusetts: The MIT Press.
Pribram, K.H. 1958. Effect of cingulectomy on social behavior in monkeys. *J. Neurophysiol.*, 20: 588–601.
Pribram, K.H. 1976. Language in a sociobiological frame. Pages 798–809 in S.R. Harnad, H.D. Steklis, and J. Lancaster, eds., *Origins and Evolution of Language and Speech.* New York: The New York Academy of Sciences.
Pribyl, T.M., C. Campagnoni, K. Kampf, V.W. Handley, and A.T. Campagnoni. 1996. The major myelin protein genes are expressed in the human thymus. *J. Neurosci. Res.*, 45: 812–819.
Price, B. 1968. *Into the Unknown.* New York: Platt and Munk.
Pugh, K.R., B.A. Shaywitz, S.E. Shaywitz, R.T. Constable, P. Sudlarski, R.K. Fulbright, R.A. Bronen, D.P. Shankweiler, L. Katz, J.M. Fletcher, and

J.C. Gore. 1996. Cerebral organization of component processes in reading. *Brain*, 119: 1221–1238.

Purpura, D.P. 1966. [Discussion]. Page 249 in R.A. Gorski and R.E. Whalen, eds., *Brain and Behavior, v. III, Brain and Gonadal Function*. Proceedings of the Third Conference, 1963. Berkeley and Los Angeles: University of California Press.

Purves, D. 1988. *Body and Brain: A Trophic Theory of Neural Connections*. Cambridge, Massachusetts: Harvard University Press.

Rambaugh, D.M., ed. 1977. *Language Learning in a Chimpanzee: the LANA Project*. New York: Academic Press.

Ramón y Cajal, S. 1894. La fine structure des centres nerveux. Croonian Lecture, March 8, 1894. *Proc. R. Soc. B.* (London), 55: 444–468.

Ramsey, E.M. 1994. George Washington Corner. December 12, 1889 September 28, 1891. *Biog. Mem. NAS.*, 65: 56–93.

Ranson, S.W. 1920. *The Anatomy of the Nervous System from the Standpoint of Development and Function. With 260 Illustrations, Some of Them in Colors*. Philadelphia: Saunders.

Ranson, S.W. 1939. Sommolence caused by hypothalamic lesions in the monkey. *Arch. Neurol. Psychiat.*, 41: 1–23.

Ranson, S.W. and Hinsey, J.C. 1930. Reflexes in the hind limbs of cats after transection of the spinal cord at various levels. *Am. J. Physiol.*, 94: 471–495.

Rasmussen, A.T., C.J. Herrick and O. Larsell. 1940. John Black Johnston 1868–1939. *Anat. Rec.*, 76: 18–21.

Rawson, M.B. 1966. *A Bibliography on the Nature, Recognition, and Treatment of Language Difficulties*. Pomfret, Connecticut: The Onton Society.

Rawson, M.B. 1968. *Developmental Language Disabilitiy; Adult Accomplishments of Dyslexic Boys*. Baltimore: Johns Hopkins Press.

Rechtschaffen, A. and A. Kales, eds. 1968. *A Manual of Standardized Terminology. Techiniques and Scoring System for Sleep Stages of Human Subjects*. Los Angeles: Brain Information Service/Brain Research Institute, University of California. Reprinted 1973, 1977.

Reichlin, S. 1966. Neuroendocrine aspects of reproductive physiology. Pages 3–34 in R.A. Gorski and R.E. Whalen, eds., *The Brain and Gonadal Function. Brain and Behavior*, v. III. Proceedings of the Third Conference, 1963. Berkeley and Los Angeles: University of California Press.

Renshaw, B. 1940. Activity in the simplest spinal reflex pathways. *J. Neurophysiol.*, 3: 373–387.

Renshaw, B. 1941. Influence of discharge of motoneurons upon excitation of neighboring motoneurons. *J. Neurophysiol.*, 4: 167–183.

Renshaw, B. 1946. Central effects of centripetal impulses in axons of spinal ventral roots. *J. Neurophysiol.*, 9: 191–204.

Renshaw, B., A. Forbes, and B.R. Morison. 1940. Activity of isocortex and hippocampus: electrical studies with micro-electrodes. *J. Neurophysiol.*, 3: 74–105.

Resek, Carl. 1960. *Lewis Henry Morgan, American Scholar*, Chicago: University of Chicago Press.

Reymert, M.L. 1928. Why feelings and emotions? Appendix B. in M.L. Reymert, ed., *Feelings and Emotions. The Wittenberg Symposium*. Worchester, Massachusetts: Clark University Press. Reprinted 1973. in *Classics in Psychology*.

Richter, C.P. 1956. Rats, Man, and the welfare state. *Am. Psychol.*, 14: 18–28.

Richter, C.P. 1968. Experiences of a reluctant rat-catcher: the common Norway rat – friend or enemy? *Proc. Am. Phil. Soc.*, 112: 403–415.
Rioch, D.M. 1929a. Studies on the diencephalon of Carnivora. I. The nuclear configurations of the thalamus, epithalamus, and hypothalamus of the dog and cat. *J. Comp. Neurol.*, 49: 1–119.
Rioch, D.M. 1929b. [Studies on the diencephalon of Carnivora] II. Certain nuclear configurations and fiber connections of the subthalamus of the dog and cat. *J. Comp. Neurol.*, 49: 121–153.
Rioch, D.M. 1931. [Studies on the diencephalon of Carnivora]. III. Certain myelinated-fiber connections of the dicephalon of the dog (Canis familiaris), cat (Felis domestica) and aevisa (Crossarchus doscurus). *J. Comp. Neurol.*, 53: 319–388.
Rioch, D. McK. 1975. Adolf Meyer 1866–1950. Pages 153–159 in D. Denny-Brown, ed., *Centennial Anniversary Volume of the American Neurological Association 1875–1975.* New York: Springer.
Ristau, C.A. and D. Robbins. 1982. Language in the great apes. *Adv. Study Behav.*, 12: 141–255.
Robertson, J.D. 1957. New observations on the ultrastructure of the membranes of frog peripheral nerve fibers. *J. Biophys. Biochem. Cytol.*, 3(2): 1043–1047.
Robinson, B.W. 1976. Limbic influences on human speech. Pages 761–771 in S.P. Harnad, H.D. Steklis, and J. Lancaster, eds., *Origins and Evolution of Language and Speech.* New York: The New York Academy of Sciences.
Rogers, F.T. 1922. A note on the excitable areas of the cerebral hemisphere of the pigeon. *J. Comp. Neurol.*, 35: 61–65.
Rogers, F.T. 1924. An experimental study of the cerebral physiology of the virginian opossum. *J. Comp. Neurol.*, 37: 265–315.
Romanes, G.J. 1882. *Animal Intelligence*. London: Paul Trench.
Rose, J. 1951. Adolf Meyer's contributions to neuroanatomy. *Bull. Johns Hopkins Hosp.*, 89: 56–63.
Ross, D. 1972. *G. Stanley Hall: The Psychologist as Prophet.* Chicago: University of Chicago Press.
Rothmann, M. 1912. Über die Errichtung einer Station zur psychologischen und hirnphysiologischen Erforschung der Menschaffen. *Berliner Klin. Wochenschr.*, 42: 1981–1985.
Routtenberg, A., ed. 1980. *Biology of Reinforcement*. New York: Academic Press.
Rumbaugh, D.M., E.S. Savage-Rumbaugh, and R.A. Sevcik. 1994. Biobehavioral roots of language: a comparative perspective of chimpanzee, child, and culture. Pages 319–334 in R.W. Wrangham, W.C. McGrew, F.B.M. de Waal, and P.G. Hetne, eds., *Chimpanzee Cultures.* Cambridge, Massachusetts: Harvard University Press.
Rundles, R.W. and J.W. Papez. 1938. Fiber and cellular degeneration following temporal lobectomy in the monkey. *J. Comp. Neurol.*, 68: 267–296.
Ruch, T.C. 1941. *Bibliographia Primatologica. Part I. Anatomy, Embryology, and Quantitative Morphology, Physiology, Pharmacology and Psychobiology, Primate Phylogeny and Miscellanea.* Springfield, Illinois: C.C Thomas.
Sabin, F.R. 1934. *Franklin Paine Mall: the Story of a Mind.* Baltimore: Johns Hopkins Press.
Sabin, F.R. 1945. Stephen Walter Ranson 1880–1942. Pages 365–397 in *Biog. Mem.*, v. 23. Washington, DC. National Academy of Sciences.

Saffran, M., A.V. Schally, and B.G. Benfey. 1955. Stimulation of the release of corticotropin from the adenohypophysis by a neurohypophysial factor. *Endocrinology*, 57: 439–444.

Sahakian, W.S. 1975. *History and Systems of Psychology*. New York: Wiley.

Sanders, R.J. 1985. Teaching apes to ape language: explaining the imitative and nonimitative signing of a chimpanzee (*Pan troglodytes*). *J. Comp. Psychol.*, 99: 197–210.

Savage-Rumbaugh, E.S., R.A. Sevcik, D.M. Rumbaugh, and E. Rubert. 1985. The capacity of animals to acquire language: do species differences have anything to say to us? *Phil. Trans. R. Soc.* B., 308: 177–185.

Sawyer, C.H., J.W. Everett, and J.E. Markee. 1949. A neural factor in the mechanism by which estrogen induces the release of luteinizing hormone in the rat. *Endocrinology*, 41: 218–233.

Schaller, G.B. 1963. *The Mountain Gorilla: Ecology and Behavior.* Chicago: University of Chicago Press.

Schally, A.V. 1957. *In vitro* studies on the control of the release of ACTH. Ph.D. Thesis, McGill University.

Schally, Andrew V. 1978. Aspects of hypothalamic regulation of the puitary gland with major emphasis on its implications for the control of the reproductive processes. Pages 201–234 in *Les Prix Nobel*. Stockholm: Almgrist & Wiksell International.

Schaltenbrand, G. and S. Cobb. 1931. Clinical and anatomical studies on two cats without neocontex. *Brain*, 53: 449–491.

Scharrer, B. 1975. The concept of neurosecretion and its place in neurobiology. Pages 230–243 in F.G. Worden, J.P. Swazey, and G. Adelman, eds., *The Neurosciences: Paths of Discovery*. Cambridge, Massachusetts: The MIT Press.

Scharrer, E. 1928. Die Lichtempfindlichkeit blinder Elritzen (Untersuchungen über das Zwischenhirn der Fische. I.). *Z. Verge. Physiol.*, 7: 1–38.

Scharrer, E. and B. Scharrer. 1963. *Neuroendocrinology*. New York: Columbia University Press.

Schevill, W.E. and A.F. McBride. 1956. Evidence for echolocation by cetaceans. *Deep-sea Res.*, 3: 153–154.

Schiller, F. 1952. Consciousness reconsidered. *Arch. Neurol. Psychiat.*, 67: 199–227.

Schlag, J. and A.B. Scheibel, eds. 1967. Special Issue. Forebrain inhibitory mechanisms. *Brain Res.*, 6: 1–200.

Schmitt, F.O. and R.S. Bear. 1939. The ultrastructure of the nerve axon sheath. *Biol. Rev.*, 14: 27–50.

Schreiner, L. and A. Kling. 1953. Behavioral changes following rhinencephalic injury in cat. *J. Neurophysiol.*, 16: 643–659.

Schreiner, L. and A. Kling. 1954. Effects of castratlion on hypersexual behavior induced by rhinencephalic injury in cat. *Arch. Neurol. Psychiat.*, 72: 180–186.

Schrödinger, E. 1944. *What is Life*? Cambridge: Cambridge University Press.

Searcy, W.A. and P. Marler. 1981. A test for responsiveness to song structure and programming in female sparrows. *Science*, 213: 926–928.

Searle, J.R. 1979. Chairman's opening remarks. Pages 1–4 in G. Wolstenholme and M. O'Connor, eds., *Brain and Mind*. Ciba Foundation Symposium 69 (new series). Amsterdam: Excerpta Media.

Searle, J.R. 1995. The mystery of consciousness: Part II. *The New York Review*, November 16, 1995, pp. 54–61.

Searle, J.R. 2000. Consciousness. *Annu. Rev. Neurosci.*, 23: 557–578.
Sebeok, T.A., 1972. *Perspectives in Zoosemiotics.* The Hague: Mouton.
Sebeok, T.A., ed. 1977. *How Animals Communicate.* Bloomington: Indiana University Press.
Sebeok, T.A. and D.J. Umiker-Sebeok, eds. 1980. *Speaking of Apes: A Critical Anthology of Two-way Communication with Man.* New York; Plenum Press.
Seller, H. 1991. In memoriam: Chandler McCuskey Brooks. *J. Autonomic Nerv. Sys.*, 34: vii–x.
Semendeferi, K., H. Damasio, R. Frank, and G.W. Van Hoesen. 1997. The evolution of the frontal lobes: a volumetric analysis based on three-dimensional reconstructions of magnetic resonance scans of human and ape brains. *J. Human Ecol.*, 32: 375–388.
Seyfarth, R.M., D.L. Cheney, and P. Marler. 1980. Monkey response to three different alarm calls: Evidence for predator classification and semantic communication. *Science*, 210: 801–803.
Shannon, C.E. 1948. A mathematical theory of communication. *Bell Sys. Tech. J.*, 27: 379–423; 623–656.
Shapiro, H.L. 1974. *Peking Man.* New York, Simon & Schuster.
Shaplen, R. 1964. *Toward the Well-being of Mankind: Fifty Years of the Rockefeller Foundation.* Garden City, New York: Donbleday.
Shaywitz, S.E. 1995. Dyslexia. *Sci. Am.*, 275: 98–104.
Shaywitz, S.E., B.A. Shaywitz, J.M. Fletcher, and M.D. Escobar. 1990. Prevalence of reading disability in boys and girls. Results of the Connecticut Longitudinal Study. Comments. *J. Am. Med. Assoc.*, 264: 998–1002.
Shelton, G.A.B., ed. 1982. *Electrical Conduction and Behavior in "Simple" Invertebrates.* Oxford: Clarendon Press.
Sherrington, C.S. 1896. Cataleptoid reflexes in the monkey. *Proc. R. Soc. B.* (London), 60: 411–414.
Sherrington, C.S. 1898. Decerebrate rigidity and reflex coordination of movement. *J. Physiol.* (London), 22: 319–332.
Sherrington, C.S. 1900. Experiments on the value of vascular and visceral factors for the genesis of emotion. *Proc. R. Soc. B.* (London), 66: 390–403.
Sherrington, C.S. 1906. *The Integrative Action of the Nervous System.* New York: Scribner's Sons. Reprinted Cambridge University Press, 1947.
Sherrington, C.S. 1909. A mammalian spinal preparation. *J. Physiol.* (London), 38: 375–383.
Sherrington, C.S. 1941. *Man on His Nature.* [The Gifford Lectures Edinburgh 1937–8.] New York: The Macmillan Co.
Sherrington, C.S. and R.S. Woodworth. 1904. A pseudoaffective reflex and its spinal path. *J. Physiol.* (London), 31: 234–243.
Shorey, M.L. 1909. The effect of the destruction of peripheral areas on the differentiation of the neuroblasts. *J. Exp. Zool.*, 7: 25–63.
Shorey, M.L. 1911. A study of the differentiation of neuroblasts in artificial culture media. *J. Exp. Zool.*, 10: 85–93.
Simmons, J.A. 1973. The resolution of target range by echolocating bats. *J. Acoust. Soc. Am.*, 54: 157–173.
Simmons, J.A. and A.D. Grinnell, 1988. The performance of echolocation: Acoustic images perceived by echolocating bats. Pages 353–385 in P.E. Nachtigall and

P.W.B. Moore, eds., *Animal Sonar: Processes and Performance.* New York: Plenum Press.
Skinner, B.F. 1931. The concept of the reflex in the description of behavior. *J. Gen. Psychol.*, 5: 427–458.
Skinner, B.F. 1935. Two types of conditioned reflex and a pseudo type. *J. Genet. Psychol.*, 12: 66–77.
Skinner, B.F. 1956. A case history in scientific method. *Am. Physiol.*, 11: 221–233. Also in *Cumulative Record*, 1961.
Skinner, B.F. 1959. A case history of scientific method. Pages 359–379 in S. Koch, ed., *Psychology: A Study of a Science.* New York: McGraw-Hill Book Co.
Skinner, B.F. 1966. Some responses to the stimulus "Pavlov". *Conditional Reflex.* 1: 74–78. Also in *Cumulative Record*, 3rd ed., 1972, pp. 592–596.
Skinner, B.F. 1967. B.F. Skinner. Pages 387–413 in E.G. Boring and G. Lindzey, eds, *A History of Psychology in Antobiography.* v. 5. New York: Appleton-Century-Crofts.
Skinner, B.F. 1979. *The Shaping of a Behaviorist. Part Two of an Autobiography.* New York: Knopf.
Sloan, N. and H. Jasper. 1950. Identity of spreading depression and "suppression." *Electroenceph. Clin. Neurophysiol.*, 2: 59–78.
Smalheiser, N.R. 2000. Walter Pitts. *Persp. Biol. Med.*, 43: 217–226.
Small, W.S. 1900. An experimental study of the mental processes of the rat. *Am. J. Psychol.*, 11: 133–165.
Small, W.S. 1900–1901. Experimental study of the mental process of the rat. II. *Am. J. Psychol.*, 12: 206–239.
Smith, J.M. 1996. [Reply to Noam Chomsky] Letters, *The New York Review of Books*, February 1, 1996, p. 41.
Smith, J.R. 1938. The electroencephalogram during normal infancy and childhood. I, II, and III. *J. Genet. Psychol.*, 53: 431–453; 455–469; 471–482.
Smith, K.W. and A.J. Atelaitis. 1942. Studies on the corpus callosum. I. Laterality in behavior and bilateral motor organization in man before and after section of the corpus callosum. *Arch. Neurol. Psychiat.*, 47: 519–543.
Smith, P.E. 1916. Experimental ablation of the hypophysis in the frog embryo. *Science*, 44: 280–282.
Smith, P.E. and A. Elwyn. 1951. Oliver S. Strong 1864–1951. *J. Comp. Neurol.*, 94: 179–180.
Smith, P.E. and E.T. Engle. 1927. Experimental evidence regarding the role of the anterior pituitary in the development and regulation of the genital system. *Am. J. Anat.*, 40: 159–217.
Smith, P.E., A.T. Walker, and J.B. Graeser. 1923–1924. The production of the adiposogenital syndrome in the rat, with preliminary notes upon the effects of a replacement therapy. *Proc. Soc. Exper. Biol. Med.*, 21: 204–206.
Smith, W.J. 1991. Singing is based on two markedly different kinds of signalling. *J. Theor. Biol.*, 152: 241–251.
Sokal, M.M. 1971. The unpublished autobiography of James McKeen Cattell. *Am. Psychol.*, 26: 626–635.
Sourkes, T.L. 1966. *Nobel Prize Winners in Medicine and Physiology 1901–1965.* London: Abelard-Schuman.
Sperry, R.W. 1951. Regulative factor in the orderly growth of neural circuits. *Growth Sym.*, 10: 63–87.

Sperry, R.W. 1966. Brain bisection and mechanisms of consciousness. Pages 298–313 in J.C. Eccles, ed., *Brain and Conscious Experience*. New York: Springer-Verlag.

Sperry, W.M. 1966. Heinrich B. Waelsch. *J. Neurochem.*, 13: 1261–1263.

Spiegel, E.A., H.R. Miller, and M.J. Oppenheimer. 1940a. Forebrain and rage reactions. *Trans. Am. Neurol. Assoc.*, 66: 127–131.

Spiegel, E.A., H.R. Miller, and M.J. Oppenheimer. 1940b. Forebrain and rage reactions. *J. Neurophysiol.*, 3: 538–547.

Spitzka, E.A. 1905. Study of the brains of six eminent scientists and scholars belonging to the American Anthropometric Society, together with a description of the skull of professor E.D. Cope. *Trans. Am. Phil. Soc.*, 21: 175–308.

Steffens, J.L. 1931. *The Autobiography of Lincoln Steffens*. New York: Harcourt, Brace and Company.

Stein, Gertrude. 1898. Cultivated motor automatism; a study of character in relation to attention. *Psychol. Rev.*, 5: 295–306.

Stein, Gertrude. 1933. *The Autobiography of Alice B. Toklas*. New York: Literary Guild.

Stephan, F.K. and I. Zucker. 1972. Circadian rhythms in drinking behavior and locomotor activity of rats are eliminated by hypothalamic lesions. *Proc. Nat. Acad. Sci. USA*, 69: 1583–1586.

Stern, M.B. 1971. *Heads & Headlines, the Phrenological Fowlers*. Norman: University of Oklahoma Press.

Stewart, C.C. 1898. Variations in daily activity produced by alcohol and by changes in barometric pressure and diet, with a description of recording methods. *Am. J. Physiol.*, 1: 40–56.

Stone, C.P. 1939. Sex drive. Pages 1213–1262 in E. Allen, ed., *Sex and Internal Secretions*. Baltimore: Williams & Wilkins.

Story, W.E. 1899. *Clark University 1889–1899 Decennial Celebration*. Worchester, Massachusetts: Printed for the University.

Strong, O.S. 1895. The cranial nerves of Amphibia. A contribution to the morphology of the vertebrate nervous system. *J. Morphol.*, 10: 101–230.

Struhsaker, T.T. 1967. *The Red Colobus Monkey*. Chicago: University of Chicago Press.

Swazey, J.F. 1974. *Chlorpromazine in Psychiatry. A Study of Therapeutic Innovation*. Cambridge, Massachusetts: The MIT Press.

Tasabi, I. 1974. Energy transduction in the nerve membrane and studies of excitation processes with extrinsic fluorescence probes. *Ann. New York Acad. Sci.*, 227: 247–267.

Temkin, O. 1947. Gall and the phrenological movement. *Bull. Hist. Med.*, 21: 275–321.

Terrace, H.S. 1979. *Nim*. New York: A.A. Knopf.

Terry, R.D. 1965. The development of morphogenic concepts of myelin: a brief review. Pages 3–5 in L. Goldman and J. Weber, eds., The Laser. *Ann. N.Y. Acad. Sci.*, v. 122.

Terzian, H. and G.D. Ore. 1955. Syndrome of Klüver and Bucy. Reproduced in man by bilateral removal of the temporal lobes. *Neurology*, 5: 373–380.

Thomson, E.H. 1950. *Harvey Cushing. Surgeon, Author, Artist*. New York: Henry Schuman.

Thorndike, E.L. 1898. Animal intelligence: an experimental study of the associative processes in animals. *Psychol. Monogr.*, 2, No. 8.

Tolman, E.C. 1949. There is more than one kind of learning. *Psychol. Rev.*, 56: 144–155.

Tourney, G. 1969. History of biological psychiatry in America. *Am. J. Psychiat.*, 126: 29–42.

Tower, D.B. 1958. Origins and development of neurochemistry. *Neurology*, 8, suppl. 1, 3–31.

Tower, S. and M. Hines. 1935. Dissociation of the pyramidal and extrapyramidal functions of the frontal lobe. *Science*, 82: 376.

Travis, L.E. and R.Y. Herren. 1930. Action currents in the cerebral cortex of the dog and the rat during reflex activity. *Am. J. Physiol.*, 93: 693.

Travis, L.E. and J.R. Knott. 1936. Brain potentials from normal speakers and stutterers. *J. Psychol.*, 2: 137–150.

Turner, C.H. 1891. Morphology of the avian brain. Parts I–VI. *J. Comp. Neurol.*, 1: 39–92; 107–133; 265–286.

Turner, C.H. 1907. The homing of ants: an experimental study of ant behavior. *J. Comp. Neurol.*, 17: 367–434.

Tuttle, R.H. 1986. *Apes of the World: Their Social Behavior, Communication, Mentality, Ecology.* Ridge Park, New Jersey: Noyes Publications.

Tyack, P. 1986. Whistle repertoires of two bottlenose dolphins, *Tursiops truncatus*: mimicry of signature whistles? *Behav. Ecol. Sociobiol.*, 18: 251–257.

Uzman, [Geren] B. and G. Nogueir-Graf. 1957. Electron microscope studies of the formation of nodes of Ranvier in mouse sciatic nerves. *J. Biophys. Biochem. Cytol.*, 3: 589–598.

Valenstein, E.S. 1973. *Brain Stimulation and Motivation. Research and Commentary.* Glenview, Illinois: Scott Foresman.

Van Wagenen, W.P. and R.Y. Herren. 1940. Surgical division of commissural pathways in the corpus callosum: relation to spread of an epileptic attack. *Arch. Neurol. Psychiat.*, 44: 740–759.

Velde, T.H. van de. 1930. *Ideal Marriage; Its Physiology and Technique.* Trans. by S. Browne. New York: Random House.

Voneida, T.J. 1997. Roger Wolcott Sperry 1913–1994. *Biograph. Mem. Natl. Acad. Sci., USA*, 71: 315–331.

Wada, J. and T.B. Rasmussen. 1960. Intracarotid injection of sodium amytal for the lateralization of cerebral speech dominance. *J. Neurosurg.*, 17: 266–282.

Wade, N. 1981. *The Nobel Duel. Two Scientists' 21-year Race to Win the World's Most Coveted Research Prize.* Garden City, New York: Anchor Press/Doulleday.

Walker, A.E. 1938. *The Primate Thalamus.* Chicago: University of Chicago Press.

Walker, A.E., ed. 1951. *History of Neurological Surgery.* Baltimore: William and Wilkins.

Walker, A.E. 1960. John Farquhar Fulton, 1899–1960. *J. Neurophysiol.*, 23: 346–349.

Walker, A.E. 1998. *The Genesis of Neuroscience.* Park Ridge, Illinois: American Association of Neurological Surgeons.

Waller, W.H. 1940. Progressive movements elicited by subthalamic stimulation. *J. Neurophysiol.*, 3: 300–307.

Wallman, J. 1992. *Aping Language.* Cambridge: Cambridge University Press.

Walshe, F.M.R. 1957. The brain-stem conceived as the "highest level" of function in the nervous system; with particular reference to the "automatic apparatus" of Carpenter (1850) and to the "centrencephalic integrating system" of Penfield. *Brain*, 80: 510–539.

Walter, R.D. 1970. *S. Weir Mitchell, M.D. – Neurologist; a Medical Biography*. Springfield, Illinois: C.C Thomas.

Wang, S.-C. 1981. A tribute to the late Stephen Walter Ranson. Pages 165–167 in W.F. Windle, ed., *Stephen Walter Ranson; Ground-breaking Neuroscientist*. Los Angeles: Brain Research Institute, University of California.

Wang, S.-C. and S.W. Ranson. 1939. Autonomic responses to electrical stimulation of the lower brain stem. *J. Comp. Neurol.*, 71: 437–455.

Ward, A.A. 1948. The anterior cingulate gyrus and personality. *Res. Pub. Assoc. Nerv. Ment. Dis.*, 27: 438–445.

Washburn, M.F. 1895. Über den Einfluss der Gesichtsassociationen auf die Raumwahrnehmungen der Haut. *Phil. Stud.*, 11: 190–225.

Washburn, M.F. 1908. *The Animal Mind: A Textbook of Comparative Psychology*. New York: Macmillan.

Washburn, M.F. 1922. Introspection as an objective method. (Address of the president before the American Psychological Association, 1921.) *Psychol. Rev.*, 29: 89–112.

Washburn, M.F. 1932. Some recollections. Pages 333–358 in C. Murchison, ed., *A History of Psychology in Autobiography*, v. 2. Worchester, Massachusetts: Clark University Press.

Waters, R.H. 1934. The law of effect as a principle of learning. *Psychol. Bull.*, 31: 408–425.

Watson, J.B. 1903. *Animal Education; An Experimental Study on the Psychical Development of the White Rat. Correlated with the Growth of Its Nervous System*. Chicago: University of Chicago Press.

Watson, J.B. 1906. The need of an experimental station for the study of certain problems in animal behavior. *Psychol. Bull.*, 3: 149–156.

Watson, J.B. 1907. Kinaesthetic and organic sensations: their role in the reactions of the white rat to the maze. *Psychol. Rev. Monog. Suppl.* v. 8, 2: 1–100.

Watson, J.B. 1913. Psychology as the behaviorist views it. *Psychol. Rev.*, 20: 158–177.

Watson, J.B. 1914a. A circular maze with camera lucida attachment. *J. Anim. Behav.*, 4: 56–59.

Watson, J.B. 1914b. *Behavior: An Introduction to Comparative Psychology*. New York: Holts.

Watson, J.B. 1961. John Broadus Watson. Pages 271–281 in C. Murchison, ed., *History of Psychology in Autobiography*, v. III. Worchester, Massachusetts: Clark University Press.

Watson, J.B. and H.A. Carr. 1908. Orientation of the white rat. *J. Comp. Neurol. Psychol.*, 18: 27–44.

Watson, J.B. and R. Raynor. 1920. Conditioned emotional reactions. *J. Exp. Psychol.*, 3: 1–14. Reprinted *Am. Psychologist*, 55: 313–317 (2000).

Weiss, P. 1952. Survey of genetic neurology. Pages 27–40 in National Research Council Committee on Neurobiology, *Survey of Neurobiology*. Washington, DC: National Academy of Sciences-National Research Council.

Weiss, P.A. and H.B. Hiscoe. 1948. Experiments on the mechanism of nerve growth. *J. Exp. Zool.*, 107: 315–395.
Weissenberg, T.H. 1931. Charls Karsner Mills, M.D. 1845–1931. *Arch. Neurol. Psychiat.*, 26: 170–178.
Wells, F.L. 1947. James McKeen Cattell: 1860–1944. Pages 1–6 in A.T. Poffenberger, ed., *James McKeen Cattell: Man of Science*, v. 2. Lancaster, Pennsylvania: The Science Press. Also in *Am. J. Psychol.*, 57: 270–275, 1944.
Wernicke, Carl. 1874. *Der aphasische Symptomencomplex: Eine psycholigische Studie auf anatomischer Basis.* Breslau, Cohn and Weigert. Reprinted 1974, Berlin, Springer-Verlag.
Wheatley, M.D. 1944. The hypothalamus and affective behavior in cats: a study of the effects of experimental lessions, with anatomic correlations. *Arch. Neurol. Psychiat.*, 52: 296–316.
White, W.A. 1907. *Report of the Government Hospital for the Insane to the Secretary of the Interior for the Fiscal Year ended June 30, 1907.* The Psychological Laboratory. Pages 24–27. Washington, DC: Government Printing Office.
Whitehorn, J.C. 1944. Review of psychiatric progress 1943: biochemistry, endocrinology and neuropathology. *Am. J. Psychiat.* 100: 550–551.
Wiersma, C.A.G., ed. 1967. *Conference on Invertebrate Nervous Systems. Their Significance for Mammalian Neurophysiology*. Chicago: University of Chicago Press.
Wiersma, C.A.G., ed. 1975. *Invertebrate Neurons and Behavior*. Cambridge, Massachusetts: The MIT Press.
Wijngaard, M. v.d. 1997. *Reinventing the Sexes: The Biomedical Construction of Femininity and Masculinity.* Bloomington: Indiana University Press.
Williams, L.W. 1912. *The Anatomy of the Common Squid,* Loligapealii, Lesueur. Leiden: Library and Printing Office.
Wilson, E.O. 1978. Book Reviews. Amimal communication: a summing up. [Review of T.A. Sebeok, ed., *How Animals Communicate*.] *Science*, 199: 1058–1059.
Windle, W.F. 1931. The neurofibrillar structure of the spinal cord of cat embryos correlated with the appearance of early somatic movements. *J. Comp. Neurol.*, 53: 71–113.
Windle, W.F. 1979. *The Pioneering Role of Clarence Luther Herrick in American Neuroscience*. Hicksville, New York: Exposition Press.
Windle, W.F. 1980. The Cayo Santiago primate colony: its relationship to establishment of regional primate centers in the U.S.. *Science*, 209: 1486–1491.
Windle, W.F., ed. 1981. *Stephen Walter Ranson: Groundbreaking Neuroscientist. Memoirs of Students and Colleagues at His Centenary.* Los Angeles: Brain Research Institute, University of California.
Windle, W.F. and A.M. Griffin. 1931. Observations on embroynic and fetal movements of the cat. *J. Comp. Neurol.*, 52: 149–188.
Windle, W.F., J.E. O'Donnell, and E.E. Glasshagle. 1933. The early development of spontaneous reflex behavior in cat embryos and fetuses. *Physiol. Zool.*, 6: 521–541.
Winters, E.E., ed. 1950–1952. *The Collected Papers of Adolf Meyer.* vols. 1–4. Baltimore: Johns Hopkins Press.
Wislocki, G.B. and L.S. King. 1936. The permeability of the hypophysis and hypothalamus to vital dyes, with a study of the hypophyseal vascular supply. *Am. J. Anat.*, 58: 421–472.

Witelson, S.F., M.P. Bryden, and M.B. Bulman-Fleming. 1999. Handedness. Pages 855–856 in G. Adelman and B.H. Smith, eds., *Encyclopedia of Neuroscience.* Amsterdam: Elsevier.

Woodburne, R.T. 1959. Elizabeth C. Crosby. A biographical sketch. *J. Comp. Neurol.*, 112: 19–29.

Woods, James V. 1964. Behavior of chronic decerebrate rats. *J. Neurophysiol.*, 27: 635–644.

Woodworth, R.S. 1931. *Contemporary Schools of Psychology.* New York: Ronald Press.

Woodworth, R.S. 1943. The adolescense of American psychology. *Psychol. Rev.*, 50: 10–32.

Woodworth, R.S. 1944. James McKeen Cattell – In memoriam. Some personal characteristics. *Science*, 99: 160–161.

Woolsey, C.N. and E.M. Walzl. 1942. Topical projection of nerve fibers from local regions of the cochlea to the cerebral cortex of the cat. *Bull. Johns Hopkins Hosp.*, 71: 315–344.

Woolsey, T.A. and H. Van Der Loos. 1970. The structural organization of layer IV in the somatosensory region (SI) of mouse cerebral cortex. *Brain Res.*, 17: 205–242.

Wortis, J. 1994. In memoriam. Williamina Elizabeth Armstrong Himwich 1912–1993. *Biol. Psychiat.*, 35: 291–292.

Wundt, W. 1874. *Grundzüge der physiologischen Psychologie.* Leipzig: W. Engelman.

Wyman, J. 1847. Notice of the external characteristics and habits of Troglodytes Gorilla .... By T.S. Savage. Osteology of the same. By J. Wyman. *Boston J. Nat. Hist.*, 5: 417–442.

Wyman, J. 1853. Anatomy of the nervous system of *Rana pipiens. Smithsonian Contributions to Knowledge*, 5: 1–51.

Wyman, J. 1868. Review of the American Beaver and His Works. *Atlantic Monthly*, 21: 512.

Yasukawa, K. 1981. Song repertoires in the red-winged blackbird (*Agelaius phoeniceus*): a test of the Beau Geste hypothesis. *Anim. Behav.*, 29: 114–125.

Yerkes, R.M. 1902a. A contribution to the physiology of the nervous system of the medusa Gonionemus murbachii. Part I. – The sensory reaction of gonionemus. *Am. J. Physiol.*, 6: 434–449.

Yerkes, R.M. 1902b. A contribution to the physiology of the nervous system of the medusa Gonionema murbachii. Part II. – The physiology of the nervous system. *Am. J. Physiol.*, 7: 181–198.

Yerkes, R.M. 1907. *The Dancing Mouse: A Study in Animal Behavior.* New York: The Macmillan Co. Reprinted in "Classics in Psychology", 1973.

Yerkes, R.M. 1914a. The study of human behavior. *Science*, 39: 625–633.

Yerkes, R.M. 1914b. The Harvard Laboratory of animal psychology and Franklin field station. *J. Anim. Behav.*, 4: 176–184.

Yerkes, R.M. 1915. A study of the behavior of the crow *Corvus Americanus Aud.* by the multiple choice method. *J. Anim. Behav.*, 5: 75–113.

Yerkes, R.M. 1916a. The mental life of monkeys and apes: a study of ideational behavior. *Behav. Monogr.*, 3(1): 1–145.

Yerkes, R.M. 1916b. Provision for a station or institute for the study of monkeys and apes. *Science*, 43: 231–234.

Yerkes, R.M. 1925. *Almost Human.* New York: Century.

Yerkes, R.M. 1927. The mind of a gorilla. Part II. Mental Development. *Genetic Psychol. Monogr.*, 2: No. 6: 377–551.

Yerkes, R.M. 1940. Laboratory chimpanzees. *Science*, 91: 336–337.

Yerkes, R.M. 1943a. Early days of comparative psychology. *Psychol. Rev.*, 50: 74–76.

Yerkes, R.M. 1943b. *Chimpanzees: a Laboratory Colony.* New Haven: Yale University Press.

Yerkes, R.M. and C.A. Coburn. 1915. A study of the behavior of the pig *Sus scrofa* by the multiple choice method. *J. Anim. Behav.*, 5: 185–225.

Yerkes, R.M. and C.E. Kellogg. 1914. A graphic method of recording maze-reactions. *J. Anim. Behav.*, 4: 50–55.

Yerkes, R.M. and B.W. Learned. 1925. *Chimpanzee Intelligence and Its Vocal Expression*. Baltimore: Williams and Wilkins.

Yerkes, R.M. and J.B. Watson. 1911. Methods of studying vision in animals. *Behav. Monogr.*, 1(2): 1–90.

Yerkes, R.M. and A.W. Yerkes. 1929. *The Great Apes, a Study of Anthropoid Life.* New Haven, Connecticutt: Yale University Press.

Young, F.A. and D.B. Lindsley, eds. 1970. *Early Experience and Visual Information Processing in Perceptual and Reading Disorders.* Washington, DC: National Academy of Sciences Press.

Young, J.Z. 1975. Sources of discovery in neuroscience. Pages 14–46 in F.G. Worden, J.P. Swazey, and G. Adelman, eds., *The Neurosciences: Paths of Discovery*. Cambridge, Massachusettes: The MIT Press.

Yu, A.C. and D. Margoliash, 1996. Temporal hierarchical control of singing in birds. *Science*, 273: 1871–1875.

Zenderland, L. 1998. *Measuring Minds: Herbert Henry Goddard and the Origins of American Intelligence Testing.* Cambridge: University of Cambridge.

Zuckerman, Solly. 1970. *Beyond the Ivory Tower: the Frontiers of Public and Private Science.* London: Weidenfeld & Nicolson.

Name and Subject Index

Page numbers in italics indicate name is noted without elaboration